LANDMARK COLLECTOR'S LIBRARY

JAMES WATT

Volume 2:
The Years of Toil, 1775-1785

Rev. Dr. Richard L. Hills

Watt, James, a Philosopher,
Mechanician and Civil Engineer,
whose inventive talents, extensive knowledge
of the Sciences and Arts and practical application
of them to the purposes of life, place him the
foremost rank of those illustrious men
whose discoveries have influenced the
state of Society and conferred
distinction upon the are
and country in which
they lived.

James Watt junior,
Encyclopaedia Britannica, 1824.

LANDMARK COLLECTOR'S LIBRARY

JAMES WATT

Volume 2:
The Years of Toil, 1775-1785

Rev. Dr. Richard L. Hills

Landmark Publishing

Published by

LANDMARK
Publishing Ltd

Ashbourne Hall, Cokayne Ave
Ashbourne, Derbyshire DE6 1EJ England
Tel: (01335) 347349 Fax: (01335) 347303
e-mail: landmark@clara.net
web site: www.landmarkpublishing.co.uk

1st edition

ISBN 1 84306 046-9

British Library Cataloguing in Publication Data: a catalogue
record for this book is available from the British Library.

Printed by Cromwell Press Ltd, Trowbridge

Design & reproduction by James Allsopp

Cover captions:

Front cover & page 3: James Watt, F.R.S., 1736-1819

Back cover: Watt's Single-Acting Pumping Engine. *Farey (1827)*

Contents

Watt's Life	6
Illustrations	9
Introduction	11
1. From Glasgow to Birmingham	12
2. Prototype Engines and Patent Act of 1775	30
3. Continuing Trials and the First Engines	50
4. Family and Partners, 1774 – 1780s	76
5. Fair Prospects in Cornwall	92
6. Development of the Single-Acting Pumping Engines and Foreign Patents	122
7. The Financial Struggle	142
8. Problems in Cornwall	160
9. Natural Philosopher	182
Epilogue	237
Appendix 1 Specification of Patent 913, 1769	239
Appendix 2 An Account of James Watt's Improvements upon the Steam or Fire Engine, 1775	240
Appendix 3 Proposals to the Adventurers	244
Short Bibliography	250
Index – People	254
Index – Places & Subjects	255

Abbreviations

A.o.S.	Archives of Soho at Birmingham Central Library.
B.C.L.	Birmingham Central Library.
Boulton	Boulton Papers at Birmingham Central Library.
B. & W. Col.	Boulton and Watt Collection at Birmingham Central Library.
C.C.R.O.	Cornwall County Record Office, Truro.
JWP	James Watt Papers at Birmingham Central Library.
M	Muirhead Papers at Birmingham Central Library.
N.L.S.	National Library of Scotland, Edinburgh.
S.R.A.	Strathclyde Regional Archives, Glasgow.
S.R.O.	Scottish Record Office, Edinburgh.
R.I.C.	Royal Institution of Cornwall, Truro.
T.N.S.	Transactions of the Newcomen Society.

Wilson Papers at Cornwall County Record Office.

Watt's Life

1774 May 31; Watt arrived in Birmingham.
December; failure of tin cylinder.

1775 January; Watt in London about extending 1769 patent.
February; Petition presented in House of Commons.
March 9; Bill presented.
May 22; Royal Assent given to Act extending patent to 1800. Roebuck relinquished his two-thirds of patent to Boulton.
July; Draft of partnership between Boulton and Watt.
October; Watt in Scotland.

1776 March; Bloomfield Colliery engine began work.
April; New Willey engine started. Watt moved to No. 1, Newhall Walk, Birmingham.
July; Watt in Scotland married Ann McGrigor.
November; Order received for Ting Tang engine, Cornwall.

1777 March; Watt moved to Regent's Place, Harpur Hill, Birmingham.
April; Experiments on expansive engine.
September; Wheal Bussy engine working, first in Cornwall.
October 11; Birth of Gregory Watt.

1778 April; Provisional grant of French patent.
July 30; Ting Tang engine set to work. Watt in Cornwall.
September; Chacewater engine working.

1779 May 5; Birth of Jessy Watt.
May; Watt started experiments on copying machine. Wrote directions for engine erectors.
September to February 1780, Watt in Cornwall.

1780 February 27; Copying machine patent 1,244 sealed.
June and August; Watt in London over alkali patent.
Wheal Virgin, Cornwall, ordered five engines.
R. Cartwright stole Watt's idea of crank for rotative engines.

1781 February; Watt sent J. McGrigor plans for steam-heated drying cylinders.
May; Watt started work on rotative engines.
June to April 1782; Watt in Cornwall.
July; J. Hornblower junior patented his double-cylinder engine, No. 1,298.

1782 Death of J. Fothergill.
February 23; Patent 1,306 granted for rotative engines, sun and planet.
April; Watt experimenting on cement for steam engine joints. Watt calculated horse power as 33,000 pounds raised one foot high with one bushel of coal.
July 4; Patent 1,321 for expansive engine.
August 23; Death of J. Watt senior.
November; Work started on double-acting engine at Soho.
December 1; Letter from J. Priestley about converting water into air.

1783 March; Proposals for steam-driven corn mill in London.
April 26; Watt wrote letter to J. Priestley about composition of water to be submitted to the Royal Society but not read.
November; Watt suggested ideas for metric measurements.

1784 February; Watt elected Fellow of Royal Society of Edinburgh.
May; Watt's paper on composition of water read to Royal Society of London.
June; Watt paid off last debts to J. Black.
August 25; Patent 1,432 for three bar motion and steam carriage.

1785 April & June; Watt in London over trial of Arkwright's patent.
July 9; Patent 1,485 for smoke consuming furnace.
September 1; Cornish Metal Company launched.
November 24; Watt elected Fellow of Royal Society of London.

1786 February; First rotative engine in a textile mill set to work.
March; Albion corn mill engine set to work.
June; Dutch patent granted.
November to January 1787; Boulton and Watt in Paris.

1788 Watt started experiments to produce chlorine in late winter.
May; King George III visited Boulton and Watt engine at Whitbread's brewery.
Boulton saw centrifugal governor which Watt adapted for the steam engine.
July 16; Watt elected corresponding member of the Batavian Society.

1788 August; American patent.

1789 May; Lease purchased of house on Handsworth Heath.

1790 September; Watt moved into Heathfield.

1791 March 3; Albion Mill destroyed by fire.
June 6; Marriage of Margaret Watt to James Miller.
July 14; Riots in Birmingham, Priestley's house destroyed.

1792 January & August; Watt in London opposing Hornblower junior's bill to extend patent.
March 29; Margaret Miller born, Watt's first grandchild.

1793 May 6; Watt elected member of the Society of Civil Engineers.
May 14; Birth of grandson.
June; Trial of E. Bull starts in Court of Common Pleas.

1794 March; Injunctions obtained against 'pirate' engines.
June; Trial of Bull continued.
June 6; Death of Jessy Watt.
July to December; Watt carried out experiments on medicinal gases.
October; Partnership for Soho Foundry settled.

1795 January to February; Trial of Boulton v. Bull.
May 16; Judgement suspended.
September; J. Wilkinson's dispute with his brother exposed his building engines with Watt's condenser.

1796 January 30; Boulton formally opened Soho Foundry.
June; Margaret Watt died.
Steam engine indicator developed.

December to February 1797; Boulton & Watt v. Hornblower & Maberley in Court of Common Pleas.

1797 February; Arguments about 'Writ of Error' not resolved.

1798 May; Watt purchased Manor of Gledestry, Kington, Radnorshire, first of his 'Welsh' estates.
November; Boulton & Watt v. Hornblower & Maberley in Court of King's Bench.

1799 January; Judgement given in favour of Boulton & Watt against Hornblower & Maberley.
December 6; Death of Joseph Black.

1800 June; Termination of 1775 Act, end of Boulton and Watt partnership.
Watt helped Robison edit Black's lectures.

1801 Peace of Amiens, Soho Manufactory illuminated with gas light.

1802 Portrait painted by Sir W. Beechey.
September to October; Watt toured the Continent seeking Gregory.

1803 Purchase of Doldowlod estate in Wales.
February 6; Death of Joseph Priestley.

1804 Watt turned down appointment as Sheriff of Staffordshire.
Greenock property sold.
Watt started experiments with sculpturing machines.
October 16; Death of Gregory Watt.

1805 January 30; Death of John Robison.

1806 Watt honoured by Glasgow University with award of Doctor of Laws.

1808 Watt elected corresponding member of the Académie des Sciences in Paris.
Watt established prize at Glasgow University for an essay.
December; Deaths of Gilbert Hamilton and Thomas Beddoes.

1809 Watt drew up a scheme for heating the Hunterian Museum, Glasgow.
August 17; Death of Matthew Boulton.

1810 January; Death of Ambrose Weston.
Watt's scheme for Glasgow Waterworks finished.

1812 Watt advised about Sheerness Dockyard for Navy Office.

1813 Watt's portrait painted by Sir Thomas Lawrence.

1814 Watt elected as Foreign Associate by Académie des Sciences, Paris.
Watt's bust carved by Sir Francis Chantrey.
Watt wrote articles for *Edinburgh Encyclopaedia* and started correcting Robison's articles on 'Steam' and 'Steam Engine'.

1815 Death of John Southern.

1816 Watt declined being made Sheriff of Radnorshire.
Watt made gift to provide books for Greenock Library.

1819 James Watt died August 25.

Illustrations

James Watt, F.R.S., 1736 – 1819 (Williamson, 1856), Frontispiece.

From Glasgow to Birmingham

Matthew Boulton, F.R.S., 1728 – 1809. (Smiles, (1878), p. 12.

Soho Manufactory, c. 1800, p. 19.

Sketch map of north west Birmingham, p. 22.

Illustrations, Prototype Engines and the 1775 Act

Diagram of Watt's single-acting pumping engine, p. 32.

The steam-wheel, 1775. (By permission of Birmingham Central Library), p. 35.

Drawing of Watt's single-acting pumping engine, 1775. (By permission of Birmingham Central Library), p. 37.

Memorial to William Small, 1734 – 1755. (Muirhead, (1858), p. 253), p. 41.

More Engine Trials and the First Engines

Side elevation, Penrhyn Du engine, 1779. (Redrawn from B. & W. Col., PF 545. By permission of Birmingham Central Library), p. 53.

Watt's slide valve, 1775. (By permission of Birmingham Central Library), p. 54.

New Willey Blowing Engine, p. 61.

Single-acting pumping engine, Stratford-Le-Bow, 1775. (B. & W. Col., PF 1072. By permission of Birmingham Central Library), p. 65.

Boulton and Watt engines in 1777. (Lord, (1923), p. 154), p. 65.

Exhaust valve for Broseley and Bradley engines, 1777. (Dickinson and Jenkins, (1927), plate XLII. By permission of Birmingham Central Library), p. 67.

Twin air pumps and hot water pump for Bedworth condenser, 1776. (Redrawn from Dickinson and Jenkins, (1927), plate LXXIV. By permission of Birmingham Central Library), p. 69.

Single-acting pumping engine for Peter Colville, Torryburn, Fife, 1776. (B. & W. Col., PF 624. By permission of Birmingham Central Library), p. 70.

Plan of Torryburn engine and boilers, 1776. (B. & W. Col., PF 624. By permission of Birmingham Central Library), p. 70.

Cataract mechanism, injection and blowing-through valve, 1779. (Redrawn from Dickinson and Jenkins, (1927), plate LIII. By permission of Birmingham Central Library), p. 72.

Family and Partners, 1774–1780s

Watt's family by his first marriage, p. 79.

Regent's Place, Harpur Hill. (Smiles, (1878), p. 160), p. 81.

Watt's family by his second marriage, p. 84.

Fair Prospects in Cornwall

Smeaton's engine for Chacewater mine, 1776. (Smeaton, (1812), Vol. II, plate VI), p. 95.

Watt's engine for Ting Tang mine, 1778. (B. & W. Col., PF 1288. By permission of Birmingham Central Library), p. 102.

Watt's cylinder inside Smeaton's for Chacewater mine engine, 1778. (B. & W. Col., PF 1315. By permission of Birmingham Central Library), p. 104.

Cosgarne House, Cornwall. (Smiles, (1878), p. 215), p. 107.

William Murdock, 1754 – 1839. (Smiles, (1878), p. 343), p. 109.

Improvements to Single-acting Pumping Engines

Watt's single-acting pumping engine for Ales and Cakes, 1779. (B. & W. Col., PF 1293. By permission of Birmingham Central Library), p. 122.

Early layout of two conical valves. (Farey, (1827), Vol. I, p. 322), p. 124.

Valve operating mechanism. (Farey, (1827), Vol. I, p.373), p. 125.

Sizes of Valves. (Watt's Note Book, in possession of the late N. Bertenshaw), p. 126.

Front elevation of Penrhyn Du engine and boiler, 1779. (Redrawn from B. & W. Col., PF 545. By permission of Birmingham Central Library), p. 127.

Cylinder with three valves and altered steam casing. (Farey, (1827), Vol. I., p.333), p. 129.

Items destroyed in Soho fire, 1778. (Boulton 348.119, no date), p. 129.

Diagram of expansion after cut-off, 1780. (Watt's Note Book, in possession of the late N. Bertenshaw), p. 132.

Plan of Penrhyn Du engine, 1779. (Redrawn from B. & W. Col., PF 545. By permission of Birmingham Central Library), p. 135.

The Financial Struggle

Premiums expected in 1779. (B. & W. Col., M. Boulton to C. Vere, April 1779), p. 146.

State of Accounts on 18 December 1780. (JWP, 6/21.25, 18 December 1780), p. 148.

Problems in Cornwall

Average output of copper in Cornwall and Devon, 1770 – 1800. (Barton, (1961), p. 94), p. 161.

Average premiums received, 1782. (Lord, (1923), pp. 155 – 57), p. 167.

Royalties received by Boulton and Watt from Cornish engines, 1778 – 1785. (Wilson Papers DDX 318/11, Mine Accounts), p. 167.

Performance of Newcomen and Watt engines compared. (Farey, (1827), Vol. I., p. 368), p. 169.

Duty of Wheal Maid engine, 1788 – 1791. (JWP, 3/18, J. Watt to M. Prony, 30 April 1803), p. 170.

J. Hornblower junior's Radstock engine. (B. & W. Col., PF 1379. By permission of Birmingham Central Library), p. 174.

Drawing of improved version of Hornblower's engine, 1788. (Howard's *New Royal Cyclopaedia*, 1788), p. 178.

Natural Philosopher

Watt's steam heated drying cylinders. (*Edinburgh Encyclopaedia*, (1830)), p. 186.

Calculating device in garret workshop. (R.L. Hills), p. 189.

Watt's copying presses. (Patent 1,244, 1780), p. 205.

Jospeh Priestley, F.R.S., 1733 – 1804. (Uglow (2002), p. 439), p. 217.

Priestley's apparatus for converting water into air. (Gibbs, (1965), p. 159), p. 221.

Lavoisier's apparatus for decomposing water into hydrogen. (Muirhead, (1846)), p. 223.

Introduction

In the first volume of this new biography of James Watt, we covered many of his interests and activities in Scotland never included before in accounts of his life. This was possible through studying the James Watt Papers recently purchased from the late Lord Gibson Watt. These papers opened up a whole new field of study which has thrown fresh light on many aspects of Watt's later career in England as well. In this volume, we look at arguments appertaining to his acquiring the extension to his 1769 patent by the 1775 Act of Parliament, so crucial to the success of the partnership of Boulton and Watt. We learn how his single-acting steam engine was turned into a standard production design. More surprisingly, we see how, in order to attain financial stability and viability, Watt took over the management of the accountancy as well as the technical development of the partnership. In addition, the James Watt Papers throw new light on Watt's career as a natural philosopher which has been largely forgotten. For example, his development of his copying press was based largely on his chemical experiments to concoct a suitable ink. His earlier studies on the properties of steam led him to submit a paper to the Royal Society of London on the composition of water which played an important role in his being elected to that prestigious Society, a recognition of his achievements through his remarkable combination of scientific knowledge and practical application.

This new biography could not have been written without recourse to the publications of other scholars such as H.W. Dickinson, Rhys Jenkins, D. King-Hele, J. Lord, D. McKie, A.E. Musson, E. Robinson, E. Roll, R.E. Schofield, J. Tann and many others. The fresh material in the James Watt Papers has often led me to give different interpretations from theirs and to show Watt in another light. Errors in what I have written are mine and not theirs. Over the many years of research and writing, I have received so much help and encouragement from so many people that, while wishing to thank them all, inevitably I will leave some out. That the research was possible at all has been due to travel grants from the Royal Society. These enabled me to go, for example, to the National Library of Wales at Aberystwyth to study the microfilms of Watt's Letter Books by permission of Lord Gibson Watt. I wish to thank the Librarians there for their help as well as Librarians and Archivists at Cornwall Country Record Office, Edinburgh University Library, the National Library of Scotland, Manchester Central Library, the Royal Institution of Cornwall, the Royal Society and what was the University of Manchester Institute of Science and Technology. But above all I wish to thank the Archivists and members of the team cataloguing the Archives of Soho at Birmingham Central Library for their forbearance in handling all my requests without whom this new biography would not have been possible. Then there have been the many friends whom have helped me and given advice. Among them have been Kenneth Brown, Martin Clagett, Jur Kingma, Joe Marsh, Fred Page, Adam Reid, Bill Slatcher, Jan Verbruggen and Raj Williamson. In addition, this study would not have seen the light of day without Lindsey Porter and his team at Landmark Publishing through their tireless efforts converting my manuscripts into something presentable. To them all I owe a great debt.

Richard L. Hills.

1
FROM GLASGOW TO BIRMINGHAM

Matthew Boulton, F.R.S., 1728 – 1809.

The spring of 1774 saw James Watt set out on a venture which was to prove a new start in life for him and helped to create a new era in the development of industrial technology. His decision to journey from Glasgow to Birmingham became the most important in his career because it was in Birmingham where he was able to turn his 'perfect' steam engine into one that worked successfully, first a reciprocating type for raising water and then into a rotative version for driving machinery directly. In 1763, he had been asked by the University of Glasgow to repair a model of the type of steam engine invented by Thomas Newcomen. Its high consumption of fuel and steam caused him to initiate a series of scientific investigations which led to his studying the relationships between temperature and steam pressures, the value of latent heat in steam and other factors. But he could not discover how to improve the performance and make the engine more economical until inspiration came in the spring of 1765. This was when he realised that the steam could be condensed in a separate vessel from the working cylinder. From this he developed his concept of a perfect steam engine in which the working cylinder could be kept as hot as the steam entering it, thus avoiding wasteful condensation, while his separate condenser could remain cold, enabling a complete vacuum to be generated. He patented these principles in 1769 but did not succeed in constructing a working engine while he remained in Scotland. At Birmingham, the various threads of his earlier tortuous existence were drawn together so that he rose from virtual obscurity to become one of the foremost engineers of all time. In his own day, he received international recognition and his renown has continued down the years. The various elements of his early career will be reviewed briefly to see how they formed a basis for his later life as well as seeing why Birmingham presented better prospects which of course partly revolved round the personality of Matthew Boulton.

The last entry in the Journal which Watt kept for the early part of 1774 was for 31 May when he noted, 'as approach Birmingham stone grows softer and gravel harder'. [1] He had travelled with his friend, the geologist James Hutton, from Edinburgh where he had tried to wind up his Scottish business connected with surveying canals. He must have recognised that he would have to remain in the south for some time in order to put his steam-wheel and his reciprocating steam pumping engine into proper working order with the assistance of Boulton at the Soho Manufactory, but it is unlikely that he ever thought that his stay in Birmingham would extend for the rest of his life and that he would never live in Scotland again. Watt could play his cards close to his chest and did not reveal his intentions since his friends in Scotland were expecting his early return while his friends in Birmingham were pressing him to stay there.

We can see good reasons for Watt deciding to leave Scotland, a principal one being his bereavement through the death of his wife, Margaret usually called Peggy, on 24 September 1773. His grief was profound and he wrote at the time, 'In her I lost the comfort of my life, a dear friend and a faithful wife'. [2] (See Vol. I, p. 127) Then he had sought consolation from his close acquaintance in Birmingham, William Small, a medical doctor, and wrote to him:

> You are happy, Small, that have no such connection. Yet this misfortune might have fallen upon me when I had less ability to bear it, and my children might have been left suppliants to the mercy of the wide world. I know that grief has its period; but I have much to suffer first. I grieve for myself, not for my friend; for if probity, charity, and duty to her family can entitle her to a better state, she enjoys it. I am left to mourn. [3]

Their marriage had been a true love-match. Small replied, 'I most deeply condole with you, and wish I could comfort you... Come to me as soon as you can'. [4] In Small, Watt knew he would find a sympathetic person. So the children, Margaret and James Watt junior, were left at Greenock in the care of their grandfather, James Watt senior, when Watt set out for Birmingham.

Watt was probably aware that he had reached a crucial turning point in his life, and, if he failed to make a success of his steam engine, his other career prospects were problematic. A brief examination of his earlier business connections shows that they were all virtually finished and it is doubtful whether he could have returned to them. He had started out as a maker of and dealer in scientific instruments, particularly those connected with navigation. Through repairing instruments for the observatory at Glasgow University, he made the acquaintance of professors such as John Anderson and Joseph Black from whom he learnt a great deal about natural philosophy. He set up a workshop in a room in the precincts of Glasgow University, out of which he expanded into a shop in the Saltmarket assisted by the financial backing of John Craig. He also had a workshop and employed a maximum of fourteen apprentices and journeymen working for him in June 1763. [5] (See Vol. I, p. 106) Yet this business declined from around that time and may have been sold off to one of his apprentices, John Gardner, in December 1770 when Watt's Journal records that he spent three days 'sorting John Gardner's account', finishing on Christmas Day. [6]

By 1774, Gardner was well established in Glasgow in the scientific instrument trade. Another of Watt's apprentices, Murray Osborne, also set himself up as a mathematical instrument maker so that, had Watt attempted to start up again in this business there, he would have faced severe competition from these two and others. Yet, although he did not engage in scientific instruments commercially while at Birmingham, this background of competence in making small instruments proved useful for other ventures and he retained an interest in them for the rest of his life. His garret workshop at his later Birmingham home, Heathfield House, contained a metal-working lathe and a variety of hand tools. He used slide rules as calculating devices in his engineering designs. For his steam engines, he adapted a device to count the number of strokes as well as an early form of indicator to show the pressure in the cylinder. His experience of working with glass may have helped to fashion the apparatus for producing chlorine. And there was also the apparatus used for pneumatic medicine in conjunction with Dr. Thomas Beddoes at the Pneumatic Institution, Clifton, near Bristol. Watt's sculpture copying machines with which he amused himself during the last years of his life could be considered as forms of precision instruments. The scientific instruments business had taught Watt the importance of accuracy in manufacture, something which he carried into his steam engine business. Also the neatness of his handwriting was notable, showing his dexterity. If James Watt, scientific instrument maker, saw no future in 1774 in trying to pursue that career again, he also realised that another early career, that of retail merchant, was equally dead. In conjunction with his partner, Craig, and additional financial backing from Black, Watt had branched out into an ambitious shop in the Trongate selling a wide range of merchandise. When Craig died in December 1765, Watt's total indebtedness to him was £757.4s.7d. [7] (See Vol. I, p. 104) This debt was taken over by Dr. John Roebuck when he became a partner to exploit Watt's steam engine patent of 1769 for the separate condenser. This debt, incurred principally in purchasing stock for the shop as well as supporting the scientific instrument business, formed a large part of the financial settlement between Roebuck and Boulton when Boulton took over Roebuck's two-thirds share in the patent.

Watt's debts to Black were additional to this sum. Watt was still owing Black £239.2s. in 1772. [8] (See Vol. I, p. 80) This account was not finally cleared until the outstanding balance of £18.8s.6d. was paid on 7 June 1784. [9] He had been reluctant to borrow from one friend to repay another and apologised to Black for the length of time this debt had been outstanding. [10] This shows for how long the debts of Watt's speculative merchanting venture hung round his neck and could well have forced him into bankruptcy without the generous forbearance of his friends.

We can be certain that the shop itself was closed before Watt accepted the post of resident engineer for the construction of the Monkland Canal in June 1770, [11] because he noted in his Journal for 24 May 1770, 'Finished packing up of shop goods'. [12] These goods still had a value which could be set against the debt Roebuck had taken over. They were included in the discussions concerning the transfer of a share in Watt's steam engine patent of 1769 to Boulton [13] and still remained stored when Watt made his journey to Birmingham in 1774. They were not finally disposed of until it was obvious that he would remain in England for a considerable time. In November 1775, Watt's close friend in Glasgow, Gilbert Hamilton, sold Watt's household furniture but took the shop goods to his own apartments, saying, 'I intend to have all the shop things cleaned up & if the Hardware people will not give a tolerable offer for them to sell them by auction'. [14] It certainly could not be considered as a successful business and was not one to be embarked on again. The career of James Watt, merchant, was definitely closed with no prospect of resumption. But both the scientific instrument and merchanting businesses gave Watt experience of accountancy and management. Watt's interest in chemistry literally continued to pay dividends. He had helped to improve the kilns and production at the Delftfield Pottery in Glasgow so that it became a profitable concern. He had purchased part of a share which made him a partner, a position he held for the rest of his life. [15] His share gave him a steady income but nothing is known about how important this was as a contribution towards his finances.

Watt's other main chemical venture in Scotland, together with Black and Roebuck, trying to manufacture alkali from sea salt, was a failure. But he continued with chemical experiments in England where his most successful one was his method of making a copy of a letter or a drawing. While he pioneered the manufacture of chlorine on an industrial scale, others soon overtook him in this business because he had to concentrate on his steam engines. He also acted as chemical adviser to Beddoes in producing medicinal gases for patients. His paper about the composition of water was influential in procuring his membership of the Royal Society, London. Watt's interests in chemistry were supported by other members of the Lunar Society in Birmingham, such as Boulton and Dr. Joseph Priestley.

Since taking up the post of resident engineer for the Monkland Canal, Watt had carried out no further experiments on his steam engine at Kinneil but had concentrated on his surveying commissions. The seven mile level stretch of that canal had been finished during the summer of 1773 but there was no money to complete any form of link down to Glasgow itself. (See Vol. I, p. 210 f.) There it languished and, in 1777, Gibson, Glasgow's historian, wrote, 'The Monkland Canal, which was undertaken with a view to reduce the price of coals, has not answered the end of which it was intended'. [16] Out of all the other surveys undertaken by Watt, there resulted only improvements to Port Glasgow harbour and its dry dock, improvements to Greenock harbour, Greenock water supply and a few small undertakings, principally involving water power. (See Vol. I, p. 279 f.) Yet he had shown his competence in surveying as well as preparing estimates and reports. He had been employed by Lord Cathcart for the Commissioners of Police in Scotland as well as the Board of the Forfeited Estates who may have favoured him because he was a Scot. Through these connections, as well as travelling around parts of Scotland, he had met numerous influential people and his surveying career gave him much broader experience in many fields. But Cathcart was far from well and may have lost confidence in Watt's abilities so that Watt no longer had the backing of his most influential supporter. Neither Watt's concept of cheap level canals linked to other stretches by railway inclined planes to avoid locks nor his wooden aqueducts were taken up elsewhere.

When he left Edinburgh in May 1774 to set out for Birmingham, Watt had no more surveying commissions to undertake. He had finished his reports for Cathcart on the Great, or what would become the Caledonian, Canal and the River Devon behind Alloa. He may have realised that his stay in Birmingham would last for at least a few months until his steam engines were running properly and therefore deliberately ran down his surveying business. But it is also evident that he was tempted by the better prospects in England for the type of surveying work in which he had gained competence. While the lack of suitable work in Scotland may have been due to the economic downturn resulting partly from the failure of the Ayr Bank in the summer of 1772, the slump affected England as well and was prolonged by the American War of Independence. Watt might never have returned to his steam engines but for this slump. At the same time, Boulton weathered the failure of his bankers, Neal, Fordice and Downe.

Watt reported to Small in December 1773:

> The engineering business is not a vigorous plant here; we are in general very poorly paid... in short I must, as far as I can see, change my abode. There are two things which occur to me, either to try England, or endeavour to get some lucrative place abroad; but I doubt my interest for the latter. What I am fittest for is a surveying engineer. Is there any business in that way? [17]

It may have been Small who earlier had put the idea into Watt's head of a surveying career in England, for, in April 1773, he had written, 'Dr. [Erasmus] Darwin and I, for Mr. Boulton is still in London, are endeavouring to persuade the Coventry Canal Company to be so much their own friends as to send for you to help them out of great difficulties'. [18] The Birmingham Canal, for which the Act had been passed in 1768, reached Birmingham from Wolverhampton and the north in September 1772. The Coventry Canal, authorised in the same year, still awaited completion to form a route from the south via the Birmingham and Fazeley Canal. [19] Part of the problem here may have been the death of their engineer, James Brindley, in 1772. Small replied to Watt's later plea for help, encouraging him that he would find work in the south:

> Had you been here, in all probability you would have been consulted about a difficult canal, and should you come soon you may still; but we cannot persuade the people to send into Scotland for you, or for anything else...
>
> I have no doubt that after a little time you might have good business in England as an engineer, could you once find an opportunity of giving a specimen of your talents...
>
> Should you come hither soon, pray consider aqueducts in the meantime, and the conveyance of water in pipes, and the expenses of both... Also consider the price of lockage very minutely. I long to see you. [20]

Watt took these possibilities seriously and told Small at the end of April 1774 as he was making his preparations to leave Scotland, 'I begin now to see daylight through the affairs that have detained me so long, and think of setting out for you in a fortnight at furthest... I intend to bring copies of several of my reports and plans with me to show to you and friends'. [21]

Whatever Watt may have hoped, no surveying career emerged at Birmingham, and there are no suggestions that he tried to initiate one, possibly because he quickly found that he was fully occupied with developing his steam engines. However there were definitely a few times when serious requests were made for him to resume connections with earlier undertakings. The first was in the summer of 1781 when new proprietors had purchased the shares of the

Although Watt had visited Birmingham earlier to purchase stock for his shops, it was not until May 1767 that he went to the Soho Manufactory as he was returning from London after the failure to secure an Act for their 'ditch' canal to link Glasgow with the Firth of Forth. Boulton was away so he was shown round by Small and Boulton's partner, John Fothergill. Watt was duly impressed but he only recorded his impressions many years later on Boulton's death in 1809:

Soho Manufactory, by Francis Eginton, c. 1800.

> The goods then manufactured there were steel gilt & fancy buttons, steel watch chains & sword hilts, plated wares, ornamental works in Or moulu, Tortoise shell snuff boxes, Bath metal buttons inlaid with steel & various other articles which I have now forgot. A mill with a water wheel was employed in Laminating metal for the buttons, plated goods etc and to turn Laps for grinding & polishing steel work & I was informed that Mr Boulton was the first inventor of the inlaid buttons, & the first who had applied a mill to turn the Laps. Mr. B. at that [time] also carried on a very considerable trade in the manufacture of buckle chaps, in the making of which he had made several very ingenious improvements. Besides the Laps in the mill, I saw an ingenious lap turned by a hand wheel for cutting & polishing the steel studs for ornamenting buttons, chains, and sword hilts etc, and a shaking box put in motion by the mill for scouring button blanks & other small pieces of metal which was also a thought of Mr B: - there was also a steelhouse for converting iron into steel, which was frequently employed to convert the cuttings & scraps of the chapes & other small iron wares into steel which was afterwards melted & made into cast steel for various uses. [35]

While Watt's memory was slightly at fault for it is very doubtful whether Boulton had started to manufacture ormolu as early as 1767, [36] he obviously was impressed by the scale of the Soho Manufactory and the equipment Boulton had designed for it. In 1770, Boulton had:

> ... almost every machine that is applicable to those arts; I have two water-mills employed in rolling polishing, grinding and turning various sorts of lathes. I have trained up many, and am training up more plain country lads into good workmen; and wherever I find indications of skill and engineering, I encourage them. [37]

The Soho Manufactory was on a scale not seen in Birmingham before and was once called the eighth wonder of the world. It brought many parts of the manufacturing processes under one roof instead of being scattered among a large number of small workshops. It was designed on the lines of a grand country house, was equipped with the best tools and provided a much higher standard of working conditions. This enabled Boulton to attract outstanding workmen. James Boswell was suitably impressed by 'the vastness and the contrivance of the machinery' as well as Boulton's remark, 'I sell here, Sir, what all the world desires to have – Power'. Boulton's creation of the Soho Manufactory was a major contribution to the development of factory organisation and was the forerunner of mass-production. [38] In addition to Boulton's competence in mechanical engineering and his success in business, Watt would have been impressed by Boulton's connections with people in positions of power and influence whom he recognised might be able to support sales of his steam engines. But the Soho Manufactory was not equipped for building steam engines as it had no foundry, smith's shop or suitable machine tools.

Watt continued to correspond with Small, a role crucial for the future of the steam engine. After arranging for the initial application of the patent with his agent in London during August 1768, Watt stayed for a couple of weeks with Boulton in Birmingham. The two struck up a friendship which survived until Boulton's death when Watt described him as:

> ... not only an ingenious mechanick, well skilled in all the practices of the Birmingham manufacturers but possessed in a high degree the faculty of rendering any new invention of his own or others useful to the public by organizing & arranging the processes by which it could be carried on, as well as of promoting the sale by his own exertions & by his numerous friends & correspondents.
>
> His conception of the nature of any invention was quick & he was not less quick in perceiving the uses to which it might be applied & the profits which might accrue from it.
>
> When he took any scheme in hand, he was rapid in executing it & on these occasions neither spared his own trouble [n]or expence. He was a liberal encourager of merit in others & to him the country is indebted for various improvements brought forward under his auspices. [39]

Watt certainly found in Boulton a fellow enthusiast for experimenting in natural philosophy, in which neither had any formal training. James Keir commented,

> Mr. Boulton is a proof how much scientific knowledge may be acquired without much regular study, by means of a quick and just apprehension, much practical application, and nice mechanical feelings. He had very correct notions of the several branches of natural philosophy, and was master of every metallic art, and possessed all the chemistry that had any relation to the objects of his various manufactures. Electricity and astronomy were at one time among his favourite amusements. [40]

Keir came from a wealthy Scottish family near Edinburgh. His father died when he was very young. He attended Edinburgh University without taking a degree and served in the army for eleven years before retiring as a captain and settling near Birmingham to become a pioneer of industrial chemistry. He became involved with both Boulton and Watt in many of their projects and was a member of the dining group later called the Lunar Society. Boulton was a leading member and other members gave Watt assistance in many of his activities. With the departure from Glasgow of his close friends, Black and John Robison, Watt may have felt isolated with few others to share his interests in natural philosophy except Hamilton. His university connections seem to have diminished, certainly no references have been found to any in the final years of his residence in Scotland. In contrast, he received a warm welcome from Darwin who, having met Watt during his return from his London visit in 1767, wrote enthusiastically, 'Now my dear new Friend... I wish the Lord would send you to pass a week with me, and Mrs. Watt along with you; – a week, a month, a year'. [41] Darwin was also interested in steam engines and was another crucial link in the chain bringing Watt to Birmingham.

It is also possible that Watt found Birmingham a more congenial city than Glasgow. Glasgow was more a commercial centre with its tobacco trade and sugar imports, while Birmingham concentrated on manufacturing. Birmingham had a long tradition of smithing. In 1586, William Camden found the place 'echoing with the noise of the anvils, for here are great numbers of smiths, and of other artificers in iron and steel, whose performances in that way are greatly admired both at home and abroad'. [42] The second half of the seventeenth century saw the introduction of brass. Birmingham metal workers quickly adapted their skills to this new material. The town had no charter, no guilds and was largely unaffected by restrictive legislation concerning religious observances. It became known as a place where people could settle without interference and take advantage of the lack of craft restrictions to set up their workshops and businesses. It attracted in particular nonconformists who tended to be strong-minded people. The historian, William Hutton, commenting about the elegance of its buildings in 1741, said that he 'was surprised at the place, but more so at the people... they possessed a vivacity I had never beheld... their very step along the street showed alacrity'. [43] The successful merchants and manufacturers of the eighteenth century held liberal views with an inquisitive interest in science and technology. The extraordinary number of patents granted to Birmingham men during that century confirmed their practical application of new ideas which were mostly concerned with metal working. [44]

From just under 25,000 in 1741, the population had doubled to over 50,000 by 1780. The people fuelling this growth had pioneered the efficient use of hand technology through division of labour, supplemented with waterpower, thus putting Birmingham ahead of its competitors. Sketchley and Adams's directory for 1770 contained thirty-three brass founders, forty-four buckle makers and forty-eight toy makers. By the 1770s, the button and toy makers had achieved a position of ascendancy in Britain and Europe and were developing trading contacts elsewhere. One of Boulton's visitors in 1775 recorded in a letter that 'Birmingham is a very large and thickly populated town, where almost everybody is very busy hammering, pounding, rubbing and chiseling'. [45] Such was the place where Boulton had become one of the leading manufacturers and where Watt was to live for the rest of his life.

While on the one hand Watt may have felt that his prospects were better in Birmingham, on the other Boulton saw potential in supporting Watt to develop his steam engines. Boulton inherited his father's business on his death in 1759. His ambitions soon outstripped the cramped premises on Snow Hill where the machinery was driven by horsepower. In 1761, he began to negotiate for a site in Handsworth, north of the town, and in the following year formed a partnership with Fothergill. The Hockley Brook was dammed to provide waterpower for this new mill and so the famous Soho Manufactory was opened in 1765. But production

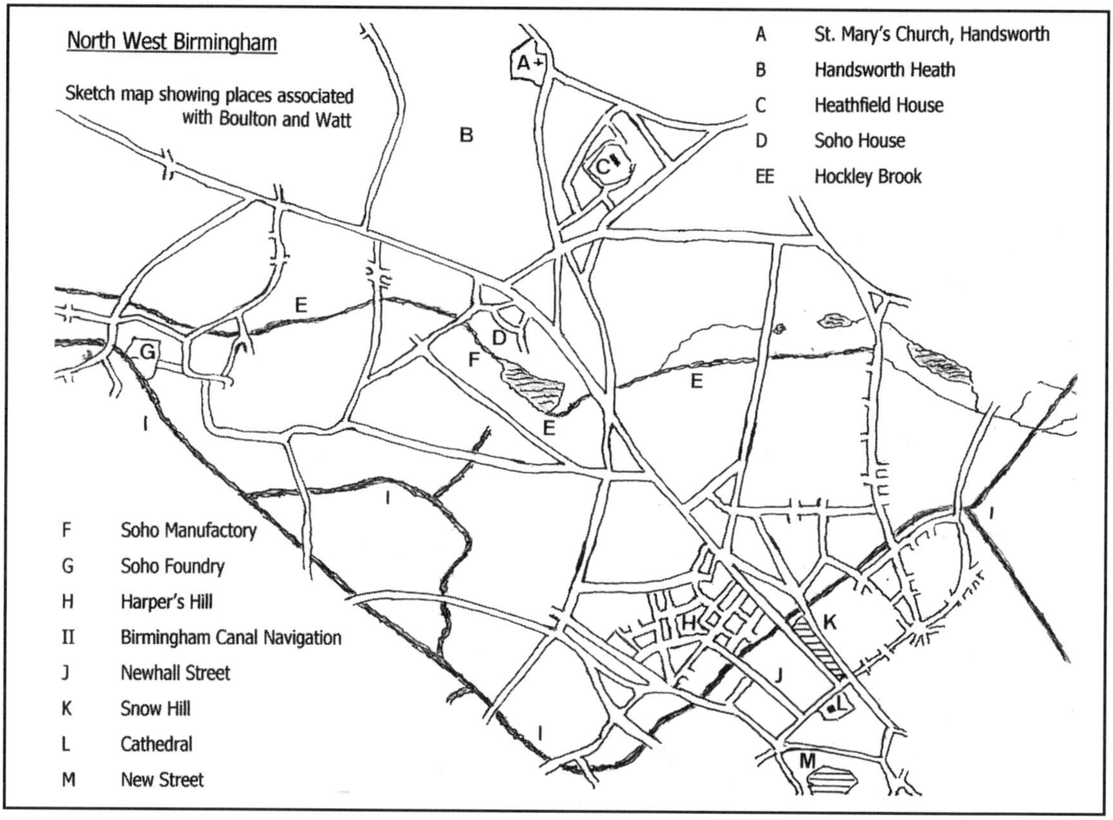

quickly outstripped the capacity of the Hockley Brook to drive the machinery. Therefore Boulton considered installing a steam engine to lift the water from the tailrace of his waterwheel back to the upper reservoir in order that the water might be used over and over again.

Boulton told a friend that 'The enormous expense of horse-power put me upon thinking of turning the mill by fire, and I made many fruitless experiments on the subject'. [46] After considering a Newcomen engine, he made a model of a Savery engine. [47] He sought advice from Benjamin Franklin in February 1766 but did not receive much help. Darwin was much more enthusiastic:

> Your Model of a Steam Engine I am told gain'd so much approbation in London, that I cannot but congratulate you on the mechanical Fame you have acquired by it; which, assure yourself, is as great a Pleasure to me, as it could possibly be to yourself; so much do I interest myself in what belongs to your Reputation. [48]

It is clear that Boulton had studied steam engines and was in a position to judge Watt's schemes. He continued his experiments possibly until he heard from Small about Watt's visit in 1767.

In the summer of 1768, Watt gave Robison an introduction to Darwin when he was travelling south. Robison called at Lichfield in early July where he was:

> ... lucky enough to meet with Dr. Small at Mr. Darwin's, and along with him another most agreeable man and your very warm friend, Mr. Boulton. Next day I went to Birmingham, and Mr. Boulton, tho lame, was so kind as to conduct me thro his vast works... I was much struck with many appearances of ingenuity in Boulton's works. It is partly at his desire that I now write to you. In hopes that you had vigorously pursued the improvements of your fire engine, and was getting your property secured, he has finished a building for one of your reciprocating Engines, and has waited delayd the erection of it on your account, because he says that he would be sorry to put the world in possession of an improvement of which you are the Author. He is however in very great want of it and [may] be obliged by necessity to finish it. He therefor earnestly begs that for his sake as well as your own you would lose no time. [49]

Robison met Boulton before Boulton had met Watt so Boulton must have been impressed with what he learnt about Watt and his two steam engines, reciprocating and rotary, from others. Darwin may have given Boulton an enthusiastic description of Watt's perfect engine which may have convinced Boulton of its scientific principles. Boulton's own opinion was strengthened when he actually met Watt in August 1768 and their lasting friendship was formed.

During the following months of 1768, Watt and Small corresponded principally over the text of the patent for the separate condenser which was sealed on 5 January 1769. [50] It was not until the beginning of that February when Boulton wrote to Watt that we can begin to learn what were his motives:

> I note what you say in respect to your connection with Dr. Roebuck, from whom I received a letter, dated the 12th December, offering me a share of his property in your engine, as far as respects the counties of Warwick, Stafford, and Derby. I am obliged to you and him for thinking of me as a partner in any degree; but the plan proposed to me is so very different from that which I had conceived at the time I talked with you upon that subject, that I cannot think it a proper one for me to meddle with, as I do not intend turning engineer. I was excited by two motives to offer you my assistance - which were, love of you, and love of a money-getting, ingenious project. I presumed that your engine would require money, very accurate workmanship, and extensive correspondence, to make it turn out to the best advantage; and that the best means of keeping up the reputation, and doing the invention justice, would be to keep the executive part out of the hands of the multitude of empirical engineers, who, from ignorance, would be very liable to produce bad and inaccurate workmanship; all which deficiencies would affect the reputation of the invention. To remedy which, and to produce the most profit, my idea was to settle a manufactory near to my own, by the side of our canal, where I would erect all the conveniences necessary for completion of engines, and from which manufactory we would serve all the world with engines of all sizes. By these means and your assistance we could engage and instruct some excellent workmen, who (with more excellent tools than would be worth any man's while to procure for one single engine) could execute the invention 20 per cent. cheaper than it would be otherwise executed, and with as great a difference of accuracy as there is between the blacksmith and the mathematical-instrument-maker. It would not be worth my while to make for three counties only; but I find it very worth my while to make for all the world.
>
> What led me to drop the hint I did to you was the possessing an idea that you

wanted a midwife to ease you of your burthen, and to introduce your brat into the world, which I should not have thought of if I had known of your pre-engagement; but as I am determined never to embark in any trade that I have not the inspection of myself, and as my engagements here will not permit me to attend any business in Scotland, and as the Doctor's engagements in Scotland will not, I presume, permit his attendance here, and as I am almost saturated with undertakings, I think I must conclude -. No, you shall draw the conclusion: yet, nevertheless, let my conclusions be what they will, nothing will alter my inclinations for being concerned with you, for rendering you all the service in my power; and although there seem to be some obstructions to our partnership in the engine trade, yet I live in hopes that you or I may hit upon some scheme or other that may associate us in this part of the world, which would render it still more agreeable to me than it is, by the acquisition of such a neighbour...

Query – is the wheel-engine invention also included in your transfer to the Doctor, or can it be separated from it with propriety? [51]

What were Watt's feelings when he received this letter? He could not fail to have been impressed with Boulton's enthusiasm and vision.

Boulton had very astutely summed up the situation. He had recognised that Watt's steam engines would need high quality craftsmanship to make them work effectively. This would need good machinery and hence capital investment in workshops. Such high standards could only be guaranteed with properly supervised production in their own premises. To receive any profit on all this outlay, they would have to capture the whole market. There is no doubt that Boulton saw such a scheme as a sound speculation, but he was also willing to incur the risks for the sake of his friendship with Watt. We must note that Boulton's assessment of the way to exploit Watt's invention was very different from the consultancy-based type that Watt had evolved in Scotland. Although Boulton cleverly rejected any partnership on these terms, he left the door open so that discussions continued on and off over the next five years.

Boulton had expressed interest in Watt's rotary engine, more usually called his steam-wheel. Boulton needed power to turn his machinery and recognised that greater economy and efficiency would be achieved if this could be done directly rather than by recirculating water over a waterwheel. He also would have known that his fellow manufacturers would be interested in a power source to replace expensive horses, both to drive machinery as well as pulling vehicles. Indeed, Darwin had corresponded with him about a 'fiery chariot' in 1764. [52] Therefore Boulton envisaged a wider market for a rotary engine over the reciprocating or pumping type and cajoled Watt into producing a rotative engine in the 1780s.

As early as February 1766, Watt told Roebuck about his simple form of a circular steam engine. [53] He made models of various types and by the beginning of 1768 was concentrating on his 'steam-wheel' which showed the most promise. [54] (See Vol. I, pp. 424 – 439) The steam-wheel consisted of a hollow tubular rim which was divided into three sections by flap valves which opened one way only. A heavy weight, such as mercury, filled part of the bottom. Steam was admitted by another valve through one of three hollow arms into the space between a flap valve and the weight, causing the weight to rise in one direction until it had reached sufficient height to overcome the resistance. Through the steam continuing to act against the weight and that closed flap valve, the wheel turned in the opposite way to the rising weight. The mercury could pass through a flap valve, which allowed admission of steam to the next section while the initial charge of steam was exhausted through another valve at the end of the appropriate arm. A detailed description of the steam-wheel formed the fifth claim in the

1769 patent although no drawing was included. Small made suggestions about this description as well as other parts of the patent text and possibly may have constructed a model for experiments. He pointed out that Richard Edgeworth was pursuing similar ideas and that a London linen draper, Moore, had taken out a patent for moving carriages by steam. It may have been this which made Watt realise that he could not develop both his reciprocating engine and the steam-wheel at the same time. Accordingly he wrote to Small at the end of April 1769:

> You want a steam-chaise; pray make one. I give you leave, and will also give you advice, – not gratis, however, but for good deeds done and to be done. Seriously, you will oblige me if you will negotiate the following affair: I find that if the engine succeeds, my whole time will be taken up in planning and erecting the reciprocating engines, and the circulator must stand still, unless I do what I have done too often, neglect certainty for hope. Now Mr. Boulton wants one or more for his own use. If he will make a model of it, of 20 inches diameter, at least, I will give him my advice and as much assistance as I can. He shall have liberty to erect one of any size for his own use, and if he should choose more, the terms will be easy, and I shall consider myself as much obliged to him. If it should answer, and he should not think himself repaid for his trouble by the use of it, he shall make and use until he is paid. If this is agreeable to him, let me know, and I will propose it to the Doctor [Roebuck], and doubt not of his consent. [55]

Watt gives us a small insight into his attitude to his career so far when he said that he had too often neglected 'certainty for hope', probably referring to his unsuccessful experiments with his steam engines.

It was only after the following Christmas that Small approached Watt:

> I am not very anxious about reciprocating engines since I have considered more carefully the circulating; one of which, equal to four horses, I can get made for fifteen guineas or less. Dr. R. is still in London, Pray let us lose as little time as possible after his return to you, for fourteen years will soon be gone. [56]

Agreement was reached quickly and construction commenced at Birmingham. Work continued on the project intermittently throughout 1770 and 1771. It was hoped to hold trials in September 1772 but probably they never took place. In February 1774, while they had still tried no actual experiments, Small reported that nearly everything was ready but nothing more would be done until Watt himself arrived. [57] There are no records to show how much the steam-wheel had cost so far but it helped to cement the links between Watt in Glasgow and his friends in Birmingham.

Now Watt's largest experimental reciprocating engine mouldering at Roebuck's house at Kinneil comes to the fore. Watt based this one on Newcomen's layout with a vertical steam cylinder below one end of a rocking beam from the far end of which were suspended the water pumps. Small had been pressing Watt to come to Birmingham for a long time. As far back as September 1770, he had written that he hoped he would soon see both Watt and Roebuck there. [58] After work had stopped on the Monkland Canal in 1772, Small wished that Watt would pass all or part of the winter with him. [59] By this time, Roebuck's financial position had worsened once more and even Boulton and Small were adversely affected. A meeting of Roebuck's creditors was called for 3 April 1773. Among them were Watt and Boulton. Boulton gave Watt the authority to act for him. Watt and Roebuck met at Kilsyth where they were able to settle their accounts on 15 May 1773 and discharged each other. [60] (See Vol. I, p. 446) Roebuck allowed the £178 which Watt had advanced him to be assumed as payment for the engine at Kinneil. So armed with his discharge, Watt set off immediately to Kinneil to rescue what was now his own property, his engine. Obviously he wanted to secure it before there could be any change of mind by Roebuck or complaints by Roebuck's creditors.

Watt was employed at Kinneil from 17 to 21 May 1773 dismantling and packing up his engine. He ordered that it should be sent by the first ship to London. [61] He wrote to Small on the twentieth:

> As I found the engine at Kinneil perishing, and as it is from circumstances highly improper that it should continue there longer, and I have nowhere else to put it, I have this week taken it in pieces and packed up the ironworks, cylinder, and pump, ready to be shipped for London in its way to Birmingham, as the only place where the experiments can be completed with propriety. I suppose the whole will not weigh above four tons. I have left the whole wood-work until we can see what we are to do, conceiving it not to be worth carriage. I would not have been in such a hurry sending it off without consulting you, but it could not be delayed, for reasons I shall advise at meeting. [62]

Watt presented not only Small but also Roebuck's creditors at their meeting on 29 May with a fait accompli. [63] At the beginning of July, Roebuck confirmed that the parts had been shipped in a dozen boxes on the *Duchess of Hamilton*. There had been some dispute about the rate for the pump and the boiler which were sent as well. [64] Later Small said that the parts ought to have been sent directly to Gainsborough on the Trent. [65] They would have to be routed from London back to Gainsborough. By this time, Watt was apprehensive about his precipitate actions and told Small, 'If you do not choose to have the engine to Birmingham, as I think you ought, I must beg you will cause care to be taken of it upon my account till I see you'. [66] Luckily for the future of the steam engine, Small did not take umbrage and accepted the situation, replying in August:

> Pray come to us and complete the engines, for they are now both at Soho; and let us endeavour to find some plan that may be beneficial to yourself and the Doctor, and just to his creditors, for it seems to be impossible to pitch upon such a plan without meeting. The wheel-engine might be tried in a very few days after your arrival, at least in the forcing way. [67]

While the engine from Kinneil had arrived at Birmingham in a remarkably short space of time, Watt followed it only after a long delay caused by his surveying commitments and the death of Peggy. Small became increasingly anxious, especially as he heard that Watt was considering spending the winter in France. He wrote in December, 'I ever since have been in daily expectation of seeing you in your journey, and now you send a letter that says nothing either of going to France or of coming hither... Shall you come soon or not?' [68] However, not only were there Watt's further experiments on steam to be completed as well as his surveying business to be wound up but Boulton's settlement with Roebuck had to be concluded. So at the end of March 1774, Small was still waiting in hopes of seeing Watt; 'If you judge the title to the engine to be clear, and your experiments to be sufficiently advanced for a reasonably good construction of it, come hither as soon as possible'. [69]

While both engines might be at Birmingham, Roebuck still retained his partnership with

Watt which entitled him to two-thirds of any profits. [70] Watt wished to see Roebuck's share transferred to Boulton but, while Roebuck's creditors valued the engine at only a farthing, [71] Roebuck himself still hoped that it might be a success and that he would receive some of the profits. Watt realised that Roebuck had borrowed money from Boulton and Fothergill jointly as well as from Boulton himself. Therefore Watt suggested to Boulton that he could trade these debts for Roebuck's share in the partnership, which caused problems with Fothergill later. Delicate negotiations followed during the summer of 1773. One of the difficulties was that Boulton did not wish to have to pay Roebuck any of the profits if this speculation succeeded. Small wrote to Watt at the end of September:

> I have pressed our friend B., until I am quite ashamed to reply to your last letter, which enclosed Dr. R.s proposal to his creditors; but he has not been able hitherto to force himself to write. Nor can I tell you what he probably will write, nor when, for this is an affair of too much delicacy for me to inquire into. [72]

The situation had not changed in December when Small told Watt, 'Dr. R. has not yet been here, so we are quite where we were about the engine'. [73] It seems that Boulton did not write to Roebuck nor did Roebuck meet Boulton before Watt left Scotland so there was no formal transfer of the partnership until after the Royal Assent had been given in May 1775 to the Act of Parliament extending Watt's 1769 patent for twenty-five years. [74] Watt arrived in Birmingham with the terms of his partnership with Boulton unresolved. All he could offer were schemes for two engines which had not yet proved capable of working, a patent which had only nine years left to run and virtually no finance with which to support himself or his schemes. Yet he maintained his faith that his invention of the separate condenser would prove to be one of the greatest advances in performance of the steam engine ever made. While this was true, he could never know that his concept of the perfect steam engine was flawed. He considered that heat was some sort of chemical substance and regarded the engine as working by the pressure of the steam. He thought it was essential to keep the cylinder as hot as the incoming steam all the time to avoid condensation and hence loss of pressure. But he could not know that heat is a form of energy which, when it is transformed into work and drives machinery, must decrease in temperature with the result that the cylinder must become colder during the working stroke. The reason for this dilemma was recognised only in the middle of the nineteenth century with the rise of thermodynamics and was resolved only partially with the uniflow engine at the beginning of the twentieth.

NOTES

1 B. & W. Col., MI, 1/21, 'Journal', 31 May 1774.

2 Ibid, MI, 1/20, 'Journal', 29 Sept. 1773.

3 Muirhead, (1854), Vol. II, p. 62, Letter 123, J. Watt to W. Small, no date; JWP 4/59.1.

4 Ibid, p. 63, Letter 124, W. Small to J. Watt, 8 Oct. 1773; Boulton 340.35.

5 Hills, (2002), p. 106.

6 B. & W. Col., MI, 1/16, 'Journal', 22 – 25 Dec. 1770.

7 Ibid, MI, 1/15, 'Journal', various dates; see Hills, (2002), p. 104.

8 Robinson & McKie, (1970), p. 28, Letter 23, J. Black to J. Watt, 22 March 1772 and Hills, (2002), p. 80.

9 JWP, C3/8, 'Diary', 8 June 1784 and B. & W. Col., MIII, 3/4, 'Personal Ledger', f.31, 'This amount was settled & Dr. Black paid in full since J.W. came to England about 1785'.

10 Robinson & McKie, (1970), p. 142, Letter 99, J. Watt to J. Black, 6 June 1784.

11 B. & W. Col., MI, 1/15, 'Journal', 27 June 1770.

12 Ibid, 24 and 28 – 9 May 1770.

13 Ibid, MI, 1/19, 'Journal', 13 April 1773.

14 JWP, 4/20.21, G. Hamilton to J. Watt, 15 Nov. 1775.

15 For a full account, see *Proceedings of the Society of Antiquaries of Scotland*, Vol. 131, 2001, pp. 375 – 412, Hills, R.L., 'James Watt and the Delftfield Pottery, Glasgow'.

16 *Scottish Historical Review*, Vol. XXIX, No. 108, October 1950, p. 131, Thomson, George, 'James Watt and the Monkland Canal'.

17 Muirhead, (1854), Vol. II, p. 67, No. 127, J. Watt to J. Small, 11 Dec. 1773; Boulton 348.38.

18 Ibid, p. 45, No. 110, W. Small to J. Watt, 3 April 1773; Boulton 340.30.

19 For the various Acts, see Priestley, (1969), p. 63 for Birmingham Canal Navigations and p. 165 for the Coventry Canal. See also Hadfield, (1966), pp. 65 – 6 and 68.

20 Muirhead, (1854), Vol. II, p. 71, No. 129, W. Small to J. Watt, Dec. 1773; Boulton 340.38.

21 Ibid, p. 78, No. 134, J. Watt to J. Small, 29 April 1774; Boulton 348.42.

22 JWP, JW/22.43, W. Craig to J. Watt, 23 Aug. 1781.

23 JWP, 6/21.10, J. McGrigor to J. Watt, 8 Nov. 1784.

24 JWP, Letter Book 1, p. 86, J. Watt to J. McGrigor, 30 Oct. 1784.

25 Ibid, p. 95, J. Watt to J. McGrigor, 21 Nov. 1784.

26 Hills, (2002), p. 268.

27 Ibid, p. 245.

28 Muirhead, (1854), Vol. II, p. 34, Letter 102, J. Watt to W. Small, 24 Nov. 1772; Boulton 348.25.

29 JWP, 4/50, J. Watt to Ld. Provost, 17 Aug. 1771.

30 SRA, B10/15/7640, 'Contract betwixt the City of Glasgow and William Anderson and Robert Adam', reg. 3 Aug. 1773 and SRA, B/10/15/7590, 'Contract betwixt the City of Glasgow and James Johnston', reg. 5 April 1773.

31 Hills, (2002), p. 273.

32 JWP, 4/19.8, J. Montgomerie to J. Watt, 25 Dec. 1770.

33 Smiles, (1878), p. 98.

34 Hills, (2002), p. 359.

35 Dickinson, (1936), p. 203.

36 Goodison, (2002), p. 44.

37 Smiles, (1878), p. 127.

38 King-Hele, (1999), p. 49 and Crowther, (1962), pp. 145 – 8.

39 Dickinson, (1936), p. 208 and King-Hele, (1999), p. 49.

40 Goodison, (2002), p. 16, from James Keir, 'Memorandum of Matthew Boulton', 3 Dec. 1809, p. 1.

41 King-Hele, (1981), p. 46, Letter 67D, E. Darwin, to J. Watt, 18 Aug. 1767.

42 Goodison, (2002), p. 12, from Camden, W., *Britannia*, 1586, Vol. I, p. 609.

43 Schofield, (1963), p. 16, from Hutton, W., *A History of Birmingham to the End of the Year 1780*, Birmingham, 1781, p. 63.

44 Goodison, (2002), p. 13, from Prosser, R.B., *Birmingham Inventors and Inventions*, Birmingham, 1881, p. 3.

45 Ibid, p. 15 from Mare, M.L. & Quarrell, W.H., *Lichtenberg's Visits to England as described in his Letters and Diaries*, Oxford, 1938, pp. 98 – 9.

46 Smiles, (1878), p. 130.

47 See *Transactions of the Newcomen Society*, Vol. 58, 1986 – 87, p. 34, Hills, R.L., 'A Steam Chimera: A Review of the History of the Savery Engine'.

48 King-Hele, (1981), p. 38, Letter 66B, E. Darwin to M. Boulton, 11 March 1766.

49 Robison & McKie, (1970), p. 13, Letter 6, J. Robison to J. Watt, 8 July 1768.

50 Patent 913, 5 Jan. 1769.

51 Muirhead, (1854), Vol. I, pp. 41 – 3, Letter 43, M. Boulton to J. Watt, 7 Feb. 1769; JWP, 6/19.23.

52 King-Hele, (1981), p. 27, Letter 64E, E. Darwin to M. Boulton, [1764].

53 Muirhead, (1854), Vol. I, p. 113, Letter 15, J. Watt to J. Roebuck, 19 Feb. 1766; JWP, 4/32.30.

54 Hills, (2002), pp. 424 – 439 for a shorter version of the paper in *Transactions of the Newcomen Society*, Vol. 70, No. 1, 998 – 99, pp. 89 – 108, Hills, R.L., 'James Watt and his Rotary Engines'.

55 Muirhead, (1854), Vol. I, p. 55, Letter 54, J. Watt to W. Small, 28 April 1769; Boulton 348.8.

56 Ibid, p. 84, Letter 71, W. Small to J. Watt, 26 Dec. 1769; Boulton 340.11.

57 Ibid, Vol. II, p. 74, Letter 131, W. Small to J. Watt, 23 Feb. 1774; Boulton 340.39.

58 Ibid, p. 4, Letter 84, W. Small to J. Watt, 17 Sept. 1770; Boulton 340.14.

59 Ibid, p. 27, Letter 99, W. Small to J. Watt, 29 Oct. 1772; Boulton 340.23.

60 Hills, (2002), p. 446.

61 B. & W. Col., MI, 1/20, 'Journal', 17 – 21 May 1773.

62 Muirhead, (1854), Vol. II, p. 49, Letter 114, J. Watt to W. Small, 20 May 1773; Boulton 348.31.

63 B. & W. Col., MI, 1/20, 'Journal', 29 May 1773.

64 B. & W. Col., JWP, 4/57.22, J. Roebuck to J. Watt, 6 July 1773.

65 Muirhead, (1854), Vol. II, p. 50, Letter 115, W. Small to J. Watt, 2 June 1773; Boulton 340.32.

66 Ibid, p. 56, Letter 118, J. Watt to W. Small, 25 July 1773; Boulton 348.35.

67 Ibid, p. 57, Letter 119, W. Small to J. Watt, 8 Aug. 1773; Boulton 340.33.

68 Ibid, p. 71, Letter 129, W. Small to J. Watt, Dec. 1773; Boulton 340.38.

69 Ibid, p. 76, Letter 133, W. Small to J. Watt, 29 March 1774; Boulton 340.40.

70 Ibid, pp. 47 – 8, Letter 113, 'Discharge by Mr. Watt to Dr. Roebuck', 17 May 1773; JWP, 4/57.21 and B. & W. Col., MI, 1/19, 'Journal', 13 April 1773.

71 Muirhead, (1854), Vol. II, p. 56, Letter 118, J. Watt to W. Small, 25 July 1773; Boulton 348.35.

72 Ibid, p. 60, Letter 122, W. Small to J. Watt, 29 Sept. 1773; Boulton 340.34.

73 Ibid, p. 72, Letter 129, W. Small to J. Watt, Dec. 1773; Boulton 340.38.

74 Tann, (1981), p. 34, Doc. 4, 'Copy of Award between J. Roebuck and M. Boulton', 29 May 1775.

2
Prototype Engines and the Patent Act of 1775

Freed from the constraints of his civil engineering career, Watt could return to his first love, his steam engines. Boulton's support at last gave him some security so, on his arrival in Birmingham, his immediate task was to make both his engines operate successfully. He also needed to test other engines to confirm the claims he was making about the performance of his own. It was soon realised that it would take time and money to complete his engines and hold trials so that the 1769 patent might have expired before there would be any profit on the costs and capital expended. Small had pointed this out as far back as December 1769. [1] Therefore a way had to be found to prolong the patent. In 1775, Boulton and Watt decided to petition Parliament for a special Act to protect Watt's invention of the separate condenser in order that they could profit from it.

Watt's Perfect Engine

Watt claimed that he had invented the perfect engine and that the chief attraction for anyone using his patent ideas would be greater economy in steam and fuel. The principles on which these claims were based were stated in his 'Plea Before Parliament' in 1775:

> Watt's improvement of the steam engine consists of two leading principles & some other subsidiary ones. These two principles which constitute its greatest excellence & distinguish it in a striking manner from all other steam engines whatsoever, are, first the keeping the vessel or great cylinder wherein the powers of steam are to be exerted to work the machine always as hot as the steam itself, and suffering no cold water to enter it. Secondly, the procuring a vacuum in the cylinder by opening a passage from it into another vessel previously exhausted of air and kept very cold into which the steam rushes by its elasticity or expansive quality, and is instantly by the cold of that vessel reduced into water, and the cylinder is left empty without being cooled. [2]

He had arrived at these principles through a series of experiments starting in 1764 and carried on intermittently until he left for Birmingham ten years later. The equipment he used was quite simple. With a thermometer he could measure the temperature of the steam and that of the condensate. The tables he drew up relating the pressure of steam to its temperature enabled him to deduce the pressures in his experimental engines which showed him how inadequate was the vacuum being achieved on condensation in the Newcomen engine. A pressure gauge in the form of either a closed mercurial barometer or an open manometer would show the highest and lowest pressures but not easily intermediate ones. A balance or scale beam could weigh measured amounts of water or fuel to determine consumption. So his theories were based on various experiments. The concept of his perfect engine came as he was walking in the Green of Glasgow one Sunday afternoon in the spring of 1765 with his inspiration of condensing the steam in a separate vessel from the main steam cylinder. His experimental engines in Scotland had shown that application of this principle was achieving considerable economy but he had been unable to construct an engine which worked to his satisfaction. Yet these trials revealed methods which were failures, leaving him to pursue those which showed more promise in Birmingham.

We have a good summary of progress in applying these principles up to July 1775 in a description of his engine sent to J.D.H. van Liender in the Netherlands. Watt started by describing the steam cylinder of his reciprocating engine.

My engine consists of a cylinder and piston and works the pumps by a great lever as usual. The improvement consists in keeping the cylinder always of the same heat with steam from the Boiler; by covering it with materials which transmit heat slowly and by using no injection of water into the Cylinder by these means the Cyl is filled by less than half the quantity of steam necessary in common engines. For them the cylinder is cooled by the water thrown in to condense the steam and w[hen] fresh steam enters it destroys or reduces to water the greatest part of what is inte[nded] to fill it; but in this new engine the Cylinder is never cooled so it can conden[se] no part of the steam and is thereby more easily filled. [3]

Diagram of Watt's Single-Acting Pumping Engine.

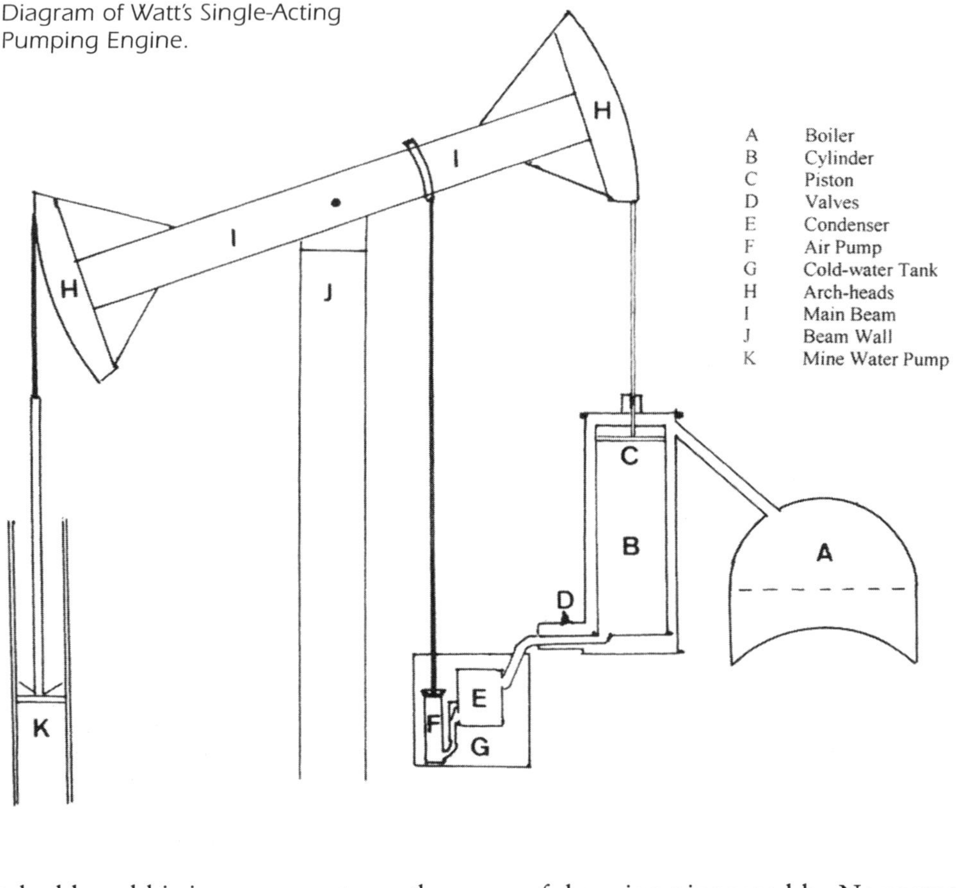

Watt had based his improvements on the successful engine pioneered by Newcomen which worked with two phases, or cycles, the steam phase when the piston was ascending and the vacuum phase when it was descending and raising water in the pumps. For the first phase, Watt had determined in his early experiments the volume of steam produced from a given amount of water at atmospheric pressure, a ratio not known accurately before. He had constructed a boiler which showed how much water had been evaporated. This enabled him to calculate how much steam had been used for a given number of strokes. He had found that this was much greater than the volume of one cylinder-full for each stroke, so he conceived his perfect engine as one which would need only one cylinder-full per stroke. Having determined the relationship between the pressure of steam and the temperature, he realised that pressure increased faster than temperature which might give greater economy if higher pressures could be used. If he knew either the pressure or the temperature, he could determine how his engine was performing.

To achieve his perfect engine, he encased in a steam jacket as much as possible of the cylinder and valve gear, or regulators as he called the valves, to keep these parts as hot as the incoming steam. He placed a cover over the top of the cylinder through which the piston rod passed so that no cold air could enter the cylinder as the piston descended. The jacket was fed with steam by a pipe from the boiler and it communicated directly with the top of the cylinder and piston under the cover. Watt had seen the advantage of using the pressure of the steam to push the piston down, both to avoid ingress of cold air and to take advantage of steam at higher pressures to give a more powerful engine. Hence he regarded his engine as working by the pressure and not the heat of the steam. On his first experimental engine in Birmingham, the jacket was made from wood which insulated the steam on the inside from the cold air on the outside. Watt even tried a steam casing below the cylinder base. In these ways he avoided any condensation of steam, and hence waste, when the steam entered the cylinder and performed the working stroke. This was where his greatest saving lay.

The piston was pushed down the cylinder by the steam against a vacuum. Watt pointed out to van Liender:

> Secondly in common fire engines the vacuum is not made perfect, because were [they] to throw in so much cold water as would compleately condense the Steam they wo[uld] lose so much more steam when they filled the cylinder over again that they would find themselves greater losers in point of Steam than they were gainers in point of power. They therefore generally confine their power to 7 lb of a column of water for each square Inch of the Piston and even that is more than they ought to use. To remedy this inconvenience I condense the Steam in a vessel appropriated to th[at] purpose which communicates with the Cylinder during the time of condensation only. This vessel I call a condenser the engine keeps it perpetually exhausted of air & wa[ter] by a pump and the steam rushes from the Cylinder into it with great violence upon opening a valve for the purpose when the condensation is wanted and the steam is there made as cold as is necessary to make the vacuum compleate. An engine I have made here is loaded to 13 lb pr. Inch and uses only one third of the fuel of a common engine. [4]

It had been known that hot water continued to boil when placed in a partial vacuum but this had never been quantified until Watt carried out experiments. Using the figures he had determined for the relationship between steam pressure and temperature above atmospheric pressure, he extrapolated these to pressures below atmospheric level and was surprised to discover that he had to cool the condenser to well below 100°F. To begin with at Birmingham, he continued to use surface condensers surrounded by cold water because he recognised that they would not require so much condensate and water to be pumped out of them as well as less air. By condensing in a separate vessel which could be kept cold all the time, not only could he achieve a near perfect vacuum, but he avoided heating and cooling these parts, so saving steam once again. He had determined the specific heats of various metals so he could calculate the amount of steam required to heat them. This showed him yet another source of waste incurred in alternately heating and cooling the cylinder of a Newcomen engine. However this information became redundant for his own engine where the various parts did not vary in temperature. (See Vol. I, pp. 303 – 75 for Watt's experiments)

Watt had also stumbled across the phenomenon of latent heat which Black explained to him. This pointed to the enormous amount of heat or fuel necessary to generate steam, so a reduction in steam consumption would effect a proportionate reduction in fuel consumption. Knowing the value of latent heat enabled Watt to calculate the amount of cooling water necessary for condensation. With all these figures, he could follow the passage of steam through his engine so that his calculations showed where it was failing to achieve perfection. Through his theoretical knowledge, he could continue to improve his 'perfect' engine by trial and error during the next few years but he kept the results of his experiments to himself for his figures were not published until after his death when they appeared in Robison's *Mechanical Philosophy* in 1822. [5] Armed with his scientific knowledge of steam and his theory of how his steam engines should perform, Watt had to improve not only the accuracy of the bore of the cylinder of the reciprocating engine but also to devise a piston that would be steam tight. A practical method of condensing had to be found as well as satisfactory air pumps to remove the condensate. So his 'perfect' engine still needed a great deal of improvement when he arrived in Birmingham.

The Steam-Wheel

Little information has survived about the period from June 1774 to the beginning of 1775 because Watt, Boulton and Small were together in Birmingham and Watt discontinued his Journal. The steam-wheel was most likely to have been ready first for Small had written in February 1774:

> We have tried no experiment with the circular engine. The engine is mounted, the boiler is ready, the piece that connects the boiler to the steam-pipes is also ready, but not ground to exclude air; lastly, the quicksilver is ready. The valves were ground very carefully, but whether they are steam-tight we know not.
>
> Probably nothing more will be done before you come, for neither Mr. B. nor I have leisure to attend the trials or the workmen; and to employ workmen on articles of this kind, without super-intendence, we have experienced to be too expensive. Besides, you ought to see all the facts that may occur. [6]

Watt quickly carried out some trials. (See Vol. I, pp. 424 – 36 for a full account of the steam-wheel) In an undated letter ascribed to July 1774, Hutton told George Clerk that Watt 'has brought his curious wheel to go, it works by steam - he says it appears to be right in all essentials and goes with an equable motion & great power'. [7] Then in November Roebuck asked to be informed of the experiment when Watt tried the 'fusible metal in your circular machine'. [8] Trials continued with some success. In the publicity pamphlet prepared for the Act of Parliament, Watt described his steam-wheel. (See p. 44):

> The second species of the improved engines is a machine which produces a circulating or rotative motion by the immediate action of steam; it consists of a wheel placed vertically, which has a hollow channel round its circumference, as if a tube were bent into a circle and its two ends were joined together...
>
> The advantages of this particular construction are evident. It gives the means of producing rotative motions of all kinds by steam, and by the equality of its velocity is better adapted to many purposes than those engines which have a reciprocating or alternate motion. In point of power gained from a given quantity of steam, it has all the advantages of the reciprocating species, as the steam vessel is kept uniformly hot, and the steam is condensed in a distinct vessel in the same manner as the other. [9]

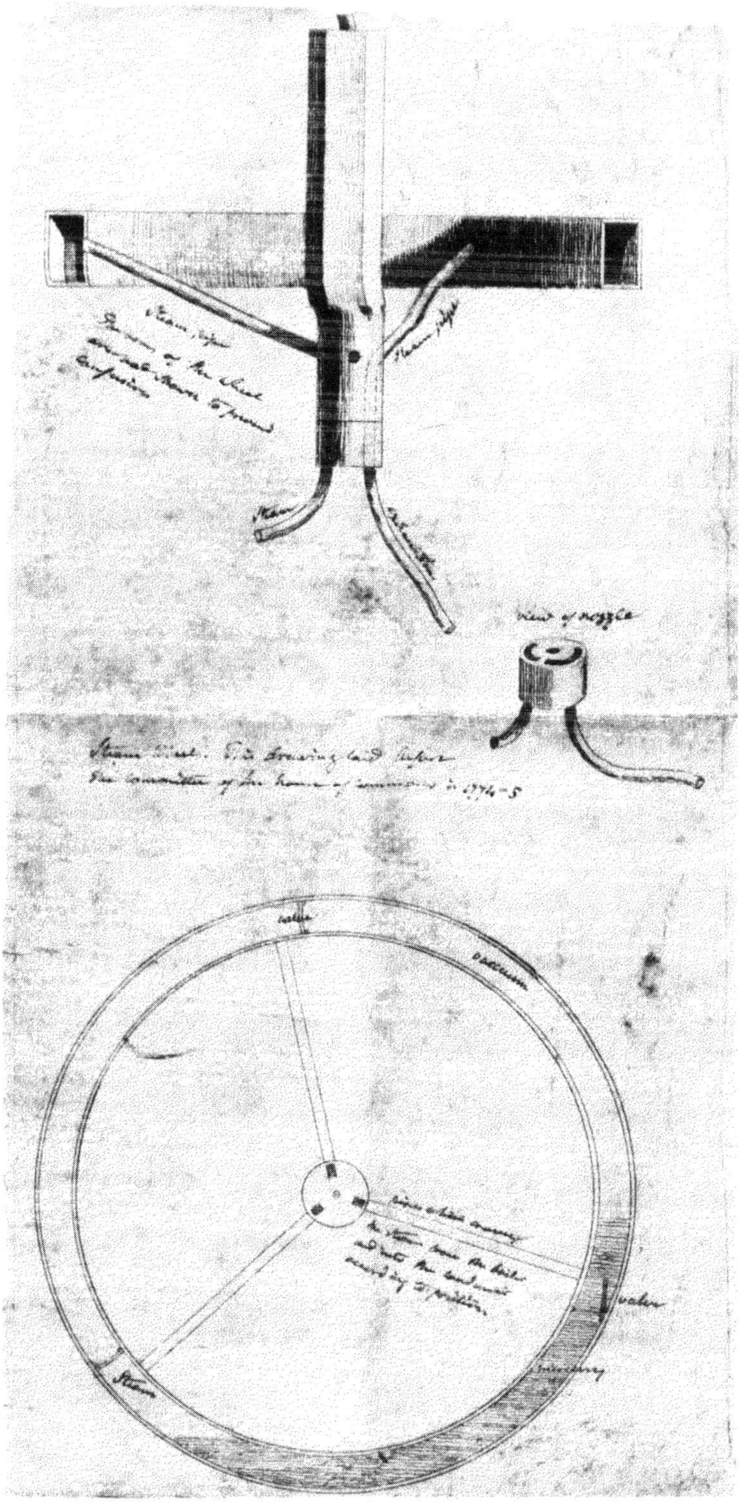

The steam-wheel shown in a drawing laid before Parliament, 1775.

At the time, it appeared to promise better prospects than the reciprocating engine.

During the questioning by the second Parliamentary Committee in 1775, Joseph Harrison, one of the smiths at the Soho Manufactory, gave evidence that he had seen a six foot diameter steam-wheel moving with great velocity. He and another man could not hold it when starting from rest. It weighed over a ton. [10] His evidence was confirmed by Samuel Garbett who had gone out of curiosity to see it. [11] Neither mentioned any condensing apparatus, so steam at a fairly high pressure must have been used. Their accounts were sufficiently convincing for the steam-wheel to be included in the Act.

The Reciprocating Engine

Watt started to erect the reciprocating steam engine which he had despatched from Kinneil so that it would act as a water-returning engine for the waterwheels at the Soho Manufactory. Since he had left all the wooden parts behind, these had to be manufactured at Birmingham so some time elapsed before it was ready. Some dimensions were reported by Harrison to the Parliamentary Committee which presumably were the same as at Kinneil. The steam cylinder was 18 ins. diameter with 5 ft. stroke. The pump was 18½ ins. diameter with 6 ft. stroke, having a lift of 44 ft. This is the figure quoted but the more likely one should be 24 ft. Therefore the pivot of the main beam was not placed centrally. It consumed half a hundredweight of coal in an hour, when making 800 strokes recorded on a special counter. [12] The cylinder, made from a sheet of tin bent to shape, was surrounded by a wooden casing which acted as a steam jacket. The combined steam and exhaust valve was a form of sliding box. Although Watt had experimented at Kinneil with a jet of cold water into the eduction pipe to condense the exhaust steam, he found that a great deal of air was drawn in owing to faulty construction so in Birmingham he used surface condensers. [13] He fitted two air pumps to remove the condensate and air from the condenser.

Unfortunately Watt's log book, 'Experiments on ye first engine at Soho' seen by H.W. Dickinson, has not been traced. In it were records of number of strokes made, weight of coal burnt, water evaporated and so on. [14] Trials had started before November 1774 because Watt had been using one of his early plate surface condensers and Roebuck expressed surprise 'that it should work at all, considering the slightness of the materials and its long exposure to the injuries of the weather'. [15] He also wondered whether the willow piston was keeping tight after it had worked for some weeks. Watt had been trying various types of piston and packings.Watt was able to send a good report on progress to his father in early December; 'The business I am here about has turned out rather successful: that is to say, that the fire-engine I have invented is now going, and answers much better than any other that has yet been made; and I expect that the invention will be very beneficial to me'. [16] On 14 December, the condenser was in good order and was being worked 'by a beam by the engine'. Earlier ones at Kinneil had been worked by hand or a waterwheel. On 23 December, the 'woollen piston being found wore out & torn by the baked oil sticking to the cylinder, it was taken out and a new wooden one put in... the piston was covered with horse muck'. [17] Perhaps it came as no surprise that that failed and the horse muck was replaced with paper pulp mixed with flour paste. These were only a few of the packing materials tried. On 27 December, a serious failure occurred when 'the tin cylinder had been forced in by the steam'. In 1769, Watt had told Small that this cylinder 'was a full quarter inch thick and 18 inch diameter. At the worst place the long diameter exceeded the short about ⅜'. [18] Ovality of this dimension would have been almost impossible to correct with the types of packing they were using.

A new cast iron cylinder was quickly ordered from the iron founder John Wilkinson. In 1774, Wilkinson had patented a method for casting cannon solid and then machining the

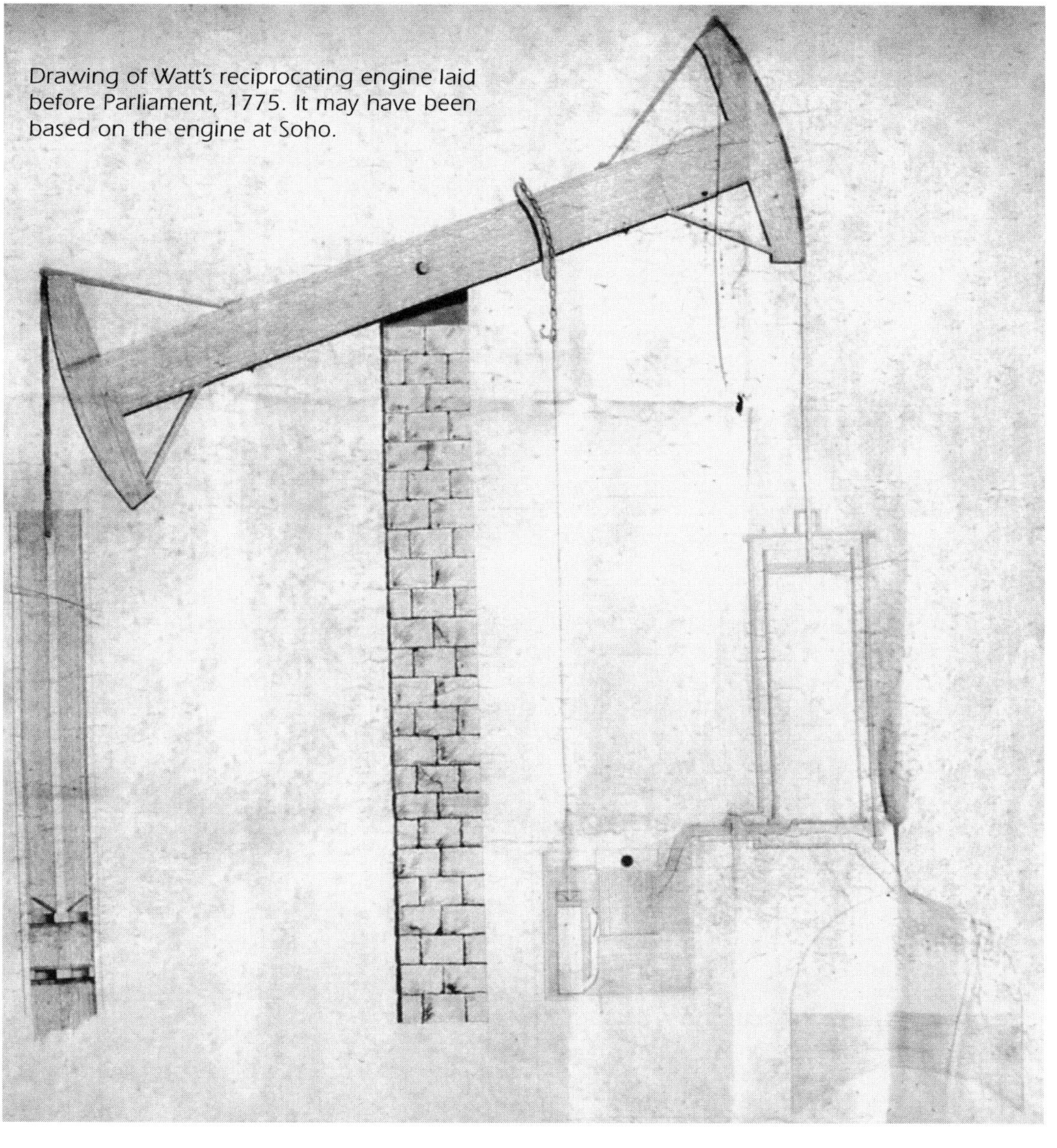

Drawing of Watt's reciprocating engine laid before Parliament, 1775. It may have been based on the engine at Soho.

bore. [19] He quickly developed a different machine for boring the large hollow cylinders for steam engines. In these, a boring bar to hold the cutting head was passed through the cylinder and mounted steadily in bearings at either end. As the cutting head rotated, it was advanced along the cylinder bore. This machine was far more accurate than that designed by Smeaton for the Carron Foundry. (See Vol. I, p. 325) Watt pointed out to Smeaton that Wilkinson had 'improved the art of boring cylinders so that I promise upon a 72-inch cylinder being not farther distant from the absolute truth than the thickness of a thin six pence at the worst part'. [20] Wilkinson's boring machine proved to be the answer to Watt's quest for an accurately bored cylinder, so much so that Watt insisted on his customers ordering their cylinders from him until Bersham Ironworks was closed in 1795.

Boulton had heard nothing by 26 January 1775 about the cylinder from Wilkinson when he promised Watt he would 'send him a line of stimulation'. [21] It was not until 24 April that Boulton wrote to Watt in London to tell him, 'The cast-iron cylinder is arrived; it seems tolerably true, but is an inch thick and weighs about 10 cwt.' [22] It was considerably more substantial than the tin one and continued in use for several years. Some trials continued in the intervening period with the tin cylinder until the cast iron one arrived but Watt was in London for most of that time. The fact that this engine was not working could explain Boulton's comment to Watt in February, 'I note what you say about fire-engines, and only regret that we have not one of each species ready to work, as two working models would clear their way into the House of Commons better than blind words can do'. [23]

While in his case before Parliament, Watt based his claims principally on his scientific experiments showing that he had invented the perfect engine, the engine with the tin cylinder had worked sufficiently well to convince the Parliamentary Committee that the reciprocating engine was viable. The cast iron cylinder arrived too late to do this. It may have been the failure of the tin one which emphasised the need for more trials, alterations and expense before either type would become a commercial proposition that turned the minds of Boulton and Watt to seek a way of prolonging the patent. They may have taken up a suggestion made by Hutton in 1774:

> Invention is too great a work to be well paid in a state where the general system is to be best paid for the thing that is easiest done... I admire so much your reciprocating engine. I dare say Mr. Boulton's mill will never want water in the dryest summer... I am sincerely of opinion that a short account of the machine should be made out, with a distinct estimate of its value compared with the common one, and then represent that besides the invention of this uncommon advantage, it has cost many years' labour and expense to bring it to perfection of utility, and that such is the difficulty attending this operation, that without an accurate description of the method necessary to render this invention practicable, it may remain as useless as in the head of the inventor. Therefore application should be made to Parliament and give the choice either of buying the public use and description of the invention, or of giving an exclusive privilege for a sufficient length of time, as it cannot be expected that anything worth the value can be got whilst the present engines subsist. [24]

Some of this advice was followed.

Prolongation of the Patent

Watt went to London in the middle of January 1775, primarily to discuss prolonging his patent but also to help Boulton and Fothergill in various ways, such as adjusting and marking weights they had manufactured, finding workmen and other jobs. [25] He took the opportunity of testing Newcomen engines in the London area. He had already recorded the performance of the Ranter engine in Wednesbury and found that it was loaded to 7.2 p.s.i. [26] He noted how much steam was wasted at the New River engine which had been installed by Smeaton and told Boulton that it did '1/3 in point of power & 1/2 of perfection'. [27] The engines at York Buildings were also tested at various times, the largest being loaded at only 5.73 p.s.i. However some tests on the Chelsea engine were a failure because he could not try the quantity of the injection water and 'the fellow told me lies about the height of the col[umn] of water'. [28] Yet he did record a load of 6.2 p.s.i. A later test on the engine at Soho with the cast iron cylinder gave nearly 13 p.s.i. [29] which shows the great improvement introduced by Watt.

In London, Watt sought the advice of Alexander Wedderburn, then Solicitor-General, over his patent. This was to surrender it and apply for a new one. [30] Boulton approved of this, [31] but Watt continued with his enquiries. 'I have taken advice of several people whom I could trust about the patent; they all agree than an Act would be much better and cheaper, a patent now being £130; and the Act, if obtainable, £110'. [32] Having ascertained this, Watt returned to Birmingham in early February for a short while. [33]

After consultation with Boulton and Small, it was decided to try to obtain a new Act. So, with Small's help, Watt wrote an eight page pamphlet, *An Account of James Watt's Improvements upon the Steam or Fire Engine*, which was printed and distributed to gain support. [34] In it, Watt commenced by outlining his philosophical investigations on the nature and properties of steam and how earlier inventors of the fire engine had little knowledge of this. He stated the principles of the Savery and Newcomen engines, emphasising their deficiencies. Although Newcomen's engine had been invented seventy years ago and was superior to Thomas Savery's, it had barely been improved since. Few experiments upon the generation of steam and its condensation had been made before Watt started his in 1764. These pointed out the defects of the common Newcomen engine which he, Watt, had been able to overcome. He described his separate condenser as applied to his reciprocating engine and how this resulted in a much more powerful machine. He stated the advantages of his engine:

> First, that the cylinder being surrounded with steam from the boiler, is kept always uniformly hot as the steam itself, and is therefore incapable of destroying any part of the steam which should fill it, as the common engines do. Secondly, the condenser being kept always as cold as water can be procured, and colder than the point at which it boils in vacuo, the steam is perfectly condensed, and does not oppose the descent of the piston, which is therefore forced down by the full power of the steam from the boiler, which is somewhat greater than that of the atmosphere. [35]

This gave Watt his claim to have invented the perfect steam engine and he pointed out that the pressure on his pistons was 12 p.s.i. compared with that in the common engines of 7 p.s.i. He concluded with the account of his steam-wheel covered above. He hoped to convince people of the very real merit and value of his invention. To whom it was distributed is not known but N.D. Falck, a Dutch engineer, obtained a copy and criticised Watt's engine when describing his own. [36]

Also a *Petition to Parliament for Engine Act* was prepared as the start of lengthy proceedings at Westminster. [37] This would be submitted to the House of Commons for leave to bring in a Bill. If accepted the Bill would receive a first formal reading followed by receipt of more petitions for and against. After a second reading, the Bill would be sent to a committee for a report which would form the basis for debate at the third reading. In Watt's case, it was referred back to the committee before final acceptance. In his petition, Watt pointed out how he had spent many years and a considerable part of his fortune in making experiments to improve steam engines. But he had encountered many difficulties constructing such large and complex machines so that 'he could not complete his invention before the end of the year 1774, when he finished some large engines of his construction which have fully succeeded, and confirmed your petitioner in his opinion of the expediency of his invention'. It would take not only more investment to erect suitable manufacturing facilities for these engines but also some time to convince the public of the superiority of this invention so that 'the whole term granted by the [1769] patent may probably elapse before your petitioner can receive advantage from his labours'. He hoped that his engines would furnish 'mechanical power at less expence and in more convenient forms' and would be 'of extensive utility in facilitating the operations in many great works & manufactures of this Kingdom'. He therefore sought an extension to his patent to enable him to bring his invention 'completely into execution & obtain an adequate recompense for his labour, time, & Expence'. Armed with these pamphlets, Watt returned to London.

The Death of William Small

But then came tragedy with the death of Small on 25 February. He had been ailing for some time but rallied for a short recovery. In a long letter of 2 February 1775 to Lord Dartmouth soliciting his support, Boulton wrote:

> The spirit of my politicks hath lately been absorbed by my anxiety for my dear friend Dr. Small, who is alive yet, but almost lost hopes of recovery. His virtues were more and his foibles fewer (for vices he had none) than in any man I ever knew. The public will sustain in his death a loss. I shall sustain a great and irrepareable [sic] one... I perceive I am forgetting myself. Pray my Lord forgive me. [38]

His passing came as a shock particularly to Boulton who wrote to Watt, 'The last scene is just closed; the curtain is fallen, and I have this evening bid adieu to our once good and virtuous friend for ever and ever'. [39] Now it was Watt's turn to console Boulton:

> To pretend to offer you consolation under the weight of your present sorrow, I know to be in vain. I only beg leave to repeat to you the sentiments which that dear friend we lament expressed to me on a similar occasion. It is our duty as soon as possible to drive from our minds every idea that gives us pain, particularly in cases like this, where our grief can avail nothing. Remember, my dear Sir, that our friend enjoys that repose he so much desired...
>
> Come, my dear Sir, and immerse yourself in this sea of business as soon as possible, and do not add to the griefs of your friends by giving way to sorrow. [40]

Watt must have missed Small greatly for it had been to him that he had turned over the past half a dozen years not only for advice about his engine and civil engineering businesses but also for consolation on the death of Peggy. It has been suggested that, without Small's constant intervention, Watt would have lived out a mediocre life as a surveyor in Scotland and the steam engine would have had a radically different incarnation. [41]

Small's death coincided with another shock for Boulton because he learnt that Watt had been offered a position in Russia, partly through his singing his partner's praises to the Russian ambassador. This offer of a post with a salary of £1,000 per annum came through Robison who was then Mathematical Professor at the Government Naval School at Cronstadt. [42] Such an income would have been affluence to Watt after so many years of financial struggle, but the prospect of life in that country did not appeal to him. He had probably decided to turn down the offer by the time Darwin wrote on 29 March:

> Lord, how frighten'd I was, when I heard a Russian Bear had laid hold of you with his great Paw and was dragging you to Russia. - Pray don't go, if you can help it: Russia is like the Den of Cacus, you see the Footsteps of many beasts going thither but of few returning. I hope your Fire-Machines will keep you here. [43]

Memorial to William Small, 1734 – 1775.

Obtaining the 1775 Act

In the meantime, Watt's petition had been submitted to the House of Commons and read on 23 February. It was referred to the consideration of a committee consisting of Lord Guernsey, Sir Adam Ferguson and others. [44] Although they were due to meet the following day, it was not until 3 March that Roebuck, Alexander Cumming and Boulton were called to give evidence. Roebuck, who had by no means lost interest in the engine, responded first. His evidence was typical of that presented by others:

> That to his Knowledge £2,000 at the least has been expended in bringing the said Fire Engines to their present State of Perfection; and that if an Allowance was made for the Time that Mr. Watt has already employed, in bringing the said Engines to their present State, at all proportionable to what has been paid to him when engaged in other Business, the Expence would greatly exceed £3,000; and that to carry this Invention to the Degree of Perfection, which the Witness thinks it capable of, it will be necessary to expend some Thousand pounds more.

> That the Petitioner has been employed in this Work 17 Years, and for 10 or 12 Years last past it has taken up the greatest Part of his Time.
>
> That this Engine will not consume Half the Steam that a common Fire Engine does, probably not the Third Part, and produce an equal Effect; that it will do double the Work of a common Fire Engine at the same Expence, and be of general Utility in Collieries and Mines of all Kinds; that in the Witness's Opinion, it may be applied to many mechanical Purposes with greater Advantage than the present Fire Engine, as it works by a circulating, instead of a reciprocating Motion, and that it may be applied with Advantage wherever any Kind of mechanical Power is wanted; that he is of Opinion, the Money already expended, and which he thinks may with Propriety and Prudence be laid out, together with a moderate Stock to carry this Invention into Execution, may amount to £10,000; but that it will not be prudent in the Petitioner to expend the Sum necessary for that Purpose, unless some greater Security was given for the sole making and vending the said Machines than is granted by the said Letters Patent.
>
> That he has seen One of these Engines which Mr. Watt has already erected, and which was not finished till November last, and it does double the Execution of a common Fire Engine with the same Expence of Steam and Fuel.
>
> That the Petitioner has not yet received any Advantage from the Engines he has made; that an expensive Apparatus is necessary for making these Engines, particularly boring and turning Mills for making of Cylinders; and Founderies for casting them must also be erected, in order to bore the Cylinders with sufficient Accuracy, which Mills will cost a very considerable Sum of Money; and as it must necessarily be some Time before the Public in general become acquainted with the Utility of this Machine, the Term of the Patent will probably expire before the Petitioner can receive an Advantage from his Invention. [45]

Roebuck suggested that the patent should be continued to the 'Extent of 28 years'. Whatever the exaggerations and special pleading in his evidence, it shows something of the costs already incurred and what expenditure was anticipated especially in setting up a special steam engine manufactory as Boulton had desired. It also indicates the greater economy of Watt's engine over the common type. Watt's old friend, the scientific instrument-maker and watchmaker Cumming, had not seen an engine but believed that the principles on which it functioned were a great improvement on those then in use. Boulton also confirmed that these engines were three times better than others and 'that this is not only the cheapest mechanical Power yet invented, except Wind and Water Mills, but that it may be applied to an infinite Number of Purposes to which the common Fire Engine is not at all applicable'. [46] Because it was necessary to establish a proper place where these engines could be manufactured, he asked not only that the term of the patent be extended but that it should include Scotland as well. Leave was given to bring in a Bill.

It became apparent that there would be opposition to the Bill so Watt and his friends continued to seek support. Boulton sent letters to influential people. In 1773, when campaigning for the Bill to establish the Assay Office in Birmingham, he had listed thirty-five lords and six bishops on whom he intended to wait. [47] Now he listed a dozen people whom Watt might approach. He also sent a letter to Mr. Mackworth, a colliery owner at Warrington, who had 'expended much Money in fire Engines by the assistance of the charlatan Mons. Blackey'. [48] Boulton was surprised to learn that opposition might come from 'friends & Philosophers' and he was afraid that some of it might be due to party politics against himself. The most active part in the preparation of the Bill came from Scottish and Midland members such as Sir Adam Ferguson, Lord Frederick Campbell, the Staffordshire member, Sir William Bagott and the Lichfield representative, Thomas Gilbert. There is no record of opposition or support from the many Cornish members. [49]

Roebuck proffered his help and wanted to know the day of the first reading and who were the principal opponents. He had spoken to William Adam who offered to defend the Bill in the House if necessary. [50] William Irvine, who had also offered help, saw the petition printed in the Edinburgh Courant. [51] This may have been the source from which Cochrane learnt about it; he was pleased to hear the engine had been brought to perfection and wished Watt every success 'notwithstanding that we coalmasters have no Reason to rejoice at any improvement that Diminishes the Consumption of Fewel'. He hoped to see Watt as soon as possible at Culross about an engine to drain a coal mine. [52]

On 7 March, Watt saw Wedderburn who approved the text of the proposed Bill so it went to be printed. That Thursday it was to be presented by Guernsey. Watt wrote optimistically to Boulton, 'I expect the Opposition will subside, at any rate none will probably arise untill after the second reading, which must be five days after the first. I showed it to Mr. J. Dundas yesterday who seemed satisfied and said that he would attend the Committee'. [53] A Mr. Forsyth had asked for a copy of the Bill and would use his influence with some members he knew.

The Bill was presented to the House of Commons on 9 March and read for the first time. The second reading followed on 13 March when the Bill was committed for examination to a committee under Guernsey and Edmund Burke. [54] Burke represented Bristol and opposed the Bill. He may have been supporting the coal mining interests there but it is more likely that his opposition was based on the grounds of creating a monopoly, something he hated, for this coincided with the beginning of the movement for freedom of trade and toleration. Also his opposition may have been due to Boulton and Watt seeking advice from Wedderburn with whom Burke was struggling at this time over the American question in the Commons. [55]

Then on 20 March a petition was received from a William Pengree on behalf of William Blakey who was working in Paris and had only just heard about Watt's Bill. Blakey had taken out a patent for a new design of steam engine in 1766 which he had described in the specification and provided drawings. He had patented a more complex type of Savery engine and may have been moving towards condensing the steam in a separate vessel. [56] He pointed out that Watt had not included any drawings. He would have opposed Watt's patent had he known about it and claimed that Watt's engine was only an imitation of the construction of his and contained no new principles beyond those covered in his patent, 'with very little, if any, Deviation from his Design'. He fairly pointed out that, owing to the brief time of two or three months for which Watt had been able to make his engine work, it was 'unlikely such short Experience should be satisfactory'. If Watt's Bill should pass into law, 'it will be very detrimental to the said William Blakey'. His petition was ordered to lie on the Table until the report on Watt's petition was received. [57]

Watt prepared his defence which was printed. [58] He pointed out that, in Blakey's engine, the cylinder was alternately heated and cooled, whereas his was not. Blakey condensed the steam in the cylinder with a jet of cold water, whereas he condensed it in a separate condenser. Blakey's engine was a form of Savery engine with no piston, pump, or great lever, all of which his engine had. Blakey tried to separate the steam in the cylinder from the water being raised by admitting air whereas he took the greatest care to prevent air entering his engine and had an air pump to extract any which found its way in. He also proposed 'to move engines by the alternate Expansion and Contraction of Steam, in which Cases it is not to be condensed into Water'. So his and Blakey's engines 'indeed agree in no Point except in being Steam Engines'. He had made it clear in an addendum to his own patent that he in no way intended to include the Savery type. [59]

On 29 March, Darwin expressed the hope that the Bill would meet no further opposition in the House. He also suggested that he could write to various influential people as well as actually coming to London for a few days to wait upon some others. [60] Boulton came up to London to help and Watt later recorded how the Act, 'by his [Boulton's] exertions, the assistance of Dr. R[oebuck] & all the friends we could muster was obtained after a long attendance upon Parliament'. [61] The Committee examining the Bill had made some amendments so on 7 April the resolution was passed that it be recommitted and returned to the Committee. [62]

The Committee met on 11 April. [63] Harrison was called as the first witness and asked if he had seen the engine at work, which he had. He had helped to assess the performances of common engines, the best of which was at Wednesbury with a cylinder 42 ins. diameter. He 'found that Mr. Watt's Engine did 5 Times as much Work with the same quantity of coals'. He also gave an account of Watt's steam-wheel. (See p. 36) Boulton was called next and described how Watt's engine differed from common ones with more perfect condensation in a separate condenser and that no steam was generated in the cylinder from residual hot water. Also with the use of steam instead of the atmosphere to push down the piston, Watt's engine was more powerful. He quoted the same figures as Watt for the pressure on the pistons of the two types. While the Watt engine was more expensive, it would do four times the work. Although Col. Ray, the next witness, had seen only drawings and a model of Watt's engine, his experience with common ones showed that they approached only a fourth towards perfection. He outlined the advantages of Watt's engine which would effect a great saving of steam and fuel. He had not seen any similar engines and thought that the grant of an exclusive privilege or patent would not preclude others from erecting common engines if they wanted to choose 'the less perfect engine'. This shows the importance of Watt's concept of his perfect engine.

On 12 April, Robert Mylne was called. He was engineer and architect to the City of London and engineer to the New River Company so was an important witness, even though his evidence must have been carefully prepared by Watt. [64] He described the two earlier types, the Savery and the common, which he compared with Watt's; Watt's was 'totally different' and he explained why, through condensation of steam in a separate vessel, it approached perfection:

> In the working of an Engine, were it to be quite perfect, it ought to be perfectly hot at one instant of Time - equal to steam; at another period, it ought to be perfectly cold, in order to make a compleat vacuum. This alternate Operation, being to be produced, from 16 to 20 Times in a Minute,... is in the Nature of things impossible. From this, arises the Imperfections of the Common Fire Engine - & on the contrary, the perfection of Mr. Watt's invention. Because the cylinder according to Mr. Watt, has a body of steam round it, to preserve the metal perfectly hot; it has no water over the piston, which damps the power of the steam, in the inside of the Common Engine; there is no injection of cold water into it; which, in the Common Engine is raised to a certain degree of Heat, & together with the water produced from the Steam condensed, produces Steam so as to render the Operation in part ineffectual, at the very period of Time it ought to operate - Another Merit, that Mr. Watt's invention has, by employing the pressure of the power of steam upon the piston, instead of the Atmosphere, is an Excellence, superior to the Common Engine, in the Degree of 12 to 7. [65]

He continued that, although the first cost of Watt's engine would be greater than a common one, the annual expense would be less. A common engine with a cylinder diameter 18 ins. and a pump 18½ ins. diameter, lift 32 ft., would cost £1,400 in London; 'the merit of the

Engine, indifferent'. While the Watt engine would cost twice that, it would do double the work. He pointed out that the Marquis of Worcester had been granted an exclusive right for 99 years. Boulton showed that Savery's term had been extended for a further term of 21 years, amounting in all to 35 years. Garbett was questioned about the steam-wheel and Mylne was recalled to be examined on the same subject. 'Several other respectable gentlemen' attended but were not called to give evidence and no evidence was produced against the Bill. [66]

Watt proceeded to write two further pamphlets to answer criticisms and justify his request for an extension of his patent. One was a 'Plea before Parliament' [67] and the other the 'Case of James Watt, Engineer, Inventor of the New Improvements upon the STEAM ENGINE' which was printed. [68] In his Plea, Watt again outlined the basic principles of his engine with always keeping the cylinder hot and condensing the steam in a separate vessel always kept cold. He emphasised how he could condense perfectly, which did not happen in common engines because his condenser could be kept colder. He stressed the perfection of his engine. He did not wish to 'trouble the House with any further philosophical disquisition' because he had already affirmed these facts before the Committee. So he went on to point out how his invention differed from Blakey's which was 'constructed upon principles diametrically opposed to Watt's'. Concerning priority of invention, he, Watt, could prove that he had constructed models 'so long ago as the year 1765' but difficulties had arisen when he tried 'the execution of large engines which cannot be foreseen by the trial of smaller ones'. Six years had passed since he had been granted his patent during which time Blakey had had ample opportunity to protest if he thought himself injured. Watt pointed out that he had fulfilled the requirements of a written specification and that no request had been made for drawings. How could a hot or a cold vessel be drawn? He had explained how steam and fuel could be saved. He claimed that, 'If a drawing were given a jury might judge his cause by the line of that drawing and a person leaving out some part of the machine or varying the shape of another part, though it was constructed upon the very same philosophic principles, might allege that it was not the same machine'. He had never seen an engine that worked on the principles of his and he trusted that he had proved 'sufficiently that [he] did invent or apply to use the principles in question & that he has brought the steam engine to a degree of perfection hitherto unknown, and furnished the possibility of applying it to uses of which it was formerly incapable'. He therefore hoped that 'the justice of a British parliament will secure his property and his invention'. [69]

In his 'Case', Watt pointed out the importance of the steam engine and how valuable would be any improvement which increased its power or diminished its fuel consumption. He described again his experiments to determine the properties of steam which he followed with experiments on models. These showed the 'practicability of his improved engines' so that he applied for a patent to protect his property. Had he constructed a large engine before that, he might have lost his right. It was not until 1774 that he was able to overcome the difficulties in completing large engines and had finished working engines of the two types he had invented. To achieve all this, he had had to expend several thousand pounds and, in order to assure the success of his inventions, he would incur further expenditure in erecting mills and procuring the necessary materials. Therefore he was afraid that the whole term of his patent would elapse before his engine could come into general use and so repay his costs. He showed how the Marquis of Worcester and Captain Savery had been granted longer terms for their exclusive privileges. He apologised that there had been misunderstanding about his specification which had made it necessary to recommit the Bill. The Committee had carried out a very thorough investigation and was 'satisfied with the evidence given in favour of the engine'. He then sought to show that he never wanted to establish a monopoly and the Bill was now drawn up to guard the rights of other inventors. He had 'struggled with bad health, and many other inconveniences to bring [his engine] to perfection'. The public would be the greatest gainers through the savings which would be achieved with his engine, since 'his demand for the privilege of using the invention, will amount only to a reasonable part of them', and it was 'reasonable that the author of such a contrivance should be rewarded'. It was only because he was convinced that his invention would bring considerable advantages to the public that he had been 'obliged to trouble the legislature'. He needed someone to assist him bringing his invention 'into extensive execution' but it would not be in their interest 'unless the term of his patent were prolonged'. 'To their [Parliament's] wisdom and justice he therefore recommends his Bill, and humbly hopes it will pass into an act in such a manner as may give him an opportunity of returning his otherwise lost time and expence'. [70]

Watt's pleadings and the canvassing for support by his friends achieved their hoped-for result. On 25 April Guernsey reported from the Committee that further amendments had been made so it was ordered that the Bill be engrossed. The Bill was taken to the House of Lords on 27 April where it was considered by their Committee and returned to that House on 1 May without any amendment before being passed by the Lords on the following day. It completed all its stages by 8 May and the Act received the Royal Assent on 22 May. [71] The Act started with the text of the original 1769 patent specification and followed this with most of Watt's 1775 Petition. It enlarged the area covered to include Scotland as well as extending the term of the patent for twenty-five years from the passing of the Act. By it, James Watt could enjoy the whole profit arising by reason of his inventions and no one could use these inventions without Watt's licence or consent. Watt could claim damages from such offenders. [72] Procuring the Act was in the end less costly than having the patent renewed. Incidental expenses in Watt's 'Cash Memorandum' show that he spent roughly £70 in London between January and May 1775, including costs of travel but not lodgings. These might also have been incurred in procuring a patent so were not part of the total account presented by Nathan Barwell on 9 May. Up to 7 April costs were £54.4s.4d. which covered expenses for the recommittal. At the Lords, the costs were £37.18s. Barwell's costs for 'my own trouble in solliciting the Bill through both Houses and attending at the Committee and on the Members' was £21. £5.16s. had to be added for printers' invoices, which, with other incidentals, gave a grand total of £119.8s.4d. Barwell finished his letter with, 'I most heartly wish that the profits arising from this Invention may be equal to the Abilities you have shewn and the trouble and Expences you have been at in bringing it so near to perfection'. [73]

As soon as the Act had been passed, Watt wrote on 8 May to tell his father of the joyful news. His letter reveals some of the strain and worry he had undergone:

> After a series of various and violent opposition I have at last got an Act of Parliament vesting the property of my new fire-engine in me and my assigns, throughout Great Britain and the Plantations, for twenty-five years to come, which I hope will be very beneficial to me, as there is already considerable demand for them.
>
> This affair has been attended with great expense and anxiety, and without many friends of great interest I should never have been able to carry it through, as many of the most powerful people in the House of Commons opposed it. It has been in Parliament ever since the 22nd of February, which is a very long time to be kept in suspense.
>
> I shall be obliged to stay here a few days longer, after which I return to Birmingham to set about making some engines that are ordered; after which I intend to give myself the happiness of seeing you and the dear children... My warmest wishes and affection ever attend you; may God render your age comfortable is the prayer of your ever affectionate and dutiful son. [74]

Watt was now deeply indebted to Boulton for his help in securing the Act. He recognised that 'Mr. Boulton's amiable and friendly character together with his fame as an ingenious and active manufacturer procured me many and very active friends in both houses of parliament'. [75] On his own, Watt would never have secured his Act. The good news spread quickly in Scotland. Watt's cousin, Robert Muirheid, saw in the newspapers that the Bill was to be engrossed, which he thought was very good for Watt. [76] Hamilton was:

> ... exceeding happy to hear by your last letter that your parliamentary business which has occasioned you so much trouble and fatigue & uneasiness of late, was at last brought to a conclusion & that your right had prevailed against the strong opposition mustered up to destroy it - I hope it will turn out as advantageous as you expect or could wish for - After hearing your Bill was passed I expected you would have immediately come down to Scotland. [77]

Hamilton said that a number of people there were enquiring after Watt and he hoped that business would soon keep him in Scotland where he definitely thought Watt would return to live.

Settlement with Roebuck

But Watt had one further piece of business to transact in London, which was settling the transfer of Roebuck's two-thirds of the partnership to Boulton. Accordingly William Matthews, Boulton's agent and banker in London, drew up an agreement. Since Roebuck could advance no more money to expedite the steam engine, he agreed to give up his share to Boulton. In return, Boulton gave up his claim to £584 which he personally had advanced to Roebuck as well as the £630 which Boulton and Fothergill had lent. In addition, Boulton agreed to pay Roebuck £1,000 out of the first profits arising from the engine. [78] But then Roebuck began to have second thoughts. He was trying to gain an order for an engine in Yorkshire with which Smeaton was involved. This would burn £1,200 worth of coals per annum so that Roebuck reckoned that he would be better off receiving a tenth of the savings in fuel as his share of the premium. [79] Boulton was 'vexed about the Doctor's fickleness, and... unreasonableness'. [80] When he realised that Matthews and Roebuck had different interpretations about what constituted the first £1,000 of profit, he decided it:

> ... shall be the very first £1000 profit... that arises from and after the day on which you and I commence our partnership for profit and loss... I am determined that I will not have a twenty-five years' partnership with the Doctor and his Trustees, etc., but will draw a positive line at all events. [81]

In fact, Boulton decided to pay off the bonds which had been given for this £1,000 in order to be quit of Roebuck's assignees long before any profit had been received from the engines. [82]

It had been intended to include Small in the steam engine partnership once the extension of the patent had been secured since he had given so much advice and other help at all stages of the proceedings, finishing with assisting Watt to prepare the Parliamentary petition. His sudden death prevented this. Boulton's partner in the Soho Manufactory, Fothergill, declined to join in the venture so this new firm was founded under the title of Boulton and Watt. [83] The stage was set for this firm to acquire fame and fortune in the ensuing years but this was achieved only through much labour and further development of steam engine technology by both partners.

NOTES

1 Muirhead, (1854), Vol. I, p. 84, Letter 71, W. Small to J. Watt, 26 Dec. 1769; Boulton 340.11.
2 JWP, 4/69, 'Plea Before Parliament', 1775.
3 B. & W. Col., Office Letter Book 1, J. Watt to J.D.H. van Liender, 18 July 1775.
4 Ibid.
5 Farey, (1827), p. 405.
6 Ibid, Vol. II, p. 74, Letter 131, W. Small to J. Watt, 23 Feb. 1774; Boulton 340.39.
7 SRO, GD 18/859373/3, J. Hutton to G. Clerk, ?July 1774.
8 Muirhead, (1854), Vol. II, p. 79, Letter 136, J. Roebuck to J. Watt, 12 Nov. 1774; JWP, 4/57.23.
9 B. & W. Col., MI, 8/25, 'An Account of James Watt's Improvements upon the Steam or Fire Engine', no date.
10 JWP, 4/53, 'Copy of Minutes on Recommitment of Mr. Watt's Engine Bill', J. Harrison's Evidence, 11 April 1775.
11 Ibid, S. Garbett's Evidence, 12 April 1775.
12 Ibid, J. Harrison's Evidence, 12 April 1775.
13 Farey, (1827), p. 325.
14 Dickinson, (1935), p. 87.
15 Muirhead, (1854), Vol. II, p. 79, Letter Book 136, J. Roebuck to J. Watt, 12 Nov. 1774; JWP, 4/57.23.
16 Ibid, Vol. II, p. 79, Letter 137, J. Watt to J. Watt sen., 11 Dec. 1774; JWP, 4/60.2.
17 Dickinson & Jenkins, (1927), p. 108.
18 Boulton 348.12, J. Watt to W. Small, 21 Oct. 1769.
19 Braid, D., 'John Wilkinson's First Patent', *Wilkinson Studies*, Vol. I, (1991), p. 45.
20 Dickinson, (1938), p. 74.
21 Muirhead, (1854), Vol. II, p. 80, Letter 139, M. Boulton to J. Watt, 26 Jan. 1775; JWP 6/19.21.
22 Ibid, Vol. II, p. 84, Letter 146, M. Boulton to J. Watt, 24 April 1775; JWP, 6/19.22.
23 Ibid, Vol. II, p. 81, Letter 141, M. Boulton to J. Watt, 2 Feb. 1775; JWP, 6/19.8.
24 Ibid, Vol. II, pp. 73 – 4, Letter 130, J. Hutton to J. Watt, 1774; JWP, 4/76.9.
25 Smiles, (1878), p. 151 and B. & W. Col., MI, 2/6, 'Cash Memorandum', 12 Jan. 1775, 'Paid coach to London'.
26 B. & W. Col., MI, 2/8, 'Note Book of James Watt'.
27 Muirhead, (1854), Vol. II, p. 80, Letter M. Boulton to J. Watt, 26 Jan. 1775; JWP, 6/19.21 and Boulton 345.45, J. Watt to M. Boulton, 31 Jan. 1775.
28 Boulton 345.45, J. Watt to M. Boulton, 31 Jan. 1775.
29 B. & W. Col., MI, 2/8, 'Note Book of James Watt'.
30 Muirhead, (1854), Vol. II, p. 80, Letter 138, J. Watt to M. Boulton, 13 Jan. 1775; Boulton 348.44.
31 Ibid, Vol. II, p. 80, Letter 139, M. Boulton to J. Watt, 26 Jan. 1775; JWP, 6/19.21.
32 Ibid, vol. II, p. 81, Letter 140, J. Watt to M. Boulton, 31 Jan. 1775; Boulton 348.45.
33 B. & W. Col., MI, 2/6, 'Cash Memorandum', 9 Feb. 1775.
34 B. & W. Col., MI, 8/25, and JWP, 4/69, see Appendix 2
35 Ibid.
36 Falck, (1772), see various pages.
37 Boulton 348.55 and JWP, 4/69.
38 Quoted in Schofield, (1963), p. 116.
39 Muirhead, (1854), Vol. II, p. 81, Letter 142, M. Boulton to J. Watt, [25 Feb. 1775]; JWP, 6/19.7.
40 Ibid, Vol. II, pp. 82 – 3, Letter 143, J. Watt to M. Boulton, no date; Boulton 348.46.
41 Clagett, (2003), p. 277.
42 Smiles, (1878), p. 156.
43 King-Hele, (1981), p. 71, Letter 75D, E. Darwin to J. Watt, 29 March 1775; JWP, C1/40.
44 *Journal of the House of Commons*, Vol. XXXV, 1775, p. 142.
45 Ibid, p. 168.
46 Ibid.
47 Goodison, (2002), p. 165.
48 JWP, 6/19.10, M. Boulton to J. Watt, 24 Feb. 1775.

49 Rowe, (1993), p. 77.
50 Muirhead, (1854), Vol. II, p. 83, Letter 144, J. Roebuck to J. Watt, 3 March 1775; JWP, 4/57.25.
51 JWP, 4/76.10, W. Irvine to J. Watt, 11 March 1775.
52 JWP, 4/76.12, Ld. Cochrane to J. Watt, 13 March 1775.
53 Boulton 348.47, J. Watt to M. Boulton, 7 March 1775.
54 *Journal of the House of Commons*, Vol. XXXV, 1775, pp. 185 & 191.
55 Lord, (1923), p. 102.
56 Hills, (1989), p. 39.
57 *Journal of the House of Commons*, Vol. XXXV, 1775, p. 207.
58 JWP, 4/69, 'Comparison of Mr. Blakey's and Mr. Watt's Steam Engines'.
59 See Muirhead, (1854), Vol. III, p. 15 f. for text of 1769 Patent and Appendix...
60 King-Hele, (1981), p. 72, Letter 75D, E. Darwin to J. Watt, 29 March 1775; JWP, C1/40.
61 Dickinson (1936), p. 204, 'Memoir of M. Boulton', by J. Watt.
62 *Journal of the House of Commons*, Vol. XXXV, 1775, p. 280.
63 See JWP, 4/53, 'Copy Minutes on Recommitment of Mr. Watt's Engine Bill'.
64 Richardson, (1955), p. 16.
65 JWP, 4/53, 'Copy Minutes on Recommitment of Mr. Watt's Engine Bill'.
66 Boulton 348.54, 'Case of James Watt, Engineer, Inventor of the New Improvements upon the STEAM ENGINE'.
67 JWP, 4/69, 'Plea before Parliament' [title added later].
68 Boulton 348.54, 'Case of James Watt, Engineer, Inventor of the New Improvements upon the STEAM ENGINE'.
69 Quotes from JWP, 4/69, 'Plea before Parliament'.
70 Quotes from Boulton 348.54, 'Case of James Watt, Engineer, Inventor of the New Improvements upon the STEAM ENGINE'.
71 *Journal of the House of Commons*, Vol. XXXV, 1775, pp. 313 & 387 and *Journal of the House of Lords*, Vol. XXXIV, 1775, pp. 416, 420, 422 & 427.
72 For the text of the Act, 15 Geo. III, Cap. LXI, 'An Act for Vesting in James Watt, Engineer, and his Executors, Administrators, and Assigns, the Sole Use and Property of certain Steam Engines, commonly called Fire Engines, of his Invention Described in the said Act', see Muirhead, (1854), Vol. III, pp. 17 – 26.
73 JWP, 6/42, N. Barwell to J. Watt, 9 May 1775.
74 Muirhead, (1854), Vol. II, p. 89, Letter 149, J. Watt to J. Watt sen., 8 May 1775; JWP, 4/60.4.
75 Crowther, (1962), p. 143.
76 JWP, 4/20.14, R. Muirhead to J. Watt, 3 May 1775.
77 JWP, 4/20.16, G. Hamilton to J. Watt, 22 May 1775.
78 Tann, (1981), p. 34, Doc. 4, 'Copy of Award between Dr. Roebuck and M. Boulton', 29 May 1775.
79 Boulton 348.50, J. Watt to M. Boulton, 27 May 1775.
80 Muirhead, (1854), Vol. II, p. 93, Letter 152,
M. Boulton to J. Watt, [June] 1775; JWP, 6/19.3
81 Ibid.
82 Smiles, (1878), p. 159.
83 Lord, (1923), p. 104.

3
Continuing Trails and The First Engines

After gaining their triumph with the passage of their Act through Parliament, Boulton and Watt had to concentrate on perfecting the performance of their engines. From the time Watt left Glasgow in the spring of 1774 until 1781, there is no evidence to show that he carried out any further scientific experiments on steam or latent heat. During this period, he was too busy concentrating on improving the reliability of his engines and erecting them. Watt possessed a remarkable liaison between science and technology, since, time and again in his various projects, he investigated first the theory and then persevered with the practical application until he reached a satisfactory result. Progress with the improvement of his reciprocating engine is difficult to follow because often a modification failed to stand up to the rigours of actual operation so that part needed to be modified again. We shall see how this happened on the pumping engine at Soho. Then, based on experience gained there, the first engines were supplied to customers. But the experience gained on these necessitated further modifications before a satisfactory design was achieved.

Trials at Soho

The passing of the Act to prolong the patent had been gained by stressing the philosophical principles of Watt's 'perfect' engine but, once it had been obtained, the new firm of Boulton and Watt had to concentrate on making their engines perform properly so that fresh trials had to be undertaken at the Soho Manufactory. News of the Act stirred up fresh interest in these engines so that many enquiries were soon received. Rumours resulting probably from the original 1769 patent about the much more economical performance had stimulated enquiries as early as the beginning of 1771. Small reported to Watt:

> A friend of Boulton and me in Cornwall sent us word four days ago that four or five copper-mines are just going to be abandoned, because of the high price of coals, and begs me to apply to them instantly. The York Building Company delayed rebuilding their engine, with great inconvenience to themselves, waiting for yours. Yesterday application was made to me, by a mining company in Derbyshire, to know when you are to be in England about fire-engines; because they must quit their mine if you cannot relieve them. [1]

After Watt's arrival in Birmingham in 1774, there must have been enquiries for engines because Hamilton wrote in the middle of June, 'You mention having begun to erect two Engines – I suppose this will employ you for sometime... but I hope it will defray the expense of the journey... [When] do you expect to be in Scotland again?' [2] Watt, never one to miss an opportunity, sought orders during his visit to London in January 1775. There were prospects for a couple of engines, 'one to water Picadilly, and the other to serve the south end of Blackfriars Bridge with water'. [3] While the Act was still passing through Parliament, Boulton told Watt that 'There are several engines now wanted in Cornwall; some of the proprietors of mines are impatient to know the event of our Bill and the terms we will propose'. [4] Perhaps the Cornish orders were the result of the beginning of the American War of Independence. Therefore it became imperative to make both the steam-wheel and the reciprocating engines run satisfactorily.

While attention was concentrated on the reciprocating engine, the steam-wheel was also given trials. Towards the end of February 1776, work was slack in the smith's shop at Soho so Boulton intended 'on Monday morning to divide the wheel-engine amongst them, and get it finished out of hand'. He was still optimistic for the success of this engine later that year when he told Watt, 'If we had a hundred wheels ready made, and a hundred small engines like Bow engine, and twenty large ones executed, we could readily dispose of them. Therefore let us make hay whilst the sun shines, and gather our barns full before the dark clouds of age lowers upon us'. [5] Some enquiries were received from people wanting a steam-wheel but, in November 1776, Messrs. Proctor & Beilby of Sheffield were advised that, until the rotary engine was perfected, a reciprocating engine should be used to raise the water to turn a waterwheel, as they did at Soho. [6] There were a few more enquiries in 1777 and a few more trials but then the steam-wheel was allowed to fade away quietly. Smeaton was proved right when he saw it in 1773 and told Boulton that 'the circular engine would not do... He... was sure of this'. [7] Watt did not comment about the abandonment of one of his projects. Although the rotary engine seems to us to be merely a chimera, we should note that a few years later both Jonathan Hornblower junior and William Murdock patented equally unsuccessful versions.

While Watt was in London during the first half of 1775, Boulton continued with trials of the reciprocating engine in Birmingham and considerably improved its performance. Indeed, throughout the development of both the reciprocating and later rotative engines, Boulton continued to make suggestions for technical improvements as well as giving practical help in many other ways. Watt may have been frustrated because he could not be present and learnt about progress only through correspondence. The problems they faced were well understood by Smeaton who thought that Watt's engine was far too complicated and needed such accuracy in manufacture that it would never succeed with the state of engineering at that time. Others shared in this view. Thomas Shafto wrote to Smeaton on 30 May 1779 after seeing Watt's engine at Soho:

> I think it appears much more compleat consisting of more parts to form a whole than Newcomen's which in case of any accident must require rather more ingenuity to set right than that (which with few alterations) has been in general practice in this country for near a century past, and which almost every mechanic can put together and to which the natural antipathy of most of that sort of Gentry have to the introduction of any new species of machinery, which does not coincide with their own ideas, makes them so inattentive to its success that without a director's eye being always over them they are sure to go wrong. [8]

Watt had to train not only engine erectors but also men who would operate the finished engines. M.R. Boulton pointed out in 1824 that 'a school of workmen, in every relevant branch, was speedily and successfully instituted, and the forms and construction of the machine were perfected with a skill and accuracy till then unknown in the execution of large machinery'. [9]

While Watt was away in London, Boulton had repaired the tin cylinder and set the engine to work again because Watt wrote on 29 March 1775 that he had 'expected better performance'. He gave instructions about fitting a mercurial pressure gauge to the cylinder which he had not done before because he had been unable to prevent the mercury 'jumping into the Cylinder & spoiling everything'. Was he trying to measure the terminal pressure in the cylinder to see how it differed from that in the condenser? But at this point comes a puzzle for he continued in his letter 'now that we have an Iron Cly[inde]r'. Boulton only wrote on 24 April to say that this had arrived. [10] Perhaps Watt was anticipating the event. He suggested a brass piston should be tried with iron plates to fix in the segments and the whole covered with paper pulp. He thought this would have little friction. He wanted a pair of new pumps to be made with a twelve or fifteen inch stroke for the large condenser because he suspected a lodgement of air. It was clear that the engine still needed many modifications.

Boulton wrote in April, 'The cast-iron cylinder is arrived; it seems tolerably true'. However its bore was as much larger than 18 ins. diameter as the old one was under so that Harrison had to fit a brass ring round the piston, which he promised to complete with other work in a fortnight. Owing to the cylinder being larger and the new bottom for it being very much bigger, the cooper had to rebuild the wooden steam casing. Boulton's report on the next trials was not encouraging:

> The engine goes marvellously bad. It made eight strokes per minute; but upon Joseph's [Harrison] endeavouring to mend it, it stood still. Nor do I at present see sufficient cause for its dulness [sic]. I have a few minutes ago had the top taken out, and find that I can pump down the piston, and although I can hear the air pass by it into the cylinder, yet the error is not sufficient to account for its going badly. The piston is now taken out, and although the cylinder is not perfect, yet there doth not appear any gross error. The outside of the piston is [felt] hat, filled up with paper chewed. It is nine inches thick at least, which I fear makes much friction. [11]

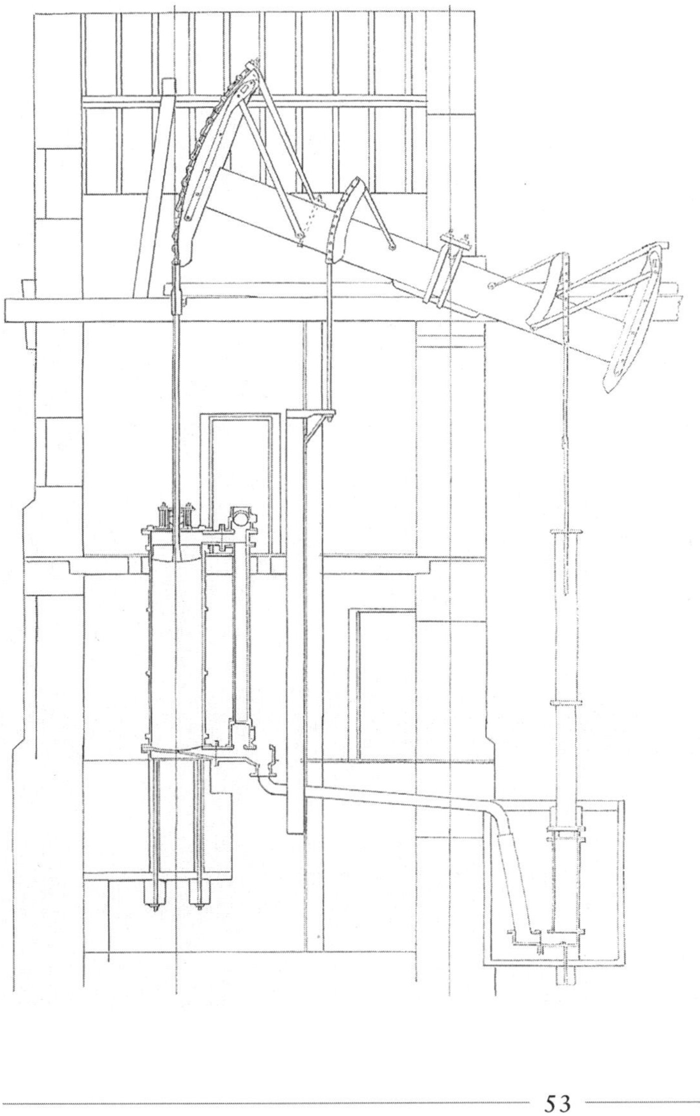

The problem of making a steam-tight seal between the piston and the cylinder walls was obviously far from solved even with the much more accurately bored cylinder from Wilkinson which did not have 'any gross error'. So it was not only the cylinder bore which had been causing problems as has been so often suggested. Boulton wished that Watt could 'learn of ye plan of Gainsborours piston as by his own acct. tis absolutely perfect for water'. [12] Humphrey Gainsborough had applied for a patent concerning a method of condensing the steam in a Newcomen engine and later wished that he had opposed Watt's patent Bill. (See Vol. I, p. 321)

Side elevation Penrhyn Du engine 1779. For front elevation see page 127 and plan page 135

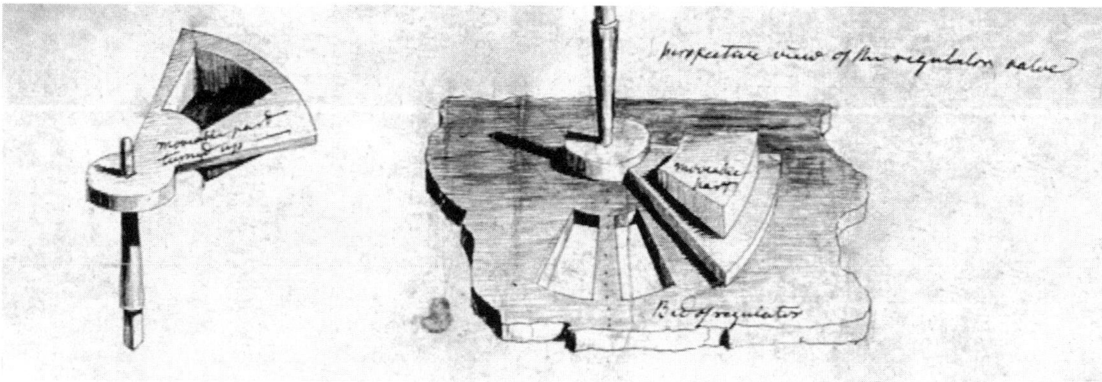

Watt's slide valve shown in the drawing laid before Parliament in 1775.

After the triumph with his patent Act, Watt was 'much mortified with the bad performance'. He suggested a method for holding the piston packing down and that the regulator valve and joints be examined for tightness. [13] Boulton continued to send reports on the trials he was carrying out. Even with a steam pressure of nine inches of mercury (4.5 p.s.i.), the engine made only nine strokes per minute. He found the slide valve very much out of true, letting in air freely. Having rectified this and parts of the valve frame which also admitted air, he thought that area of the engine 'perfect enough'. He found that there was great friction in the movement of the piston which had 90 lbs. of lead on the papier-maché packing. There was also great resistance to the piston of the water pump descending so he wondered whether a bigger barrel allowing larger valves would be an improvement. On this point he deferred to Watt's judgement, 'These are my occurring thoughts, which you will in an instant see the absurdity of, perhaps, as you have considered these subjects with more attention than I have'. He could not afford to upset Watt. He proposed to try the engine on the following day, 'as Joseph says we always had the best luck on Sundays, I will inform you of the result'. [14]

Boulton often did not date his letters, but one which may have been sent after the Sunday steaming was much more positive in tone. He had put lighter weights on the piston packing, 'which hath lessened the friction in the cylinder much'. The papier-maché and [felt] hat packing 'seems to do very well. I really could not find any fault with either piston or steam-valve. We set to work and very soon made eleven strokes per minute'. [15] He then turned his attention to the condenser. These trials were being carried out in June when the weather was much warmer than when Watt had been experimenting before Christmas. With Watt, the water coming into the condenser was 40°F but with Boulton it was '66° or 68°, which makes much difference in condensing'. Although Boulton found the water leaving the condenser was 80°, he thought the condenser should be larger. Boulton and Watt were fortunate in developing their engine in a reasonably moderate climate, where water supplies would not freeze up completely in winter, a problem faced by Marten Triewald in Sweden, [16] nor was the water generally too hot in summer to achieve reasonable condensation.

Boulton's next letter gave figures for the performance of the engine, 'No leaks; the piston and valve good', but he still thought 'the present imperfection arises chiefly from want of more cold'. So he suggested that good condensers might be made from corrugated thin iron plates as Watt had proposed earlier. [17] In his final letter of this series, Boulton found that a vacuum of 28 to 28 $^{1}/_{2}$ ins. (14 p.s.i.) was being achieved. Steam pressure was 10 to 11 ins. (5 - 6 p.s.i.), giving a pressure on the piston of 38 ins. (18 p.s.i.). He agreed with Watt that new pumps with longer strokes in the condenser would 'render the vacuum more perfect, and I likewise conclude that the condenser ought to be double the size'. He was describing other parts of the engine when Harrison came to tell him that:

> ... the engine goes very badly, although the steam is 11 inches high, owing to the valves of the condenser-pump not opening and shutting properly, which he is going to alter. Certain it is that the condenser and its pumps are the very vitals, nay, the very soul of the engine, and therefore they should be so made as not to be liable to disorder. [18]

At this point, Watt returned from London and commenced trials on 6 June. [19] He found that the paper pulp did not have 'sufficient elasticity to apply itself to the cylr.' During June, trials continued although the packing material was still giving trouble. More cow and horse muck was tried, so it seems that preventing steam passing round the piston was not immediately achieved with the new cast iron cylinder. Yet it was at this time that the engine was working with 13 p.s.i. on the piston and that leakage of steam past the piston had been reduced to acceptable levels. [20] Much later Robison commented that Watt's greatest difficulty was making the piston tight. Watt thought that the greatest waste of steam in his engine was through the unavoidable escape of steam past the sides of the piston during its descent. Luckily, in the Watt engine, this steam did not destroy the vacuum in the same way as air passing the piston of a Newcomen engine. Watt had to compromise between a piston fitting too tightly with great friction and one that moved more freely but allowed a little steam to pass. [21]

Then Watt left for Scotland where he remained until later in October. So Boulton carried on the trials. He wrote to Watt at the end of August or early September that 'The engine makes 2000 strokes with one cwt. of coals. The piston remains good, and I think lacks no improvement'. [22] At least progress was being made on that part of the engine but the condensing apparatus was still far from achieving perfection. Two small 7 ins. pumps had just arrived for the condenser but the boring head to complete them still had to be made. There was also progress elsewhere. Boulton had been extending the Soho Manufactory to cope with the steam engine business where the roof for the new forging-shop was nearly finished and two hearths built. Wilkinson had paid a visit and told him:

> ... that all his neighbours are impatient to see the event of his engine. Some, he says, have suspended their new erections until his is finished; and all of them, he is sure, will have their engines altered, which he says will be a better trade than new erections, and that work alone will be sufficient for our lives. [23]

While the steam cylinder was at last functioning reasonably, it was otherwise with the condensing apparatus. Watt had persisted with surface condensers because he thought that less air and water would have to be pumped out. But their efficiency declined since they were 'liable to be impaired by incrustations from bad water'. [24] Therefore in October 1775 he took the step of introducing condensing with a jet of cold water. [25] It is not surprising that he found this 'operates beyond my ideas in point of quickness and perfection' as Thomas Newcomen had before him. He had to increase the size of his air pumps to extract the injection water as well as the condensate and air but the action of the engine was so much improved that it developed ample power to cover the increased load. [26] He told Hamilton about the success of the new arrangement, who replied on 2 November, 'I am happy that your new condenser answers so well I expect a description of it – I hope it will answer your most sanguine expectations, & that the Engine will turn out as profitable as you could wish'. [27] Hamilton also commented just before Christmas, 'I am glad to hear your Engine Scheme is going on so well, I hope it will continue'. [28] It was only at this point in November 1775 that

Watt was able to make his engine function properly but even so the condensing apparatus still needed further modifications as well as a redesign of the valve gear. This first engine continued to be used at Soho for further experiments as Watt strove to improve the performance of those engines which he was supplying to the public. It had taken him over ten years since his first inspiration for the separate condenser to reach this point. Perhaps he might have achieved success earlier had he been able to concentrate on his steam engines without pursuing both his interests in natural philosophy such as the alkali business or his career as a civil engineer. (See Vol. I, various chapters) Yet it may also have been his search for perfection expressed in all his trials with different types of condensers while in Scotland that also took up so much of this time. These trials were based on his scientific experiments on the nature of steam but in the end experience caused him to abandon one principle, that of surface condensing, to achieve an engine which was commercially viable. (See Vol. I, p. 368)

Payment for the Engine

With the engine now a commercial proposition, Boulton and Watt had to determine how they could profit from it. Watt had been considering how he might charge users of his patent almost from the day he obtained it. In 1769, he told Small, 'Our idea at present is first to supply those places where coals are 20s. per ton and upwards, that is, provided the experiment in great answers'. [29] He intended to link the premiums to the value of coal consumed. In his 'Case before Parliament', Watt hinted at the method by which he would reap his reward for all his struggles with bad health and other problems:

> All he wishes is to be secured in the profits which he may reasonably expect from it... for he cannot expect that any person will make use of his contrivance, unless he can prove to them that savings will take place, and that his demand for the privilege of using the invention, will amount to only a reasonable part of them. [30]

Boulton explained to Mr. Glover who had enquired about engines for Cornwall:

> ... that we will undertake and contract to make an engine or engines capable of doing any quantity of work that shall be requested and described, for as little money as common engines will cost that are capable of doing as much work, and we guarantee them to do that work with half the expense of fuel that common ones will require, provided we are allowed a sum that shall be equal to its further savings over and above the said half. [31]

Watt explained their system for Britain to van Liender:

> The savings in fuel are to be estimated by comparing the effect of a Cwt of coals burnt in my engine with that of a Cwt burnt in any common engine they chuse to pitch upon, and these savings are to be divided into three parts, two whereof belong to the user of the engine and the other third part to me to be paid annually so long as they use the engine or my privilege continues, so that in places where coals are dear my profits are greater, than where they are cheap, and are always proportioned to the profits of the users. [32]

It was a neat and, on paper, deceptively simple way of assessing the premiums with the customer gaining the greater benefit. But, when it came to actually calculating what should be paid, endless controversies arose so that simpler ways of determining the savings had to be found.

Watt only rarely countenanced altering existing common engines in order to gain the maximum premiums. He always aimed for perfection which would not be achieved if his condenser were applied to an open-topped cylinder of a Newcomen engine with no steam jacket. The savings, and hence premiums for Boulton and Watt, would be less than with their fully developed engine. Also Watt never allowed anyone else to build his engines. In January 1778, Smeaton wrote to Watt asking for a licence to apply a separate condenser to an engine he was designing for pumping water at Hull. He told Watt:

> Ever since I have been acquainted with your Method of condensing out of the Cylinder, I have been much struck with the Ingenuity of the Thought; & the promising Probability of Improvement; & still more with your last Method by an Air Pump. But it always appeared to me equally applicable to the Newcommen's [sic] Engine as to your own, & even to those of the most common construction. [33]

Watt turned down the request and Smeaton recommended that Boulton and Watt should build the engine at Hull, which they did. Smeaton wrote, 'I most heartily wish you all the success that your ingenious discoverys, and indefatigable labours have deserved, and shall endeavour to promote your Interest'. [34]

The First Orders

When the Carron Company enquired in 1776 about replacing their Newcomen returning engine with a Watt patent type, he outlined his conditions and what Boulton and Watt would undertake to provide:

> We do not aim at profits in engine-building, but shall take our profits out of the saving of fuel; so that if we save nothing, we shall take nothing. Our terms are as follows: We will make all the necessary plans, sections, and elevations for the building, and for the engine, with its appurtenances, specifying all cast and forged iron work, and every other particular, relative to the engine. We will give all necessary directions to your workmen, which they must implicitly obey. We will execute, for a stipulated price, the valves, and all other parts which may require exact execution at Soho; we will see that all the parts are put together, and set to work properly; we will keep our own work in repair for one year. [35]

Boulton and Watt would act as consulting engineers, preparing the basic specifications. They did not build engines, although they would supply those smaller parts which had to be made accurately. Normally they recommended that their customers should purchase the steam cylinder from Wilkinson who could bore them with the required accuracy. Boulton and Watt would supervise the erection and oversee the commissioning. Implicit in this list is that the customer would have to find the materials for the engine house as well as acquiring almost all of the parts for the actual engine. Boulton's dream of establishing an engine manufactory on the banks of the canal in Birmingham had been quietly dropped. Watt's practice for erecting his Newcomen engines in Scotland had prevailed.

One of the points which Watt included in his 'Petition to Parliament' was that the patent might have expired so that he would not receive any advantage from it because it would require 'several years and repeated proofs... before the public can be convinced of the expedience of the invention'. [36] But he had produced virtually no evidence to support the claims he had made for the performance of his engine so that people doubted whether they were true. The criticisms which Falck expressed may have been typical even though he was trying to show the advantages of his own engine as opposed to Watt's. Falck wrote his pamphlet before Watt's engine had been really 'put into execution' [37] and he was apprehensive that it would prove to be inferior to common ones when actually working. Like Smeaton when he had seen Watt's Bow engine, [38] Falck thought 'the improvement... exceedingly complicated and expensive; and, what is worse, very subject to be out of repair'. He doubted whether steam pressure on the piston could be much greater than the atmosphere as well as the perfection of the vacuum. Direct injection of cold water would 'make a much more perfect temporary vacuum than one produced by exhausting where the cylinder retains the same degree of heat'. His argument here had some validity owing to the complexities of heat transfer across the interface of different mediums. Falck correctly recognised that, on Watt's engine, 'the exhausting of the condenser [by the air pump] must undoubtedly greatly diminish the force of the engine'. Although Watt had given no plans and Falck had not been an eye-witness to any of Watt's experiments, Falck wrote, 'at the same time I shall be very happy to be convinced... by ocular demonstration'. Watt would be able to do this shortly.

Whatever doubts other people may have had, they continued to make enquiries. From about ten enquires in 1775, the number rose to twenty in 1776. [39] Watt's Note Book for May 1776 mentions Mr. Isaac Griffiths at Swansea, Jameson Swaine at Neath Abbey, Thomas Williams also at Neath, as well as the coal owners, Mr. Arthur at Bristol and Mr. Harrington at Newton St. Loe near Bath, showing how widespread was the interest. [40] On his return from London in June 1775, Watt started to provide positive evidence for his claims with 'repeated proofs' for he was engaged on designs for two large engines, the smaller with a 38 ins. diameter steam cylinder to blow Wilkinson's blast furnaces at New Willey, near Broseley, and the other for a pumping engine with a 50 ins. cylinder at Bloomfield Colliery, near Tipton.

The Bloomfield Engine

The engine for Bloomfield Colliery was completed first and so will be described next. Boulton knew the owners of this colliery and was able to persuade them to replace the Newcomen engine which they had engaged Isaac Perrins to erect with a Boulton and Watt type. It was agreed, possibly over the dinner table, that Perrins should continue to build the engine, using some parts supplied by Soho, and that a new cylinder would be cast by Wilkinson. [41] The 72 ins. cylinder already sent by Wilkinson was too large and Watt eventually settled on 50 ins. He told the proprietors, 'If the first cost of ye new engine should exceed that of the old engine wch. it cannot do in any considerable degree, I agree that such extra expense shall be deducted out of the saving before I require any profit'. [42] He was referring to the method of charging for the use of his patent.

Watt had ordered the new cylinder from Wilkinson at the end of June:

> Messrs Bentley & Co., having agreed to erect one of my steam engines at their Colliery at Bloomfield and the cylinder they have got from you being too large for my way of working their pumps, this comes to desire you to get another cylinder cast as soon as possible of the following dimensions, viz., 9 feet long and 50 inches diameter with a flanch at each end and one upon the body of the cylinder 6 feet from the bottom end, this latter is to have no holes in it and to be quite plain one inch thick and projecting about 3 inches all round the cylr. The bottom flanch is to be pierced with 20 holes for 1¼ inch screws and the top flanch (which serves to screw on a lid) is to be divided into 16 equal parts, 14 of which are to be pierced for 1¼ inch screws and the two spaces to be left blank, at one place. I must beg of you that the cylinder may be very smooth and truely bored and got ready with all

> convenient dispatch. The drawings for the cylinder bottom, lid, and piston shall be sent to you in a post or two, meanwhile I hope you will order the cylinder to be put in hand... The thickness of Mr. Bentley's cylinder must not exceed an inch, but you may put the customary number of beads round it. There is no occasion for having so many as are upon the 6 foot one they have got. [43]

Wilkinson acted quickly so that Watt was impressed with the quality of the workmanship when the cylinder arrived in the early part of August and thought it was 'by much the best I ever saw of that dia^{tr}, and I expect it will answer well'. [44] Watt told the Carron company that the bore of this cylinder 'does not err the thickness of an old shilling in any part'. [45] The cost of the erection of the Bloomfield engine was £2,000. [46] Watt continued in his role of consulting engineer as he had done in Scotland with his Newcomen engines through preparing drawings and directions to suppliers such as Wilkinson and supervising Perrins during the erection. The colliery paid directly for the materials and the necessary labour. On those parts that were manufactured at Soho, Boulton and Watt received no profit but looked to their annual premium to reimburse them for their capital expenditure on tools and workshops. [47] On 16 February 1776, Boulton announced to Garbett that the Bloomfield engine was almost ready. [48] However, it was not until 11 March that *Aris's Birmingham Gazette* contained a report on the initial inauguration ceremony:

> On Friday last a Steam Engine constructed upon Mr. Watt's new Principles was set to work at Bloomfield Colliery, near Dudley, in the Presence of its Proprietors, Messrs. Bentley, Banner, Wallin, and Westley; and a Number of Scientific Gentlemen whose Curiosity was excited to see the first Movements of so singular and so powerful a Machine; and whose Expectations were fully gratified by the Excellence of its performance. The Workmanship of the Whole did not pass unnoticed, nor unadmired. All the Iron Foundry Parts (which are unparalleled for truth) were executed by Mr. Wilkinson; the Condenser, with the Valves, Pistons and all the small Work at Soho, by Mr. Harrison and others; and the Whole was erected by Mr. Perrins, conformable to the Plans and under the Directions of Mr. Watt. From the first moment of its setting to Work, it made about 14 to 15 Strokes per Minute, and emptied the Engine Pit (which is about 90 Feet deep, and stood 57 Feet high in Water) in less than an hour. The Gentlemen then adjourned to Dinner, which was provided in that Neighbourhood, and the Workmen followed their Example. After which, according to custom, a Name was given to the Machine, viz., PARLIAMENT ENGINE, amidst the Acclamations of a number of joyous and ingenious Workmen. This Engine is applied to the working of a Pump 14 Inches and a Half Diameter, which it is capable of doing to the Depth of 300 Feet, or even 360 if wanted, with one fourth of the Fuel that a common engine would require to produce the same Quantity of Power. The Cylinder is 50 Inches Diameter and the length of the Stroke is 7 feet. – The liberal Spirit shewn by the Proprietors of Bloomfield in ordering this, the first large Engine of the Kind that hath ever been made, and in rejecting a Common one which they had begun to erect, entitle them to the Thanks of the Public; for by this Example the Doubts of the Inexperienced are dispelled, and the Importance and Usefulness of the Invention is finally decided. – These engines are not worked by the Pressure of the Atmosphere. Their Principles are very different from all others. They were invented by Mr. Watt (late of Glasgow) after many Years Study, and a great Variety of expensive and laborious Experiments; and are now carried into Execution under his and Mr. Boulton's Directions at Boulton and

> Fothergill's Manufactory near this Town; where they have nearly finished four of them, and have established a Fabrick for them upon so extensive a Plan as to render them applicable to almost all Purposes where Mechanical Power is required, whether great or small, or where the Motion wanted is either rotatory or reciprocating. [49]

Even though the Bloomfield engine was at first a good advertisement for the Boulton and Watt engine, it was still a pioneering design and in reality still a relatively untried prototype so that design weaknesses soon became apparent and the colliery proprietors may have doubted their 'liberal Spirit... in ordering this, the first large Engine of the kind that hath ever been made'.

Watt must have been jubilant at the success of the first commercial engine with his patent condenser but we have no record of his reactions. In anticipation of this great event, the ever enthusiastic and optimistic Boulton had written to Broseley at the end of February where Watt had hoped that the New Willey engine would be running by the end of the week:

> Pray tell Mr. Wilkinson to get a dozen of cylinders cast and bored from 12 to 50 inches diameter, and as many condensers of suitable sizes; the latter must be sent here, as we will keep them ready fitted up and then an engine can be turned out of hand in two or three weeks. I have fixed my mind upon making from twelve to fifteen reciprocating and fifty rotative engines per annum.
>
> I assure you that of all the toys and trinkets which we manufacture at Soho, none shall take the place of fire-engines in respect of my attention. [50]

Boulton recognised as early as this that the potential for rotative engines would be greater than reciprocating ones.

The New Willey Engine

The next engine set to work was at the long established Willey ironworks, situated near Broseley at the end of a small coalfield where iron ore could be found close to the coal seams. A fresh partnership reorganised the works in 1757, under the name New Willey, among whom Edward Blakeway provided much of the necessary finance. In 1763, John Wilkinson married Mary Lee, Blakeway's sister-in-law, and acquired the wealth which enabled him eventually to take over the works. However the works were situated about a mile and a half from the source of the only water supply, the Dean Brook, which, although it had a reasonable fall of over 50 ft. in this distance, quickly proved to be inadequate in its flow. [51] This lack of water became more acute after Wilkinson succeeded in smelting ore with pit coal at his Bradley furnace in October 1772. He told his manager at Broseley, Gilbert Gilpin, that he was now able to produce twenty tons weekly instead of ten. [52] Coke-fired furnaces required a more powerful blast than charcoal. Also more air was needed per ton of iron poured and, while coke solved the problem of shortage of wood for charcoal, Wilkinson lacked adequate waterpower at New Willey to take full advantage of his new discovery.

One account says that Wilkinson was reluctant to use a Newcomen engine to recycle water over a waterwheel as had been employed at Coalbrookdale since 1742. [53] However another says that 'a 49 in. Newcomen engine supplied from two boilers had scarcely been able to produce enough blast for the furnace'. [54] Wilkinson decided on a Boulton and Watt engine which would drive an air-blowing cylinder directly without the intermediary of a waterwheel. Earlier, Isaac Wilkinson had a three cylinder single-acting blowing engine driven by a waterwheel and Smeaton had designed a waterpowered blowing engine with four single-acting cylinders for Carron foundry in 1769. [55] Multiple cylinders phased to work one after

New Willey Blowing Engine

Based on drawings in the Egerton Manuscripts. British Library

A Boiler
B Single-acting Steam Cylinder
C Valves
D Condenser
E Air Pump
F Main Beam
G Single-acting Blowing Cylinder
H Inlet valves
I 'Regulating Belly'
J Water
K Air Outlet

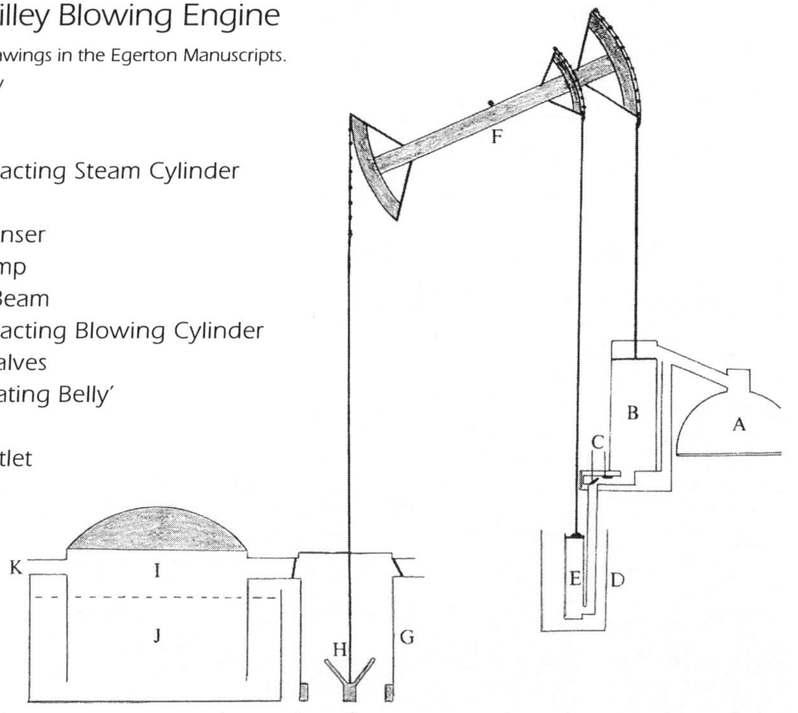

the other gave a fairly even blast. But at this period the single-acting reciprocating steam engine could operate only one single-acting blowing cylinder which gave an uneven blast with air being delivered only as the piston moved up its cylinder. Therefore it was necessary to install some form of air receiver.

While Watt must have prepared drawings for the steam engine, no evidence has survived to show who designed the blowing apparatus and air receiver, or compensation bellows, presumably Wilkinson. Although this engine was ordered in June 1775, little progress could have been made while Watt was in Scotland during that summer. Boulton reported to him at the end of August or September that, 'Wilkinson hath been here and says all his neighbours are impatient to see the event of his engine'. [56] Relations with Wilkinson were friendly at this stage but deteriorated later.

Watt went to Broseley on 8 November and again on 7 December. [57] He returned in February 1776 and was there that March to superintend the erection of the first steam-driven blowing engine for any blast furnace in the world. Samuel Smiles says that Watt took great care with the drawings and that 'the workmen did their best to execute the several parts accurately, for they understood that numerous orders would depend upon whether it worked satisfactorily or not'. [58] Boulton told Watt not to hurry but to see that anything which might hinder successful running had been ironed out 'and then, in the name of God, fall to and do your best'. [59] Watt sent favourable progress reports and, in March, Boulton replied, 'I rejoice at the well doing of Willey Engine as I now hope & flatter myself that we are at the eve of a fortune'. [60] According to van Liender who saw this engine, it was set to work on 28 April 1776. [61]

Drawings of the engine and blowing apparatus have been preserved among the Egerton Manuscripts in the British Library. [62] Although they are not drawn to scale, they show the general layout and features of Watt's design. The engine is a single-acting beam engine with the condensing apparatus situated between the cylinder and beam wall. The pistons of both steam and air-blast cylinders are suspended from chains around arch-heads. There are no additional bracings to strengthen the beam. The steam cylinder has no steam jacket which is confirmed by Boulton telling Watt in March 1776, 'Pray let the hairy coat of your cylinder be at least three inches thick'. [63] This would have been some insulating material. The top of this cylinder has direct communication with the boiler at all times so there can have been no expansive working through early cut-off. Two valves are shown at the base of the cylinder, one to let steam into the lower part of the cylinder to break the vacuum and allow the piston to rise and the other to communicate with the condenser so that this steam could rush into the vacuum there and be condensed for the downward power stroke of the engine. The condenser is shown with only one air pump which, because this is a cross-section drawing, obscures Watt's practice at this time which was to have two. Although the air pump piston here is drawn solid, the description states that it has valves. There is no foot valve between the actual condenser and air pump, nor any cover to the air pump. The house or steam cylinder was 38 ins. diameter with 7 ft. 6 ins. stroke with equal stroke for the blowing cylinder.

The blowing cylinder was 72 ins. diameter, open at the bottom to allow air in but fitted with a top cover through which the piston rod passed. As this piston descended, air passed through flap valves in it to the upper part. When the piston rose, the air above it was forced through a non-return valve into two 'Regulating Bellies'. The larger was a rectangular open box, built from bricks 33 ft. 9 ins. long by 9 ft. and nearly filled with water. Inside but taller was another brick box with domed top, constructed so that its walls reached below the water level but with passages at the bottom so that the water could flow in or out. Air from the blowing cylinder entered this second box and some passed out through another non-return valve to the furnace. But the blowing cylinder at $14^1/_2$ strokes p. min. provided more air than could immediately pass out of this valve so that it was stored in the box or 'Bellie' and forced the water down, out of the closed box and into the outer chamber. The difference in the height of water levels determined the blast pressure which was said to be 4 p.s.i. [64] This arrangement gave a more or less constant blast. The brickwork had to be strengthened because the system worked so well that the Boulton and Watt engine with one boiler provided more blast than was required and out-performed the previous Newcomen engine. The quantity of fuel consumed was very moderate and the workmen were very pleased with their efforts. [65]

The New Willey Inverted Engine

Of the early Boulton and Watt engines, least is known about one with an inverted cylinder for Wilkinson's New Willey ironworks. Increased output from his blast furnaces would have strained the capacity of finishing processes there so that Wilkinson sought to augment the waterpower for his cannon boring mill. Watt may have seen the opportunity for developing a different type of engine with a more compact layout capable of providing rotative power where a water-returning engine was needed to augment a waterwheel. This involved only a small lift so that the waterpump could be situated directly beneath an inverted steam cylinder, which was different from having to place the cylinder over a mineshaft with long heavy pump rods. Watt began investigating this in early February 1776 when he was at Broseley and wrote to Boulton, 'I have got everything my own way, but must make an inverted cylinder, which I believe I can make a good job of'. [66] Watt knew that Robison had made a model of such an engine in 1757 and he himself had built models like this as well as his first engine at Kinneil in 1765. (See Vol. I, pp. 300 & 375) This was confirmed at the patent trial against Edward Bull in 1793 when Dr. Lind stated that he had stayed with Watt for a week or ten days in 1765 and became acquainted with the principles of Watt's engine. He commented, 'I never should have put the upper ends downwards Mr Watt did himself in his first Engine and he found the inconveniency of it'. [67] History was to repeat itself.

No doubt Boulton was thinking of such an engine being placed over a mineshaft when he replied to Watt on 24 February 1776:

> I observe you are thinking of making an inverted cylinder. Pray how are you to counterbalance the descent of the piston and pump-rods, which will be a vast weight? If by a counterweight, you get nothing; but if you can employ the power that arises from the descent of that vast weight to strain a spring that will repay its debts, - if by it you can compress air in an iron cylinder, which on its return will contribute to overcome the *vis inertiae* of the column of water to be raised, you will thereby get quit of that unmechanical tax, and very much improve the reciprocating engine. But how to make your piston steam-tight requires – I know not what. [68]

Here again we find Boulton suggesting ideas for improving the steam engine, but many of his objections were invalid in the case of a water-returning engine with a low lift. Watt continued to prepare designs which Wilkinson accepted. Development and probably construction continued during the summer when Boulton suggested to Watt:

> Perhaps it might not be impolitical if you were to publish a paper in the Philosophical Transactions chiefly Elementary intimating that we have a variety of Engines invented very diff[eren]t in their construction, some where the piston is pressed upwards & without a great Beam others where there is a constant vacuum under ye piston etc etc. [69]

Watt was in Glasgow about to be married to Ann McGrigor so would have had little time to attend to the steam engine business. This may be why towards the end of July, Boulton told him that Willey was demanding pumps for this engine. [70]

Watt went to Broseley at the beginning of November 1776 when he entered in his notebook:

> New Willey Engine – goes very well but uses a great deal of steam. I measured the hot water which ran from the engine and found it 5 ?tubs? per minute had no Therm[ometer] but was as hot as I could bear the hand in it... a great deal of heat was certainly lost through the condenser sides so that the Engine uses too much steam. [71]

Unfortunately he did not record to which engine at Broseley this referred. If it were the inverted one, it might suggest why this type was discontinued. However it is more than likely that this engine was not finished by this time, since it is not included in a list Watt prepared in February 1777. Of course it may have been omitted because it was not performing well. Yet it was mentioned in a letter sent to Wilkinson in July 1777, outlining the conditions under which Wilkinson could use the patent engines:

> We will grant to you and your successors our Licenses to use the two engines now erected at New Willey and the one you propose to erect at Wilson House as well as a devil to be erected where you please during the remaining term of our exclusive privilege upon payment of one shilling yearly when demanded, provided you do not remove them to situations where coals are of much greater value. [72]

It is not clear for how long this engine worked, nor what happened to it because the records are silent until 1793 when it was mentioned during the trial against Edward Bull.

Watt asked John Southern, 'If you can find any drawing of the inverted engine made for J. Wilkinson, please bring it with you, if not it does not signify, as we have got a witness in town who saw it'. [73] This witness was Mr. More, 'Secretary for the Society for the Encouragement of Arts and Sciences', who told the Court:

> I have seen an Engine on that very Construction upside down as it was building in the year 1776 under the immediate direction of Mr. Watt and if your Lordship chooses I have thought it might be of some consequence in this Trial to produce a Manuscript written by myself at that time upon the spot where that Engine is mentioned. [74]

Unfortunately this manuscript has not survived so we have no more details about this experimental engine. It could not have worked sufficiently well since the design was not perpetuated.

The Stratford-le-Bow Engine

The partners, Messrs. Cook, Adams, Wilbie and Sager, of Cook & Company, ordered their engine also in 1775 to raise water for their distillery at Stratford-le-Bow on the outskirts of London. Wilbie, the managing partner, wanted it in a great hurry. Although this was only a small engine with steam cylinder 18 ins. diameter by 3 ft. 2 ins. stroke, it was important to Boulton and Watt because it was situated close to the capital where it could be easily observed. Indeed, in December 1776, Watt commented to Boulton that it was going well but 'a relapse of the engine would ruin our reputation here & indeed elsewhere'. [75]

Watt had journeyed to London at the beginning of December 1775 about the Bow engine. [76] The general cross section drawing is dated to this year. It shows that the single-acting beam engine with arch-heads was erected completely inside the building. In this way, it was a forerunner of later engines for waterworks where these engines were generally required to raise the water above their own level, so requiring enclosed pumps and stand-pipes. The 15 ins. diameter lift pump had a cover through which the piston rod passed with the stand-pipe rising from the top of this cylinder. The lift in this case was 38 ft. The engine made 15 strokes p. min.

This time, the steam cylinder had a steam jacket but the top still communicated directly with the boiler. Oscillating plate valves were retained for the steam inlet and condenser valves. There was a significant advance in the position of the condenser and air pump through placing this apparatus on the far side of the beam wall from the steam cylinder. In this position, the rising of the air pump piston would help to create the vacuum as the main piston was descending. This could be one reason why Boulton held high expectations for this engine and, in July 1776, told Watt who was in Glasgow, 'Bloomfield and Willey are both well & I doubt not but Bow engine will be better than any of them'. [77] This suggests that the Bow engine was not finished, which is quite likely because Watt had not sent Wilkinson a drawing of the pump until May. [78] Boulton reported some problems with it at the end of July. [79] When the engine was first set to work, probably in September, it was found that the steam could not be kept up. Its performance was improved a little by stopping the leaks and packing the piston but it still did not come up to Watt's expectations. He had to return to London at the end of November and had the machine in complete working order in December. [80] The premium was £150 per annum which Watt stated was £50 for erection and only £100 represented the actual premium. [81]

The importance of the Bow engine is shown by the visit of Smeaton to see it running in April 1777 with disastrous results. Boulton reported to Watt:

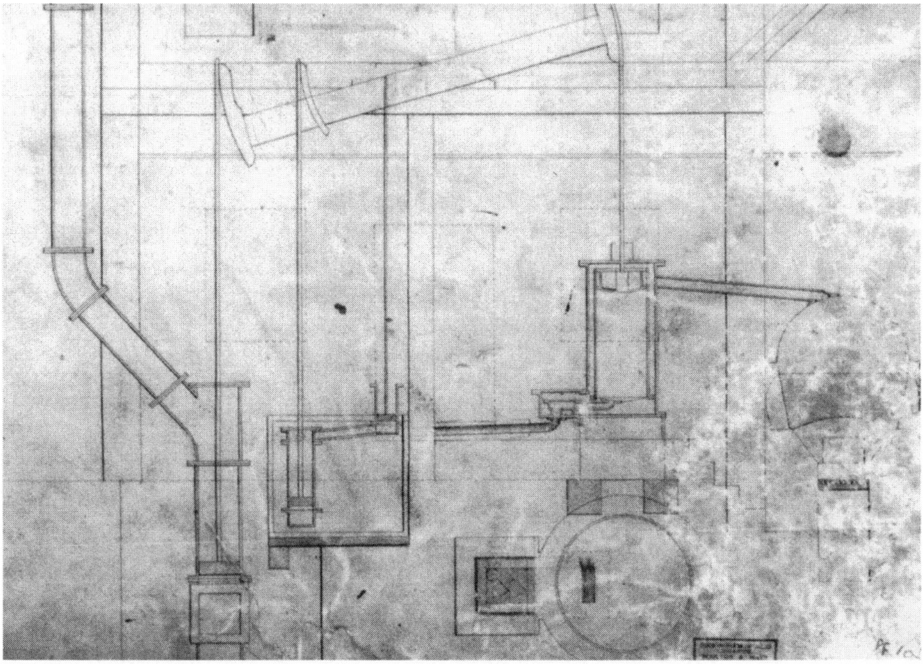

The single-acting pumping engine for Stratford-Le-Bow, 1775, was built completely inside its house.

> Hadley also told me that Mr Smeaton & Holmes called upon him to request him to attend them to Stratford le Bow: which he did. Smeaton said it was a pretty Engine but it appeared to him to be too complex but that might in some degree be owing to his not clearly under-standing all ye parts. He gave the Engineer money to drink & the consequences of that was that ye next day the Engine was almost broke to pieces. Wilbey was very angry and turned away the engineer & told Hadley the least amends he could make was to put it into order again wch he did do but was obliged to put in one new valve. [82]

Hadley was an engineer who had assisted Harrison in the erection. The repairs were satisfactory and Watt found it did 'very well' when he visited it in May 1778. [83] This was possibly in conjunction with modernising it since a new nozzle and working gear were fitted soon after. [84] The engine was still at work in September 1782. [85]

Improvements in Design

In February 1777, Watt gave a list of his engines.

Boulton and Watt Engines in 1777.

Cylinders	Place	Pump	Height of Water Raised
18 inch	Soho	18 1/2 inch	24 feet
50 inch	Bloomfield Colliery	14 1/2 inch	85 - 112 yards
38 inch	New Willey, Shropshire	Blows two large iron furnaces	
18 inch	Stratford-le-Bow, Essex	15 inch	36 - 38 feet, as tide affects it
58 inch	Bedworth, nr. Coventry	14 1/2 inch	130 yards high
	(Not finished, but will be going next week) [86]		

As might be expected, there were weaknesses in the designs of all these early engines. Instead of launching out immediately into the size of that for Bloomfield, Watt would have preferred to gain more experience by smaller incremental steps following on from the 18 ins. cylinder at Soho. There is little doubt that he would have continued with experiments to achieve his perfect engine until he had exhausted both the patience and finances of Boulton. But Boulton had no intention of going down the same road as the unfortunate Roebuck and overruled the cautious and self-deprecating Watt. Luckily for Boulton, the gamble paid off and the teething troubles were soon rectified. Wilkinson must have complained about the defects in his New Willey engine to which Watt responded in August 1779.

> I allow that it is in many respects a bad one and I should be sorry that it were in anybodies hands but your own tho' I wish you to have the best engines in the world, and that very engine may be made a very good one. I hope you will do me the justice to remember that it was against the grain with me that either Bloomfield or it were undertaken untill the matter had been more fully experimented upon at Soho and upon a larger scale many things might have been foreseen and prevented but all could not. Our business here has been pushed too hard and too much has lain upon my shoulders and a man less able to support the burden could not easily have been found. The time which should have been spent in experimenting has been lost in erecting. [87]

Watt was willing to produce improvements where experience showed there were problems but few drawings have survived at Birmingham for the first engines to show what happened. There are none for either the first engine at Soho nor its replacement as well as the one for Bloomfield Colliery. The drawings for all the engines supplied to both Wilkinson and Richard Reynolds at Ketley are missing. Another problem is to know whether the engine was erected according to the drawings. Sometimes alternatives, such as two different types of boilers, are shown. It is only when we come to the Bow engine that there are many drawings. After the Bow engine, the next two engines were to drain collieries, the earlier for Hawksbury Colliery, Bedworth near Coventry, with a 58 ins. diameter cylinder and the other in 1777 at Torryburn, Fife, on the shores of the Firth of Forth, with a 44 ins. cylinder.

Through the potential which his engine had to develop greater power than common ones, Watt found it necessary to strengthen the working beam by adding a king-post truss on his larger engines. Van Liender commented in 1786 that he had not seen this form of bracing 'because your pretty standard engine at Soho Birmingham has it not, nor that at Dudley [Bloomfield], neither that of Mr. Wilkinson at New Whilly, at least they had it not in 1776, when I have seen them'. [88] The beam for the Bedworth engine was a solid piece of timber, slung under the pivot without any bracing which first appears on that for Torryburn. In 1779, Watt carried out some experiments to determine the best height for the king-post and how much weight the diagonal stays carried. He found 'that, inversely as the height of the Kingpost above the level of the fastening of the stays was to the half beam so was the weight to the strain upon the stay so that if $^1/_2$ beam = 12 feet – KP e.g. = 6 feet the stay carried double the weight'. [89] This is but one instance of Watt employing calculations to secure the best proportions for parts of his engines.

Experience gained through working the first engines showed that the regulators or valves needed modification. One of the drawings submitted to Parliament in 1775 featured a sort of pivoted sector slide valve consisting of a box which oscillated across two openings. [90] While drawings show a valve in the top of the boiler to shut off the supply of steam, there was no other valve to regulate the steam going into the engine so that the top of the piston was open to full steam pressure all the time. The sector valve either let the steam into the bottom of the cylinder, allowing the piston to rise, or let that steam pass to the condenser, creating the vacuum, so the piston descended on the power stroke. However this did not allow any variation in the power developed. The only way to lessen the power was to reduce the boiler pressure. [91]

The drawing of the New Willey engine in the Egerton Manuscripts shows two valves, probably a sector valve for the steam inlet and a hinged flap valve for the exhaust. [92] The Bow engine most likely had two sector valves in a large valve box surrounded by a steam casing. Boulton had found this sort of valve admitted air. Certainly the flat rubbing surfaces would have been difficult to keep steam-tight and there would have been considerable friction between the sliding surfaces particularly as the engines and so valves increased in size. In

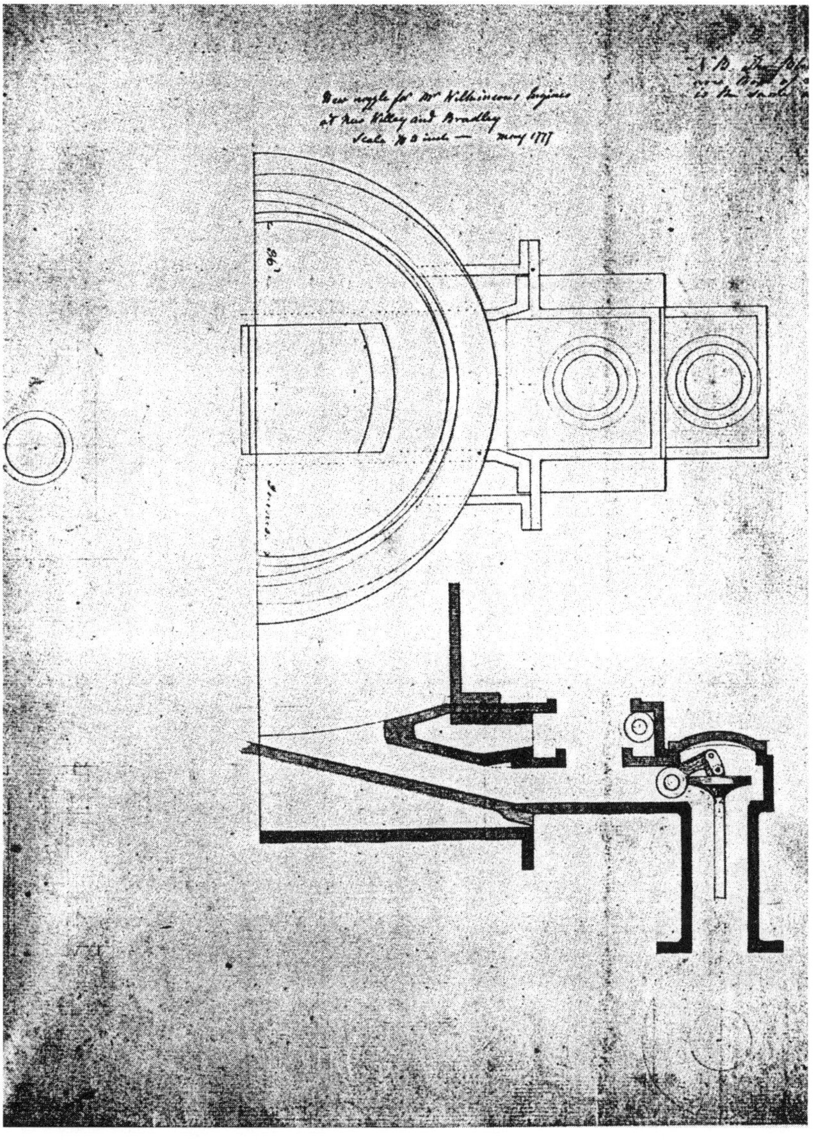

Lifting exhaust valve for Broseley and Bradley engines, 1777.

April 1776, Watt was planning conical valves closed by a weight and opened by 'the force of the steam', [93] but this idea was discarded and he designed a completely different type for the Bedworth engine in May. [94] These valves were circular copper discs which were raised vertically to fill holes in the valve box to prevent the steam passing and were lowered to open the passage. In this way Watt avoided the possibility of grit and dirt getting onto a conical seating but, in practice, it was found that the rising and falling of the discs caused grooving of the annular surfaces so that steam passed between them. A little later he introduced poppet drop valves with conical seats like those in a modern internal combustion engine. These were raised off their seats against the pressure of the steam by levers and links to open them. The Torryburn engine had disc valves when built, but New Willey was changed to drop valves in May 1777. [95]

The whole condensing apparatus was radically altered during these years. Watt had persisted with surface condensers because he recognised that less air and water would have to be removed but we have seen that he changed to jet condensing in October 1775 which he found operated much better. (See p. 55) It was only at this point in time that Watt was able to make his engine function properly but even so the air pumps needed further modifications. To understand the alterations to the air pumps, it is necessary to return to 1768 when Watt explained to Roebuck,

> In order to get quit of that small quantity of air, I propose making two exhausting barrels instead of one... I propose they should be worked by a double crank, the one ascending when the other descends... Now, suppose the communication with the cylinder opened, and the piston of one of these [air pumps] raised, the steam will rush in and be condensed, and the greatest portion of the air will be in the condenser, because, being heaviest, it will go in with the first of the steam. Now this piston, beginning to descend, shuts a valve that cuts off a communication with the cylinders; but the other piston ascending opens a fresh vacuum, in which there is no air, into which the remaining steam and air will go and be condensed, and the vacuum will be much more perfect than if there had been only one condenser having half the stroke. [96] (See Vol. I, p. 384)

The first engine at Soho had twin air pumps which presumably functioned sequentially in order to create as perfect a vacuum as possible. Watt continued to design multiple air pumps for his next engines.

Dickinson and Jenkins printed two drawings of the condensing apparatus for the Bloomfield and Broseley engines which seem not to have survived in the archives at Birmingham. [97] They show a very complex arrangement. The condenser itself was a cylinder 18 ins. diameter. At the bottom it communicated with a water pump 16 ins. diameter, which was in line with the steam cylinder and beam. It cleared out the water. Either side of the condenser was a pair of air pumps, 12 ins. diameter, joined to it through passages a little way above the bottom. The pumps were operated by chains round arcs on part of a wheel so must have worked sequentially to remove air from the condenser. This apparatus was placed below the steam cylinder between it and the beam wall. It shows Watt's concern about the amount of air which had to be removed.Two improvements followed. One was moving the whole condenser to the far side of the beam wall as at Bow. When it was situated immediately below the steam cylinder, the piston in the air pump rose, clearing out the condensate and air so improving the vacuum in the condenser, at the same time as the vacuum in the main cylinder was being destroyed by the incoming steam. By placing the air pump on the far side of the beam wall, its piston would be rising, helping to create the vacuum just when it was needed at the start of the power stroke, when the main piston in the steam cylinder was descending into the

Condenser, Air and Water Pumps For Broseley and Bloomfield Engines

Condensing apparatus for Broseley and Bloomfield engines. Water flows from condenser C through the bottom passage to the water pump cylinder E and water and air through transfer ports D to the air pumps G which operate consecutively.

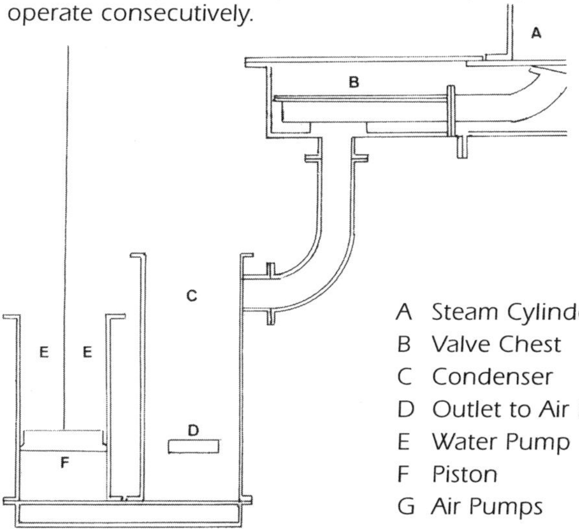

A Steam Cylinder
B Valve Chest
C Condenser
D Outlet to Air Pump
E Water Pump
F Piston
G Air Pumps

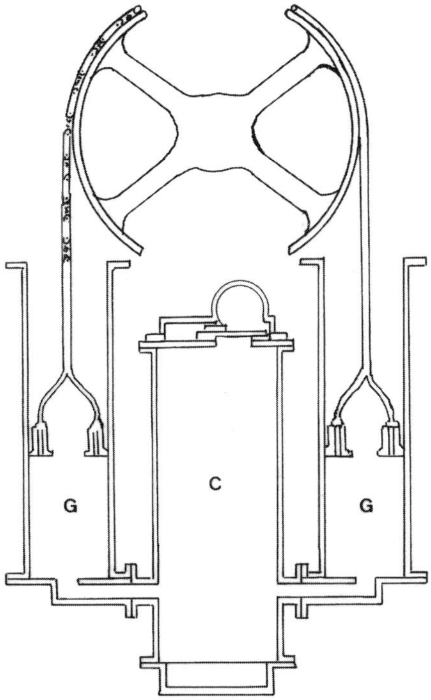

Air Pumps for Bedworth Condenser

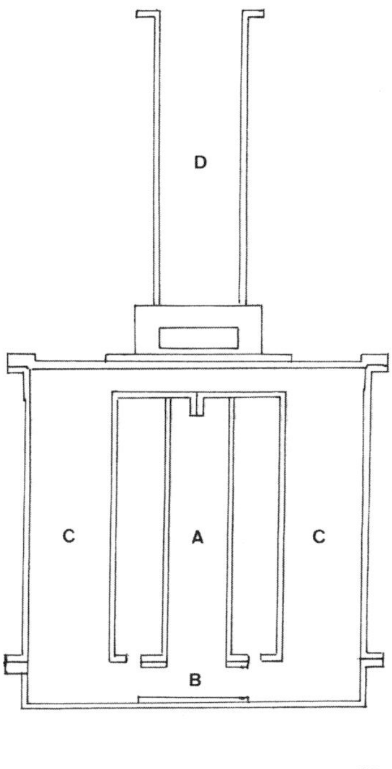

A Steam Pipe and Condenser
B Valve Box
C Air Pumps
D Water Pump

Twin air pumps feeding into the higher single hot water pump on Bedworth condenser, 1776.

Colville's Engine at Torryburn, Fife, October 1776

A Cylinder
B Valves
C Education Pipe
D Condenser
E Air and Water Pump
F Mine pumps
G Fireplace
H Boiler
I Flues
J Chimney

Single-acting pumping engine for Peter Colville, Torryburn, Fife, 1776. The steam jacket forms the main structural part of the cylinder assembly. There is a fireplace beneath the cylinder. Top: side elevation, plan bottom.

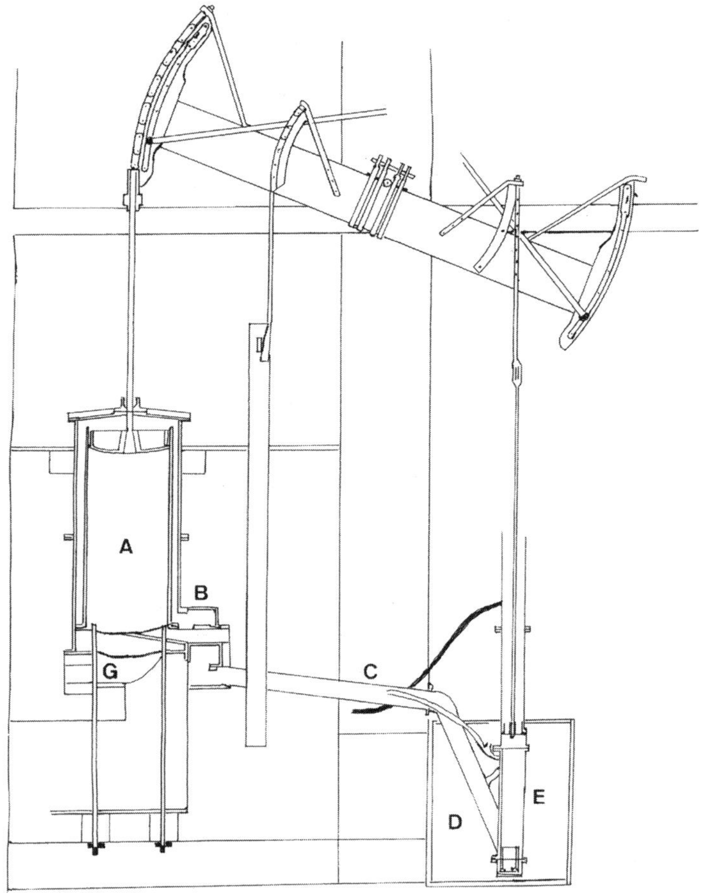

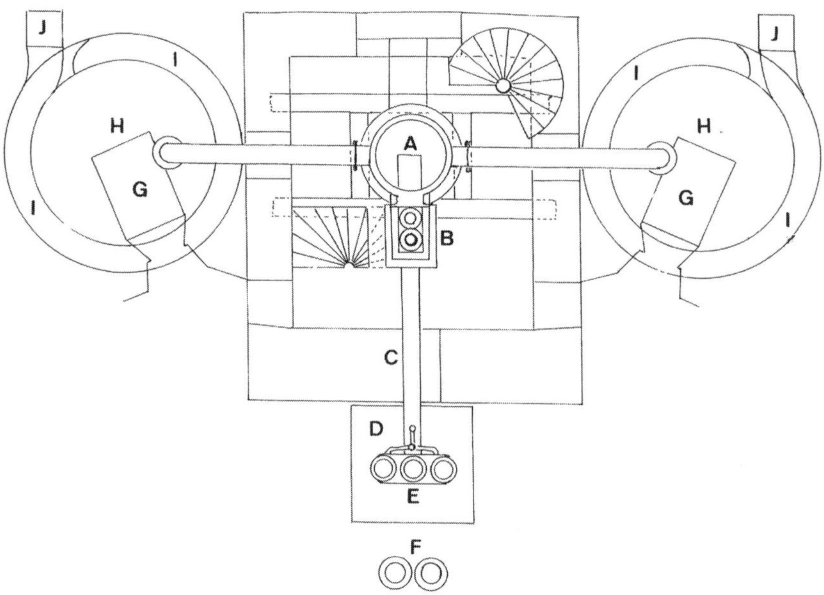

vacuum. This became the preferred position for the condenser in later single-acting pumping engines. It necessitated lengthening the eduction pipe which became the condensing vessel, dispensing with that cylinder. This pipe was sometimes jacketed with cold water to aid condensation. It sloped gently at first to pass through the beam wall but then bent to fall steeply down to the bottom of the air pumps. The nozzle of the cold water injection pipe was introduced at the bend to spray into the incoming steam.

Jet condensing was found to entrap any air so that the problem of the removal of the air was solved. The earlier complex arrangement of air pumps was discarded. A note on the drawing for the Bedworth engine reads, 'It has been necessary for several circumstances to deviate from the common form & divide the air pump into two barrels which are to be joined together at top by two side boxes cast in a piece with the pumps and forming between them the upper box'. [98] For this engine, a pair of 14½ ins. diameter air pumps lifted the water and air up to a common chamber, the upper box, from where a third water pump, also 14½ ins. diameter, lifted the hot water high enough for it to flow to replenish the boiler. All three pumps worked in unison by a yoke off the main beam. While this three pump condenser unit was installed at Torryburn, Watt quickly realised that, with all the pumps lifting at the same time, only one air pump was necessary to raise the water to the higher hot water pump which was the second improvement. This became the standard layout on later engines. At last, Watt's engine had reached a basic layout which would be followed for a while.

The Torryburn Engine.

While he was in Scotland during the summer of 1776, Watt obtained an order from Peter Colville to supply an engine for his coal mine at Torryburn in Fife. This order shows how quickly Boulton and Watt developed standard procedures and very similar designs for their engines. Watt had supplied Colville with two small Newcomen engines before 1769 and carried out repairs later. (See Vol. I, pp, 355 & 358) Colville planned to deepen his mine from twenty-five fathoms to forty-six so sought a replacement for his existing Newcomen engine. Watt visited the site in early July 1776 and accepted the order on the third. [99] Instructions for the cylinder were sent to Wilkinson at Bersham immediately. Boulton was asked to send a copy of the contract, presumably a standard form. Because Colville had no person with any experience of either erecting or running these engines, it was agreed that, 'a person is to be dispatched immediately at his [Colville's] charge to study under Messrs Perrins or Harrison the art of construction & using & to stay with us 3 months he is bred a carpenter & cabinet maker & has wrought in the latter branch at London'. [100] It is not known whether this became a common practice and was repeated for men to drive other engines.

On his return home, Watt was probably too busy catching up with work on other engines to pay any attention to this one until Hamilton reminded him early in October that Colville's stone masons were standing idle, awaiting plans. [101] Watt rapidly set to and prepared drawings for a 44 ins. diameter cylinder, coincidentally the same as the previous Newcomen engine, by 8 ft. stroke. [102] These show that the Boulton and Watt engine was settling down to a standard form for mine pumping engines. (See p135) Owing to its large size, the Torryburn engine had two haystack boilers, one either side of the engine house. The cylinder was mounted on a stone plinth with a fireplace beneath it to keep it warm, a feature later found unnecessary. The condensing apparatus was situated at the top of the mine shaft with twin air pumps.

One way of regulating the speed of these reciprocating engines may have been introduced at Torryburn by the supervising engine erector, Lt. Logan Henderson. He suggested the cataract. This consisted of a sort of bucket on a pivot which was filled gradually with water. When full, it overbalanced, tripping levers to open a valve on the injection pipe, at the same

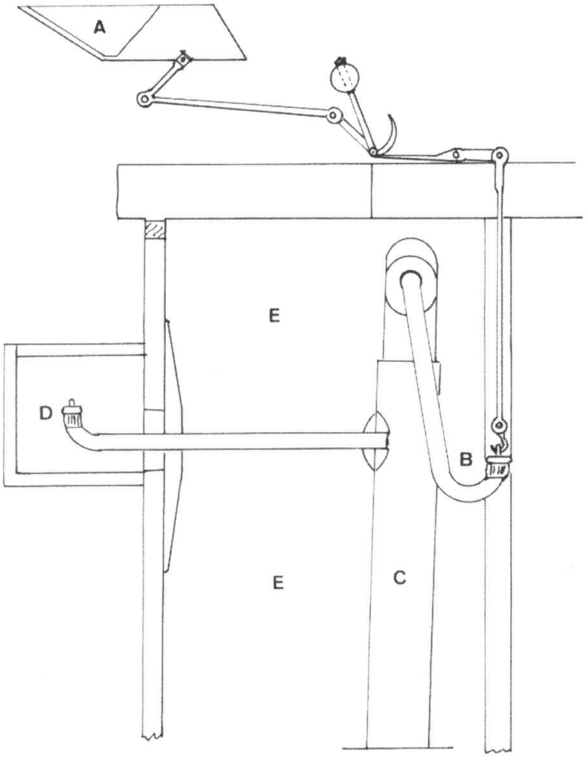

Cataract mechanism (A), injection (B) and blowing-through valve (D), 1779

time as emptying itself. Then a counterbalance weight returned it to be filled again. The rate of flow of water into the cataract bucket determined the number of strokes p. min. Watt replied to Henderson in November 1777,

> As to the cataract I have never given much consideration to the subject and if you can improve it shall be obliged to you, whatever you do with George Maclaren let it be in concert with Mr Colvile, but I think that George cannot serve us more effectually than by staying and taking care of Mr Colvile's engine for some time after sett [sic] to work. [103]

Here we find Watt willing to accept an improvement suggested by somebody else to his engine. Parts of Colville's engine were shipped to London by 3 May 1777. [104] Dickinson and Jenkins state that Henderson and William Symington the elder carried out its erection without any supervision from Watt who did not go to Scotland during this period. [105] Here again we see normal practice beginning where Watt had little involvement with actual erection. Presumably George Maclaren mentioned above was also involved and may have been the person sent to Birmingham. The engine commenced working in January 1778 but was soon stopped because Colville was declared bankrupt in 1779. It was running again by September 1782. [106] Watt had good reason to hope that a successful engine in Scotland would lead to further orders but that at Torryburn and another in 1778 for G. Meason's lead mine at Wanlock Head remained the sole examples for many years.

Progress Up To 1778

Good progress had been made so far. On Watt's shoulders had lain most of the complicated design work. The Boulton and Watt engine had proved itself to be more economical than other types and was reliable as well. It had been developed into three different versions which formed the basis for future designs up to the expiry of the patent in 1800. There was the blowing engine for Wilkinson which would be improved to give a more even blast first with a double-acting air cylinder and then a double-acting steam cylinder. Twenty-four blowing engines had been erected by 1800. Then there were the low lift pumping engines based on that for Bow. Fourteen were erected to supply water for towns or cities and a further nineteen to raise up water for re-use on canals. Most of these engines for waterworks were built entirely under covered engine houses which may have provided the model for rotative engines later. [107] So far, all these versions had limited applications. Greater potential existed in mines where the distance to which water had to be raised might be much higher. The demand here was for standard pumping engines. Coal and metalliferous mines outside Cornwall took probably forty-five in this period, spread over the rest of Britain. But the number in coalmines was not large through the way in which Watt charged for his patent by taking a proportion of the savings of the fuel. There were no savings in a coalmine when waste coal was being burnt so that coal mining was a very restricted market for Boulton and Watt engines. But this was not the case in Cornish metal mines where all coal had to be imported from South Wales and then transported to the mines. Boulton and Watt supplied forty-nine engines to this county, which was the greatest concentration of their engines anywhere. Cornwall would make or break the Boulton and Watt engine. These achievements were attained in spite of a background of serious lack of capital, financial insecurity and debt in the early years.

NOTES

1 Muirhead, (1854), Vol. II, pp. 15 – 6, Letter 89, W. Small to J. Watt, 14 Feb. 1771; Boulton 340.16.

2 JWP, 4/20.1, G. Hamilton to J. Watt, 17 June 1774.

3 Muirhead, (1854), Vol. II, p. 80, Letter 140, J. Watt to M. Boulton, 31 Jan. 1775; Boulton 348.45.

4 Ibid, Vol. II, p. 85, Letter 146, M. Boulton to J. Watt, 24 April 1775; JWP, 6/19.22.

5 Ibid, Vol. II, p. 160, Letter 160, M. Boulton to J. Watt, [1776]; Boulton 348.69.

6 Lord, (1923), p. 153.

7 Muirhead, (1854), Vol. II, p. 64, Letter 124, W. Small to J. Watt, 8 Oct. 1773; Boulton 340.35.

8 T. Shafto to J. Smeaton, 30 May 1779. I wish to thank A.P. Woolrich for drawing my attention to this letter.

9 Arago, (1839), p. 195, M.R. Boulton, Proceedings at the Freemasons' Hall, 18 June 1824.

10 Boulton 348.48, J. Watt to M. Boulton, 29 March 1775 and Muirhead, (1854), Vol. II, p. 84, Letter 146, M. Boulton to J. Watt, 24 April 1775; JWP, 6/19.22.

11 Ibid, Vol. II, pp. 86 – 7, Letter 147, M. Boulton to J. Watt, [May 1775]; JWP 6/19.6.

12 JWP, 6/19.6, M. Boulton to J. Watt, [May 1775].

13 Boulton 348.50, J. Watt to M. Boulton, 27 May 1775.

14 JWP, 6/19.5, M. Boulton to J. Watt, [May 1775]; this sentence was omitted by J.P. Muirhead in 1854, probably to avoid upsetting Victorian religious sentiment.

15 Muirhead, (1854), Vol. II, p. 90, Letter 150, M. Boulton to J. Watt, [June 1775]; JWP, 6/19.4.

16 Lindqvist, (1984), p. 294.

17 Muirhead, (1854), Vol. II, p. 91, Letter 151, M. Boulton to J. Watt, June 1775; Boulton 348.51.

18 Ibid, Vol. II, p. 93, Letter 152, M. Boulton to J. Watt, [June 1775]; JWP, 6/19.3.

19 Dickinson & Jenkins, (1927), p. 110.

20 B. & W. Col., MI, 2/8, 'Note Book, 1770 – 81'.

21 Robison, (1822), Vol. II, pp. 125 – 6.

22 Muirhead, (1854), Vol. II, p. 94, Letter 153, M. Boulton to J. Watt, [end of Aug. or Sept.] 1775; Boulton 348.52.

23 Ibid.

24 Farey, (1827), Vol. I, p. 320.

25 Dickinson & Jenkins, (1927), p. 110.

26 Farey, (1827), Vol. I, pp. 320 & 325.

27 JWP, 4/20.19, G. Hamilton to J. Watt, 2 Nov. 1775.

28 JWP, 4/20.20, G. Hamilton to J. Watt, 23 Dec. 1775.

29 Muirhead, (1854), Vol. I, pp. 63 – 4, Letter 58, J. Watt to W. Small, 28 May 1769.

30 Boulton 348.54, J. Watt, 'Case before Parliament'.

31 Muirhead, (1854), Vol. II, p. 85, Letter 146, M. Boulton to J. Watt, 24 April 1775; JWP, 6/19.22.

32 B. & W. Col., Office Letter Book 1/8-9, J. Watt to J.D.H. van Liender, 18 July 1775.

33 Ibid, MII, 4/9.6, J. Smeaton to J. Watt, 5 Jan. 1778.

34 Ibid, MII, 4/9.8, J, Smeaton to J. Watt, 30 March 1778.

35 Farey, (1827), Vol. I, pp. 327 – 28.

36 JWP, 4/69, 'Petition to Parliament', 1775.

37 Falck, (1776), pp. 18 – 20 for this and later quotes.

38 Muirhead, (1854), Vol. II, p. 103, Letter 162, M. Boulton to J. Watt, 20 April 1777; Boulton 348.71.

39 Lord, (1923), p. 153, M. Boulton to Hodgkinson, 8 May 1778; Tangye Mss.

40 B. & W. Col., MI, 2/8, 'Notebook of J. Watt'.

41 Dickinson & Jenkins, (1927), p. 113, J. Watt to J. Bentley & Co., 15 June 1775; B. & W. Col.

42 Ibid.

43 B. & W. Col., J. Watt to J. Wilkinson, 27 June 1775.

44 B. & W. Col., Letter Book, J. Watt to J. Wilkinson, 10 Aug. 1775.

45 Farey, (1827), Vol. I, p. 328.

46 Roll, (1930), p. 28.

47 Lord, (1923), p. 109.

48 Ibid, p. 108.

49 Dickinson & Jenkins, (1927), pp. 113 – 4.

50 Muirhead, (1854), Vol. II, p. 96, Letter 154, M. Boulton to J. Watt, 24 Feb. 1776; Boulton 348.57.

51 Braid, (1991), pp. 7 – 8.

52 Dickinson, (1914), p. 20.

53 Braid, (1991), p. 11.

54 Dickinson & Jenkins, (1927), p. 245 and Rolt, (1961), p. 65.

55 Dickinson, (1914), p. 16 and Smeaton, (1812), Vol. I, p. 365.

56 Muirhead, (1854), Vol. II, p. 94, Letter 153, M. Boulton to J. Watt, [end of Aug. or Sept.] 1775; Boulton 348.52.

57 B. & W. Col., MI, 2/9, 'Account Book', 8 Nov. & 7 Dec. 1775.

58 Smiles, (1878), p. 161.

59 Ibid.

60 Muirhead, (1854), Vol. II, p. 98, Letter 156, M. Boulton to J. Watt, March 1776; JWP, 6/19.1.

61 I am grateful to J. Verbruggen for pointing out J.D.H. van Liender's letter to J. Watt, B. & W. Col., Box 36, 31 May 1786.

62 British Library, Egerton Mss. 1941, ff. 5 – 20.

63 JWP, 6/19.2, M. Boulton to J. Watt, March 1776.

64 Dickinson, (1914), pp. 24 – 5.

65 Farey, (1827), Vol. I, p. 320.

66 Dickinson & Jenkins, (1927), p. 115, J. Watt to M. Boulton, 7 Feb. 1776; B. & W. Col.

67 JWP, 3/14, Boulton v. Bull, Trial in Court of Common Pleas, 22 June 1793.

68 Boulton 348.57, M. Boulton to J. Watt, 24 Feb. 1776.

69 Boulton 348.61, M. Boulton to J. Watt in Glasgow, 10 July 1776.

70 Boulton 348.62, M. Boulton to J. Watt in Glasgow, 25 July 1776.

71 B. & W. Col., MI, 2/8, 'Notebook 1770 – 1781', 2 Nov. 1776.

72 Lord, (1923), p. 158, J. Watt to J. Wilkinson, 17 July 1777.

73 Dickinson & Jenkins, (1927), p. 115.

74 JWP, 3/14, Boulton v. Bull, Trial in Court of Common Pleas, 23 June 1793.

75 Boulton 348.68, J. Watt to M. Boulton, 3 Dec. 1776.

76 B. & W. Col., MI, 2/9, 'Account Book', 7 Dec. 1775.

77 Boulton 348.61, M. Boulton to J. Watt, 10 July 1776.

78 JWP, C3/1, 'Memorandum', 3 May 1776.
79 Boulton 348.69, M. Boulton to J. Watt, [end July] 1776.
80 JWP, C3/1, 'Memorandum', 25 Nov. 1776 and B. & W. Col., MI, 2/9, 'Account Book' 9 Dec. 1776.
81 Roll, (1930), p. 33.
82 Boulton 348.71, M. Boulton to J. Watt, 20 April 1777.
83 JWP, C3/3, 'Journal', 23 May 1778.
84 Dickinson & Jenkins, (1927), p. 116.
85 Lord, (1923), p. 159.
86 Ibid, p. 154.
87 Dickinson & Jenkins, (1927), p. 114, J. Watt to J. Wilkinson, 27 Aug. 1779.
88 B. & W. Col., Box 36, J.D.H. van Liender to J. Watt, 31 May 1786.
89 JWP, C3/5, 'Journal', 3 Aug. 1779.
90 Dickinson & Jenkins, (1927), p. 117.
91 Andrews, (1988), p. 8.
92 British Library, Egerton Mss., 1941, ff. 5 – 20.
93 Farey, (1827), Vol. I, p. 320, J. Watt to J. Smeaton, April 1776.
94 B. & W. Col., Portfolio 625.
95 Dickinson & Jenkins, (1927), p. 176.
96 Muirhead, (1854), Vol. I, p. 21, Letter 27, J. Watt to J. Roebuck, 29 April 1768; JWP, 4/32.10.
97 Dickinson & Jenkins, (1927), pp. 208 – 09.
98 B. & W. Col., Portfolio 625.
99 B. & W. Col., MI, 2/8, 'Note Book', [before 8 July] 1776 and JWP, C3/1 'Memorandum', 3 - 5 July 1776.
100 Boulton 348.59, J. Watt to M. Boulton, 8 July 1776.
101 JWP, 4/20.32, G. Hamilton to J. Watt, 9 Oct. 1776.
102 Dickinson & Jenkins, (1927), p. 118, Plates XX & XXI.
103 JWP, W/5, J. Watt to L. Henderson, 9 Nov. 1777.
104 Boulton 348.73, J. Watt to M. Boulton, 3 May 1777.
105 Dickinson & Jenkins, (1927), pp. 118 – 19.
106 Lord, (1923), p. 159.
107 Andrews, (1988), pp. 23 – 5.

4
Family and Partners, 1774-1780s

When Watt wrote triumphantly from London at the beginning of May 1775 to tell his father about the successful passing of the Act to secure his patent for a further twenty-five years, it was almost exactly one year since he had left Glasgow. During that time, he had not been north of the border so had seen neither his father nor his children. He told his father that he 'intended to give myself the happiness of seeing you and the dear children'. [1] It is therefore appropriate to review what had been happening to his family during this time and in the next few years. But, before he could travel north, he had business to finish in London, such as the settlement with Roebuck. Then, when he did return to Birmingham, he was faced with the need to initiate drawings and work on orders for steam engines.

In the middle of August 1775, he left Birmingham for Sheffield, possibly in conjunction with an engine order. From there he went on to Glasgow. [2] While he took the opportunity to see his family, he was able also to settle some of the fees owed him for his canal surveys as well as discuss steam engine orders. The return to Birmingham with Watt junior followed the same route, arriving home on 21 October. [3] He was at once immersed in steam engine business, particularly with erecting the Bloomfield and New Willey engines which were finished in the spring of 1776. He was unable to set out for Scotland again until early June, this time not only to see his father and friends as well as transact more business, but also to marry his second wife, Ann McGrigor. He started his triumphant journey home with 'Annie' on 1 August, this time coming through Kendal and Warrington. He arrived home 'by miday' on the ninth. He would not return to Scotland for a few years. Once again he was quickly occupied with engine business, which included going to Bristol and Bath in September and London in December. Then orders from Cornwall began to dominate the activities of Boulton and Watt. Watt went with Annie to the West Country in early August, leaving on 28 September and arriving home on 2 October, [4] the first of many visits to that county on behalf of Boulton and Watt. These ten years up to 1785 have been described as:

> ... the most continuous and active exertion, and the most profusely fertile in mechanical invention, of the whole of his life... during which he secured by five several patents, the invention of the machine for copying letters and drawings, together with those almost innumerable improvements on the steam-engine, of infinite ingenuity and value, which he made subsequent to that of the separate condenser; besides originating other contrivances, which, although not included by him in patents, might have made both the fortune and the reputation of many a minor adventurer in that line. [5]

The Family

The letter to his father from London in May 1775 is one of the few which have survived out of Watt's correspondence with his Scottish friends from this period. He apologised to his father, 'I would have wrote to you oftener but had so little that was pleasing to communicate that I contented myself with letting you know of my welfare through the channel of our friends'. [6] The most important of these friends was Hamilton. Some of his letters to Watt have been preserved and it is mostly from these that we learn about Watt's affairs in Scotland.

Hamilton, who was always expressing his hopes that he would soon be seeing Watt back in Scotland, shared a common interest in natural philosophy. Watt kept him informed about progress with his engines, so Hamilton was pleased to learn about the success of Watt's improvements to his condensers. On his part, Hamilton enquired about such things as a triple achromatic object glass for a telescope consisting of crown and flint glass; [7] a clock in which Watt was involved with Small; barometers; a break-in at the Glasgow University observatory; increase in weight when iron was turned into steel; a press for James McGrigor and much more. Muirhead, Watt's cousin, commented, 'I expected to have heard from you before now, especially as I knew you was writing regularly to Gilb. Hamilton but I suppose you did not think my Genius was sufficiently clever, to understand the d-m'd abstruce subjects you correspond with him about'. [8]

Hamilton acted as Watt's Scottish agent for many years. He took care of Watt's finances there and tried to obtain payment of the fees owed to Watt for the surveys he had undertaken. Hamilton was in frequent contact with the officials of the Bo'ness, the Monkland as well as the Forth and Devon Canals. He was partially successful for in October 1775 Watt recorded having received £25 from the Bo'ness Canal and £50 from an unidentified one as well as some other amounts totalling over £100. [9] Hamilton told Watt that Lord Kilmaur intended to set up a small fire engine on his coal mine and wanted an estimate which 'must be for the engine compleatly finished in everything ready for work & the sooner you can send it the better as he wants to begin it as soon as he can Mr Dalyell says likewise that if this Engine should answer there are 3 or 4 more will be wanted'. [10] But of course Boulton and Watt did not supply complete engines so nothing came of this. Hamilton looked for a clerk for Boulton and Fothergill and it was most likely on his recommendation that James Pearson was offered this post. [11] He also kept an eye on Watt's family affairs in Greenock and informed Watt who had told his father, 'Please communicate what you have to say to me to Mr. Gilbert Hamilton, mercht. Glasgow who will write it to me'. [12]

By this time, Watt senior was beginning to age rapidly for he was about seventy-six. Although in Hamilton's early letters all seems well, in March 1775 he pointed out that Watt senior's memory was impaired. [13] Watt of course saw his father at Greenock when he went to Scotland that summer and had to settle a boundary dispute there. [14] However during his visit a year later, he not only arranged for an inventory to be made of his father's goods, [15] but also for his father's accounts to be examined and settlements made with the town authorities and other people. This suggests that his father was unable to handle his own affairs. This summer was possibly the last time Watt saw his father who was then being looked after by Miss Jean Cochrane. She reported in April 1779 that Watt senior's memory had quite gone. While several debts had been paid off, the opening of the Clyde navigation had taken trade away from Greenock so that the rents they received from letting houses had fallen. [16] Watt told Jean Cochrane not to worry about the low rental income since he would meet his father's and her wants as well as pay a person to attend his father. [17] Watt senior was unwell at the beginning of February 1782 [18] and took a turn for the worse in March. Thomas Cochrane wrote on 23 August 1782 that Watt's father had died in the morning of the previous day after an illness lasting four days. [19] Watt was deeply affected by the news as we see in his reply because their relationship had been close:

> Yesterday brought me your letter of the 23rd, containing the afflicting news of the death of my worthy and kind father. When we consider his death as a removal from a state of pain and disease to a state where we must hope that he will meet the reward of a well-spent and laborious life, we cannot with reason bewail his loss; yet there is something so afflicting in the thought of the final, solemn departure of a beloved friend and revered parent, that though I have been, by his long illness and declining state, prepared for the event, the account of it has given me much pain... It behoves me now, however, to lay aside unavailing regret, and to thank God that he has hitherto preserved me from signal misfortune. [20]

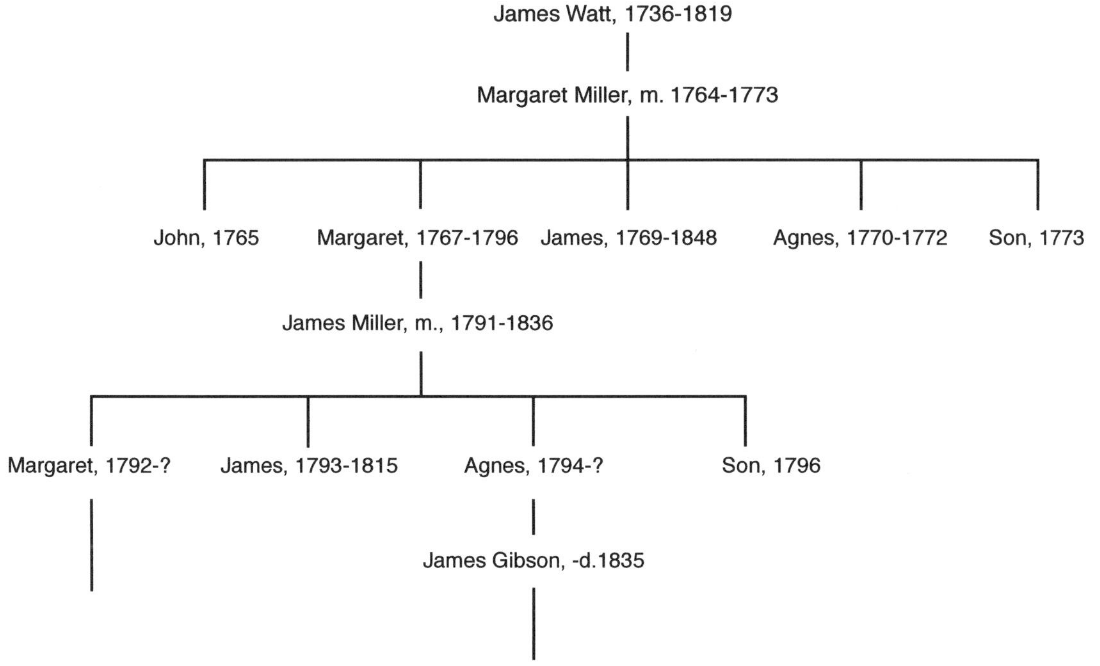

Watt could not attend the funeral but Hamilton did. [21] While Watt senior's affairs were soon settled, the house in Greenock was retained for several years.

When Watt first left Glasgow for Birmingham in May 1774, his children, Margaret and James Watt junior, were sent to Greenock, probably into the care of Watt senior, although for a time they were with Nancy Millar. Watt wanted his father to 'take care of their health and manners untill [sic] I return' and wanted them to go to a reading school. [22] In June, Hamilton wrote to Watt about looking for a boarding school, but this could have been for his own offspring. [23] Later in July, Watt senior commented that the two children and other friends were well but 'the children makes but slo progress in lerning'. [24] In September, 'Jemmy' would be going to church the following Sunday in a new set of clothes, [25] so the children appear to have been well looked after. But we have no inkling about the feelings of the children themselves, at first losing their mother through her death, then their father leaving them for an indefinite period as well as the upheaval of different people looking after them. These difficult circumstances in this period of their lives may have been the cause of the later breakdown in relations between them and their father as they grew up.

Margaret remained in Scotland when Watt returned to Birmingham in October 1775. She would have been about eight and may have felt betrayed once more by her father and so might have seen her true home with her friends in Scotland. Watt may have thought he would be unable to cope with looking after a young girl on his own. Watt junior, aged six, was taken off south and Hamilton was pleased to hear that both the James's had 'stood the journey so well'. [26] Watt junior was soon sent to school. While Watt was away at Broseley supervising the erection of Wilkinson's engine, Boulton kept an eye on the lad and took him to school. This sort of help may have led to a friendship between them. Certainly later Watt junior found he could discuss problems more easily with the avuncular Boulton than with his father. In May 1776, Watt paid £1.4s.6d. for 'Jemmy's schooling at Mr. Howels' [27] but left his son in Birmingham when he went to Glasgow.

The Lunar Society

While Watt's family remained in Glasgow, he was not left without friends in Birmingham. He had formed close relationships with Boulton and Small and through them met other people with similar interests in what became known later as the Lunar Society, perhaps better called the Lunar Circle at this time. This group may have owed its origin to Darwin for scraps of evidence point to early meetings at Lichfield rather than Birmingham. [28] Certainly Watt was received enthusiastically by Darwin as he was returning from London after failure to secure the Act for their 'Ditch' canal in 1767. (See Vol. I, p. 128) Watt had been to Birmingham, possibly to deliver a letter to Garbett from Roebuck. He saw Small at the same time who took him round the Soho Manufactory in Boulton's absence and may have sent him on to Darwin. In the following year, Small found himself in an embarrassing position after Watt had described his alkali experiments to him because Keir, who had been experimenting along the same lines, also confided in him. (See Vol. I, p. 158)

It may have been through Small that the centre of the Circle shifted to Birmingham for Small has been suggested as the true catalyst. Certainly Watt's arrival there confirmed this move. [29] It seems that it was not until after Small arrived in Birmingham that the informal Circle began to meet more regularly. [30] A letter probably written in the early 1770s by Keir to Boulton shows that meetings were being held when the moon was full:

> I had a letter yesterday from Dr. Darwin in which he says he longs for a little philosophical laughing.
> - Therefore when you are at leisure, some full moons Sunday, I hope you will indulge the Dr. and let us both know in time that we may lay aside our patients and Glass making to attend you. [31]

On his arrival in Birmingham, Watt was a prime candidate to join this dining group. An invitation is most likely to have occurred by September 1774 when Hutton sent his 'Compliments to Madam Bolton and the philosophical society'. [32] He found a group of friends with very diverse professional, business and philosophical interests. If through his personality Small was the main catalyst, then his death might have struck devastating blow. Boulton however stepped in to the rescue and told Watt in February 1776:

> Pray remember that ye celebrations of ye 3^{d} full moon will be on Sunday March 3^{d} Darwin & Keir will both be at Soho I then propose to make many motions to the Members respecting new laws & regulations such as will tend to prevent the decline of a society which I hope will be lasting. Pray bring Mr Wilkinson I think he will make a good member. [33]

These regulations have not survived but it must have been at this point when the informal Lunar Circle became the famous Lunar Society which Watt enjoyed attending for many years. Meetings were soon instituted for, while Watt was in Scotland getting married, Boulton told him that 'On Sunday next I expect Dr. G. Fordyce & his friend, also Kier [sic], Darwin, Charles & Withering'. [34] J.P. Muirhead pointed out that all the members were 'devoted to the pursuit of natural knowledge, and filled with mutual esteem and affection, who there found profitable pleasure in each other's society'. [35] It always remained a private exclusive group and never included any of the office or technical staff of Boulton and Watt. [36]

Watt's Homes in Birmingham

When he first arrived in Birmingham, Watt may have lived with Boulton at Soho House. After Boulton's help in securing the Act of Parliament for the extension of his patent, Watt felt under a moral obligation to continue the supervision and development of the engine business in Birmingham. He had probably decided to remain there before he left for Scotland in the summer of 1775 because they had agreed to form a partnership. This is evident from the letter, which Watt sent on 5 July, outlined the partnership terms. (See p. 87) Watt realised that he would have to make more permanent arrangements for living accommodation, particularly because he wanted to move his family south to be with him.

Boulton had been searching for a suitable residence while Watt was in Scotland for he wrote in August or September 1775, 'I have not been inattentive to ye procuring a House for you', [37] but the location of what he found is not known. Watt had retained his house at Calton in Glasgow, which he gave up when he left in October. Hamilton sold the household furniture for which he sent Watt the sum of £30. [38] The point of no return had been reached. After coming back from Scotland, Watt needed to furnish his new residence and recorded 'Expenses for house, £9.18s.5d.' which was spent on sheets, tablecloths, napkins, towels, chairs, coffee pot, 12 cups and saucers and £2.12s.6d. for a smoke jack. [39] At least he intended to eat well.

Watt did not remain here long because, at the beginning of April 1776, he 'Took part of Newhall house from Mr, Holloway at £16 rent, he to layout £5 upon repairs & me as much – to have 3 month's warning in case of removal'. [40] The repairs could not have taken long because Watt moved in a week later on 11 April. [41] This was 1 Newhall Walk. [42] It has been suggested that Watt moved here as soon as he arrived in Birmingham, and that this was Boulton's old residence since Boulton and his father had lived somewhere in Newhall Walk, but exactly where is uncertain. [43] At any rate, this was the house to which he brought Annie in 1776 and where the family remained until they 'flitted from Newhall to Harpur Hill', Regent's Place, on 22 March 1777. [44] It is not known where Watt established his first headquarters which were later situated at Regent's Place where he did his correspondence, drawings and calculations. His assistants worked here too so for days on end Watt might not visit Soho which was about a quarter of an hour's walk away. It was only in 1790 when Watt moved to his new home of Heathfield that the office was transferred to the Manufactory. There is no evidence that either at Newhall or Regent's Place Watt set up a workshop for experimental work with the exception of chemistry. [45]

Watt's house at Regent's Place, Harpur Hill.

Second Marriage

The first hint we have that Watt's visit to Scotland in the summer of 1775 had turned into more than just settling his affairs and securing possible engine orders comes in a letter from Hamilton that October:

> I don't know but I may turn ?bad? to you, when I consider what a sacrifice I am making of giving up the pleasure of her company altogether & being totally deprived of her unless your High Mightiness shall vouchsafe to grant us a sight of her & for all your fine promises now, things appear in a very different light after & before & you may repent at visiting these cold Bleak northerly regions so often as you intended especially as you will then have carried off what you reckon the soul & sun of the country, & there will be nobody to visit but a ?percell? of dull stupid Mortals - You will perhaps think I am mad: indeed I am very well pleased as Annie that the contents are in no ways displeasing to you - you see how disinterested I am in rejoicing at my own misfortunes. [46]

This, and the ensuing letters, are a rare instance of some light-hearted banter appearing in any of the correspondence in the Watt archives. Watt had chosen Ann McGrigor, Hamilton's sister-in-law. She was the favourite daughter of James McGrigor, the bleacher. We do not know whether the introduction came through Hamilton and his wife or through Watt's earlier business connections with McGrigor who had discounted some bills for Watt in 1764 and 1765. (See Vol. I, p. 134) Watt continued to advise McGrigor about waterwheels and other machinery in the following years so he could have been acquainted with Annie for quite a long time.

We know that Watt was an avid correspondent. He must have written to Annie during the ensuing year but no letters have survived. From the tone of Hamilton's letter, it would appear that the affair was well advanced, if not virtually settled, but Watt kept it a close secret even from Boulton and Watt junior. Watt's visit to Scotland in the summer of 1776 was, for him at any rate, primarily to marry Annie while business, such as securing the order for the Torryburn engine, took second place. It could well be that the cautious, hesitant Watt was afraid that there might be problems and the whole venture fail which was why he did not want to tell his southern friends until the knot was actually tied. However McGrigor, perhaps knowing Watt well, demanded to see the partnership articles between himself and Boulton. Therefore on 3 July, Watt had to let the cat out of the bag and wrote to Boulton:

> I have however the pleasure amidst all this to have obtained both the young lady & old Gentlemans consent to take her along with me, but shall be obliged to wait a little untill [sic] she can be ready. This gives me great pain as I am sensible I must be wanted with you, but I hope you will excuse it when you consider my situation & that it will save another journey which must otherwise have taken place soon. [47]

Then came the request that Boulton procure a legal contract for their partnership which had never been formally drawn up and signed. This is discussed in the section on the partnership. (See p. 87)

A later part of this same letter points to Watt not having revealed his new love to Boulton and being fearful of Boulton's adverse reactions. He hoped that Boulton could come to Glasgow and continued:

> ... whether a man of the world such as you look upon my present love as the folly of youth or the dotage of age - I find myself in no humour to lay it aside or to look upon it in either of these lights, but to consider it as one of the wisest of my actions & should look upon a disapointment [sic] in it as one of the greatest of my misfortunes...
>
> I have had better health since I left you than has been my lot for years and my spirits have borne me through my vexations most wonderfully. [48]

The prospect of marriage roused Watt's spirits and temporarily relieved his headaches and bouts of depression.

Watt wanted to marry as soon as possible so, hearing nothing from Boulton, wrote again five days later, stressing the need for some form of partnership contract. He also pointed out that, 'I must do him [McGrigor] the justice to say that he has behaved to me in a very open & friendly manner: and when he found that his daughters affections were engaged beyond recall, gave his consent with good grace'. [49] So Watt had found his second love, but he finished this letter with, 'I hope Jemmy is minding his school and is well You need not tell him nor anybody else that I am going to bring him home a mama'. This secrecy and sudden shock of coping with a new mother further unsettled the lad.

Watt need not have worried about Boulton's reaction. Boulton received Watt's first letter and replied on the 10th, congratulating him on his engagement and sending a pseudo-deed of partnership which he said that they must 'have it executed as soon as you return'. [50] McGrigor accepted the deed which Boulton had concocted so that the marriage arrangements went ahead. Watt made the proud announcement to Boulton on 28 July 1776:

> I reckon I have made a good one [bargain] in securing to myself a valuable friend - All is settled & upon Monday the fatal noose is to be put about my neck, or more emphatically, I am to get a lick of the eternal glue pot - At present I am something in the state of a man waiting for his dinner after the cloth is laid, comforting myself for the delay by the certainty of a good meal - I wish my appetite may be as good - I dare say I have your prayers. [51]

The happy couple left for Edinburgh the following Friday and arrived in Birmingham on 9 August. This time Margaret was taken with them, the total expenses for the three costing £18.7s. [52] The family was reunited.

Hamilton continued in his jocular vein in his first letter to Watt after the marriage:

> Poor man! How I pity you, what a sad reprobate you have got for a wife, that instead of letting you indulge yourself in your good old customs of sleeping by the fireside after dinner, indulging your favourite laziness, & being sulky or peevish, etc, rouses you up & keep you in order, & set you to that dismal work of reading, writing or drawing or that most disagreeable of all businesses, the answering of letters - Indeed, indeed, you are much to be pitied, & sorry am I that your Philosophic Spirit should be so much plagued you would not take my advice, but run into danger with your Eyes open, so you must make the best of the bargain you can - However for your comfort & consolation I think our Madder Vats will be mostly over by the end of November, when I intend to come & see how she behaves & help you to keep her in order though perhaps she may get the better of both of us. [53]

Hamilton knew Annie's sister well and perhaps realised that both had similar characters for there is some truth in this letter. Later Annie is reputed to have trained her pug dogs never to cross the hall without wiping their feet on the mat, while Watt himself had to be careful to leave his leather apron and dirt in his garret workshop for fear of upsetting her. [54]

Annie's character was very different from Peggy. She was certainly more extrovert with a more powerful personality and so perhaps was better suited as the wife of a leading industrialist. D.F.J. Arago commented, 'Her various talent, soundness of judgement, and strength of mind, rendered her a worthy companion'. [55] But there are signs that she found Watt rather dour and missed the laughter of her father's home at Clober in Scotland. J.P. Muirhead speculated on Watt's feelings concerning this second marriage:

> Yet possibly, in the long forty-three years of his second wedlock, amid all the prosperity and fame by which they were marked, there may have been moments when his heart throbbed at the retrospect of an earlier time; and of an union, in days that were no more, with one whose loving hopes had sustained him in sorrow, without being permitted to taste of his joy; who had beheld his success and renown only by anticipation, and yet with all the firm faith of undoubting affection; and who had been summoned from his side just as he was about to emerge from the comparative obscurity in which he had long so wearily pined. [56]

The Family Growing Up

The family was together for only a short time since the two children were left behind at Birmingham when Watt and Annie set out for Cornwall in August 1777, arriving home again on 2 October. Annie gave birth to their first child Gregory about one o'clock on the morning of the eleventh. The christening took place on 10 December. [57] Their second child, a daughter, was born on 5 May 1779 around nine in the evening. [58] Watt told Hamilton that Annie was 'very ill but for a much shorter time than of Gregory'. [59] Perhaps this may give a clue why we do not hear of any further children. The baby was christened on 8 June 'by the name of Jane or Janet' but later was called Jessy. [60]

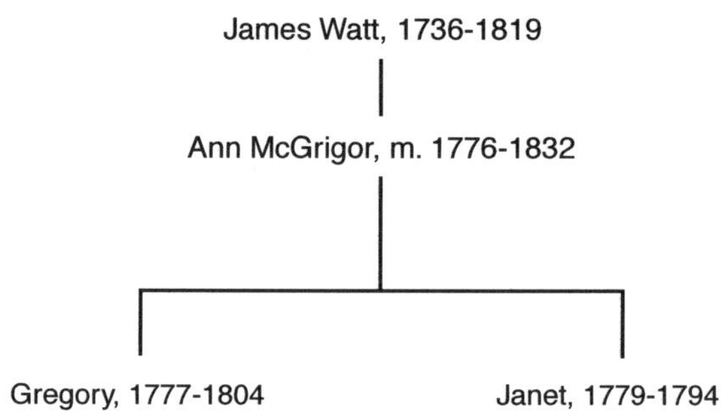

Watt's family by his second marriage.

Watt had to leave for Cornwall in September 1779 when Annie took the opportunity of going to her parents in Scotland where she remained until April 1780. Watt's journal shows that he corresponded regularly, at one time asking her to visit his relations. He obviously missed her for he wrote in March, 'This house grows more and more lonely every day for want of you and the little prattlers; I feel an excessive want of your company and advice. everybody wonders what is become of you. I believe some of them think that you are run away'. [61] Their long periods of separation, especially when Watt was in Cornwall as well as Watt's own stress through overwork, placed a great strain on their marriage. He was ill and confined to the house at the beginning of April 1780 and complained that she had delayed her journey home, telling her, 'Why should you suppose me to be grown indifferent about you, is it because I have lett you stay so long to my own vexation? I am angry at your suspicions'. [62]

Few details have survived about the family in the ensuing years. The children suffered from some ailment in the autumn of 1780 from which Gregory had recovered and Margaret, now called Peggy, was back at school again. This is all we know about her at this time. Jemmy, so called by his father and now eleven, had been causing problems at school and did not get on very well with Annie. Earlier Watt had written to Mrs. Nancy Marr in Scotland:

> As to James he possesses very considerable skills but I cannot say his word of honour is quite so much to be depended on as his sisters. [He] is a little saucy sometimes but his comrades will drub him into good manners at school and at home he dares not exert it. I am sorry his abilities do not lie in the line of my profession he seems to be best fitted for a lawyer or a merchant - but in whatever way they lie, if he prove an honest industrious man I shall be satisfied. [63]

At this time, Watt seems to have been willing to allow his son to choose his own career.

Jemmy must have been cheeky to his father who gave him 'a severe correction' and told the master at Winson Green School, Birmingham, that he was 'to be severely punished in future'. [64] Watt thought he would have to change his son's school, 'as he seems rather to be getting from under the masters command'. [65] Watt junior was obviously a high spirited lad, quite typical for his age, so that his father complained, 'I am sorry to see that he is a much greater sloven and more careless of his clothes than his father ever was'. [66]

Watt sought advice from the Rev. Mr. Deane at Shifnal over a suitable school:

> Having a son about 12 years of age at Winson green School whom I want to remove to some other school where his learning and manners may be more particularly attended to than they can be in so numerous a school my friend Doctor Withering has done me the favour to mention you as a person to whose care I may safely entrust his education...
>
> The boy is now attending writing and arithmetic in which latter he is advanced so far as to the extraction of the Cube root. He also reads latin, has read Caesar a first time and is now reading Virgils Aeneid. In a fathers opinion he is no despicable scholar for his years.
>
> I want him to be further instructed in Writing, Arithmetic, and Latin, and to be radically taught Euclids Geometry, with such of the dependent Sciences as his time admits of.
>
> But what I most desire is that a strict attention be paid to his manners and morals these being the most essential and also the prevailing reason with me to remove him from a public school
>
> When you do me the pleasure to answer this letter shall be obliged to you to lett me know if you have any persons who teach French & dancing to your Scholars. [67]

We may note Watt's concern for his son's morals as well as his ambitions for his education. Jemmy put the request to his father that he should learn French through Boulton. [68]

Jemmy was eventually sent to a Mr. Pickering to whom Watt complained in the autumn of 1781 about a drawing he had received 'which did not appear to me so well done as I wished and expected'. [69] At the same time, he wrote to Jemmy in possibly the first of many hectoring letters, complaining about the quality of this drawing and continuing:

> You may be sure that it will give me the greatest pleasure to see you superior to your contempories in every branch of learning & I must be equally mortified to see you

> inferior in any - Your childhood is now passing away and you should spare no pains to acquire manly knowledge against the time your age will entitle you to be called a man - This can only be done by serious application to the different sciences which you are learning, by suffering none of your working hours to pass away except in the pursuit of laudable knowledge, and by never desisting from the study of any thing you are sett to learn untill you make yourself master of it & then to practise it often lest you forget it. [70]

Watt had maintained cordial relations with his own father at this age so it is surprising that he failed to learn how to treat his son.

The next we know about Watt junior is that he was to learn a trade under Wilkinson with whom he was staying at the beginning of May 1784. Watt told Mrs. Campbell:

> I have sent my son to Mr. Wilkinson's ironworks, at Bersham, in Wales, where he is to study practical book-keeping, geometry, and algebra at his leisure hours; and three hours in the day he works in the carpenter's shop. I intend he should stay there a year; what I shall do with him next I know not, but I intend to fit him for some employment not so precarious as my own. He has not yet manifested anything which may be called genius, but he has abilities to learn anything, draws tolerably, and writes a letter in good style... My younger children are very thriving, very amiable, and sufficiently advanced in knowledge for their years. [71]

Watt junior had already written from Bersham, excited at having seen a cylinder being cast but added, 'it was for france although they had not any particular order for it'. [72] His time was not entirely filled with work or study for he visited Wrexham church to see the memorial for Mrs. Wilkinson and had a good view of Chester from the top of a mountain behind Minera. No doubt he found such occupations rather more to his liking than writing home so he filled the page of his next letter with large characters, only to be rebuked by his father, 'I am glad to hear of your progress in your studies which I hope you will continue apply with Spirit and attention. I remark that your writing in your letters is not improved and that in your last letter is much too large for one who is not a mere schoolboy'. [73] Watt junior's large handwriting continued to be a bone of contention for some time. His father continued,

> You should not write to me concerning Mr. Wilkinson's business it may be taken amiss and you may be looked on in the light of a tell tale I hope that you will consider you are under very great obligations to Mr. Wilkinson for permitting you to be instructed at his works and in great measure at his expense and that therefore it is your duty to do him every service in your powers and to be as useful as you can without being officious, such conduct is the sure way to be esteemed and in fact will prove the most useful to yourself. [74]

At this stage, neither Boulton nor Watt had any suspicions that Wilkinson was tricking them by casting engine parts according to their patent for other customers without permission. Watt junior would have been put in a most embarrassing position had he found this out.

Further criticism and instruction continued to pour forth from the pen of an anxious father who insisted on some manual and practical education as well as continued reading of some classics. Watt junior was to study merchants' accounts in the afternoon and arithmetic and algebra in the evening. He was advised to spend no more money than was necessary and to drink no strong liquors. [75] Perhaps if he had followed his father's injunctions to the letter, he would have had no time to have a fall from a horse, resulting in some illness. [76]

As it was, Watt junior did not remain at Bersham for the whole of the proposed year because in September 1784 Watt sought advice from Hamilton about sending his son to the college in Glasgow. But Watt changed his mind and recalled Watt junior from Bersham in October in order to send him to Geneva. [77] The reason he gave to Hamilton was rather surprising for he thought that Watt junior's 'morals are less in danger of being corrupted there and the advantage of learning french; as to the danger of learning Scotch at Glasgow that I look on as a ?nullity? as he is now too well grounded in English ever to unlearn it so far as to acquire the true Scotch tongue'. [78] Watt wrote to J.A. De Luc who was to supervise Watt junior's time in Geneva that 'My views are to breeding my son to be an Engineer (mechanician). Therefore his studies should be directed principally to that end'. [79] He had changed his views about his son's future. He pointed out how easily his son could be led astray. Perhaps he hoped that the strict Protestantism of that city would prove to be a good influence. At this point, we will leave Watt junior setting off on his first visit to the Continent and return to the story of Boulton and Watt.

The Partnership

From first meeting Watt when he was returning from London in August 1768 having initiated the start of proceedings to secure his patent, Boulton had been interested in becoming involved in the exploitation of this new form of steam engine. However he declined Roebuck's offer in December for a limited share covering three English counties only. Thir negotiations might have ceased had not Roebuck become bankrupt. Roebuck's two thirds share in the profits conditional upon his financing the development were transferred to Boulton before Watt moved to Birmingham through Boulton relinquishing claims to debts owed by Roebuck both to himself and Boulton and Fothergill. The agreement arranged by Watt and Small was that Roebuck should have a share in any eventual profits which Boulton decided to anticipate, making a settlement in June 1775, so removing Roebuck from the steam engine business. Following the unfortunate death of Small, only Boulton and Watt were left who might be involved in any partnership, Fothergill having declined to participate.

The 1775 Act mentioned only himself and his heirs and assignees. Watt was still in no position to fund his steam engines, either to continue trials for improving their performance or to supply capital for erecting any. Equally he was morally indebted to Boulton for funding the early experiments at Soho as well as securing the Act. Discussions about regularising their partnership may have begun while Watt was still in London when he was finalising the agreement with Roebuck, or when he returned to Birmingham. There had been some earlier proposals before Watt wrote from Scotland to Boulton on 5 July 1775:

> As you may have possibly mislaid my missive to you concerning our contract, I beg just to mention what I remember of the terms.
>
> 1. I assign to you 2/3 of the property of the invention on the following conditions:
> 2. You to pay all expenses of Act, or others incurred before June, 1775, and also the expense of future experiments, which money to be sunk without Interest by you being the consideration you pay for your part, but the experimental machines to be your property.
> 3. You to advance stock in trade bearing Interest, but having no claim upon me for any part of that further than my Intromissions, but the stock itself to be your security & property.

4. I to draw one-third of the profits so soon as any arise from the business, after paying the workmens wages and goods furnished, but abstract from the stock-in-trade excepting the Interest therof, which is to be deducted before a balance is struck.
5. I to make drawings, give directions, & make surveys, the company paying travelling expenses to either of us when upon engine business.
6. You to keep the books & balance them once a year.
7. A book to be kept wherin to be marked such transactions as are worthy of record, which, when signed by both to have the force of the contract.
8. Neither of us to alienate our share without consent of the other, & if either of us by death or otherwise shall be incapacitated from acting for ourselves, the other of us to be sole manager without contradiction or interference of Heirs, executors, assignees or others but the books to be subject to their inspection & the acting partner of us to be allowed a reasonable commission for extra trouble.
9 The contract to continue in force for 25 years from the first of June, 1775, when the partnership commenced, notwithstanding the contract being of later date.
10. Our Heirs, Executors & Assignees, bound to observance.
11. In case of demise of both parties, our Heirs, etc. to succeed in the same manner, & if they all please may burn the contract. [80]

Watt asked that, if Boulton found nothing 'very disagreeable in these terms', a deed should be 'drawn up by an able lawyer with expedition'. [81] The period would terminate with the expiry of the Act.

Boulton replied with the hint that he might be willing to buy Watt out. He may have been afraid that Watt would remain in Scotland, in which case he recognised that he could not continue with the steam engine business alone and would have wasted his investment:

> I know not what property you may have in Scotland but I know you have here (unincumbered from all debts whatsoever) 1/3 of Act of Parl[t] which I value at some thousands, & if I had no regard for you nor wish for you to live in this Country I would give you oppertunity [sic] of realising that property wch now perhaps may be cons[d] as a mere shadow but I say the value to ye sellor of everything is just as much as it will bring & if you was fair game I wd endeavour to make a good bargain with you & would offer you for your third share a sum equal to what I have given & what I am yet to give to Dr R[oebuck] for his 2/3 but when I am dealing with you I had rather make a bad bargain than a good one but in the present case I had rather ware out our Act of Parlt. ensemble than gain a greater fortune by being ye sole proprietor.
>
> I would have sent you the assignment & the Article of partnership had it been in my power but Mr Dadley is suddenly called to London & it cannot be done before his return but if you want to shew it to any of your friends you may give them a copy of the following wch I have extracted from our mutual missives & is to the best of my knowledge all that our articles contain. [82]

This suggests that Boulton genuinely wanted a partnership with Watt through their friendship as much as through any business attractions. But there the matter rested, probably forgotten owing to pressure of erecting the first engines, until Watt wished to marry Annie.

On his return to Scotland again in the summer of 1776, having obtained consent from both the young lady and the old gentleman to 'take her along with me' on 3 July, Watt was faced with a nasty surprise before he could marry her:

> The only disagreeable part of the business that remains to be done is the settlement, & I find that the old Gentleman wishes to see the Contract of Partnership between you & I, and as that has never been formally executed, I must beg the favour of you to get a legal contract writen [sic] & signed by yourself - sent to me by return of post or as soon as may be. Lest he should have called my prudence in question, I have been obliged to allow him to suppose such a deed did exist, but was single, so what you send must pass for a duplicate and another may be actually written which I hope you will not doubt of my readiness to execute as soon as I return, and in fact this deed should have been executed long ago, in common prudence upon both sides, particularly upon yours who have no legal assignation to the Act of Parliament - I therefore hope that you will excuse the old Gentleman's caution. [83]

Having heard nothing from Boulton, Watt wrote again on 8 July, pointing out McGrigor's fears.

> The thing which seems to stick most in his stomach is that somehow or other in the case of failure of success I may be brought into a load of debt which may totally ruin me, I hope therefore you will excuse his caution in this matter, as I do, when you consider that he is disposing of a favourite child. [84]

Perhaps if McGrigor had been able to see into the future, he might have felt that his fears were justified through the near-collapse of Boulton's empire in the early 1780s.

Boulton replied with a draft similar to that of Watt's a year earlier. He pointed out that it was difficult to value the partnership but that he would be willing to pay Watt two or three thousand pounds for his share in the Act. He hoped that they would live harmoniously for the ensuing twenty-five years 'which I had rather do than gain a Nabob's fortune by being the sole proprietor'. [85] He made the same excuse as before, that Dadley had been suddenly called to London so he could not send any formal articles of partnership but continued with a slightly modified version of the terms which Watt had sent him. The most significant addition, in view of the problems Watt had faced and his inability to deal with the contractors on the Monkland Canal, was that to the sixth article - 'I [Boulton] am also to assist in managing workmen, making bargains, or doing whatever we may jointly think is for the interest of the trade'. [86] Boulton apologised that he had 'ill expressed himself' for he was in a great hurry to catch the post. McGrigor was satisfied, never realising the machinations behind his back. Yet it is surprising that neither the cautious Watt not Boulton the businessman had formally executed any deed of partnership in over a year from its practical commencement. There is a note in Watt's Journal for 20 April 1779, 'Executed deed of partnership with Mr. Boulton' but it is more likely that this refers to a partnership with Keir. [87] No partnership deed has ever been discovered among their surviving papers, reflecting the mutual confidence the partners had in each other. Later there was at least some informal understanding of their commitment to each other which was to prove so profitable and successful. (See p. 151)

NOTES

1 Muirhead, (1854), Vol. II, p. 89, Letter 149, J. Watt to J. Watt. sen., 8 May 1775; JWP, 4/60.4.
2 B. & W. Col., MI, 2/6, 'Cash Memorandum', 17 & 21 Aug. 1775.
3 Ibid, 16 to 21 Oct. 1775.
4 JWP, C3/1, 'Memorandum', 1 – 9 Aug. 1776 and B. & W. Col., MI, 2/9, 'Account Book', 10 Aug. & 2 Oct. 1777 and JWP, C3/2, 'Journal', 28 Sept. 1777.
5 Muirhead, (1858), p. 33.
6 Muirhead, (1854), Vol. II, p. 89, Letter 149, J. Watt to J. Watt sen., 8 May 1775; JWP 4/60.4.
7 JWP, 4/20.1, G. Hamilton to J. Watt, 17 June 1774.
8 JWP, 4/20.4, R. Muirhead to J. Watt, 7 Jan. 1775.
9 B. & W. Col., MI, 2/6, 'Cash Memorandum', 10 Oct. 1775.
10 JWP, 4/20.15, G. Hamilton to J. Watt, 5 April 1775.
11 JWP, 4/20.21, G. Hamilton to J. Watt, 15 Nov. 1775.
12 JWP, 4/60.1, J. Watt to J. Watt sen., 1 June 1774.
13 JWP 4/20.12, G. Hamilton to J. Watt, 10 March 1775.
14 JWP, 4/3.1, T. Cochrane to J. Watt, 16 Sept. 1775.
15 JWP, C3/1, 'Memorandum', 22 July 1776.
16 JWP, 4/21, Miss J. Cochrane to J. Watt, 9 April 1779.
17 Ibid, J. Watt to Miss J. Cochrane, 5 March 1782.
18 JWP, JW/22.38, Miss J. Cochrane to J. Watt, 7 Feb. 1782.
19 JWP, JW/22.37, T. Cochrane to J. Watt, 23 Aug. 1783.
20 Muirhead, (1858), p. 302, J. Watt to T. Cochrane.
21 JWP, 4/43.13, G. Hamilton to J. Watt, 30 Aug. 1782.
22 JWP, 4/60.3, J. Watt to J. Watt sen., 15 July 1774.
23 JWP, 4/20.1, G. Hamilton to J. Watt, 17 June 1774.
24 JWP, 4/3.3, J. Watt sen. to J. Watt, 31 July 1774.
25 JWP, 4/3.2, J. Watt sen. to J. Watt, 28 Sept. 1774.
26 JWP, 4/20.19, G. Hamilton to J. Watt, 2 Nov. 1775.
27 JWP, C3/1, 'Memorandum', 27 May 1776.
28 Robinson, (1962 – 63), p. 155. See Schofield, (1963), for the best account of the Lunar Society. Uglow, (2002), is well written but inaccurate.
29 King-Hele, (1999), p. 120.
30 Clagett, (2003), p. 215.
31 See Robinson, (1962 – 63), p. 163.
32 Ibid.
33 Boulton 348.57, M. Boulton to J. Watt, 24 Feb. 1776. See also Schofield, (1963), pp. 141 – 42.
34 Boulton 348.62, M. Boulton to J. Watt, 25 July 1776.
35 Muirhead, (1858), p. 391.
36 Schofield, (1963), p. 152.
37 Boulton 348,52, M. Boulton to J. Watt, Aug. or Sept. 1775.
38 JWP, 4/20.21, G. Hamilton to J. Watt, 15 Nov. 1775. See also JWP, 4/20.29, 29 Nov. 1775.
39 B. & W. Col., MI, 2/6, 'Cash Memorandum', 1 Nov. 1775.
40 JWP, C3/1, 'Memorandum', 2 April 1776.
41 Ibid, 11 April 1776.
42 Dickinson & Jenkins, (1927), p. 41, note 1, Birmingham 1777 Directory.
43 Dickinson, (1936), p. 29.
44 JWP, C3/1, 'Memorandum', 22 March 1777. Dickinson & Jenkins, (1927), p. 46 gives 17 March.
45 Dickinson, (1935), p. 94.
46 JWP, 4/20.18, G. Hamilton to J. Watt, 18 Oct. 1775.
47 Boulton 348.59, J. Watt to M. Boulton, 3 July 1776.

48 Ibid.
49 Boulton 348.60, J. Watt to M. Boulton, 8 July 1776.
50 Boulton 348.61, M. Boulton to J. Watt, 10 July 1776.
51 Boulton 348.64, J. Watt to M. Boulton, 28 July 1776.
52 JWP, C3/1, 'Memorandum', 9 Aug. 1776.
53 JWP, 4/20.32, G. Hamilton to J. Watt, 9 Oct. 1776.
54 Smiles, (1878), pp. 391 – 92.
55 Muirhead, (1858), p. 260.
56 Ibid.
57 JWP, C3/2, 'Journal', 11 Oct. and 10 Dec. 1777.
58 JWP, C3/4, 'Journal', 5 May 1779.
59 JWP, W5, J. Watt to G. Hamilton, 6 May 1779.
60 JWP, C3/4, 'Journal', 8 June 1779.
61 JWP, W5, J. Watt to A. Watt, 4 March 1780.
62 Ibis, 5 April 1780.
63 JWP, W5, J. Watt to Mrs. N. Marr, April/May 1779.
64 Ibid, J. Watt to 'Dear Sir', 5 Aug. 1780.
65 Ibid, J. Watt to Jean Cochrane, 13 Aug. 1780.
66 JWP, W5, J. Watt to Mrs. N. Marr, 13 Aug. 1780.
67 Musson & Robinson, (1969), p. 203, J. Watt to the Rev. Mr. Deane, 2 Nov. 1780.
68 Tann, (1981), p. 53, Doc. 13, M. Boulton to J. Watt, 13 Jan. 1781.
69 JWP, 3/8, J. Watt to Mr. Pickering, 15 Oct. 1781.
70 Ibid, J. Watt to J. Watt jun., 15 Oct. 1781.
71 Muirhead, (1854), Vol. II, Letter 255, p. 187, J. Watt to Mrs. Campbell, 30 May 1784, JWP, Letter Book 1.
72 JWP, 4/9.39, J. Watt jun. to J. Watt, 2 May. 1784.
73 JWP, Letter Book 1, J. Watt to J. Watt jun., 29 May 1784.
74 Ibid.
75 See Musson & Robinson, (1969), p. 204 for a longer account of Watt's instructions.
76 JWP, 4/9.21, J. Watt jun. to J. Watt, 8 Oct. 1784.
77 JWP, Letter Book 1, J. Watt to G. Hamilton, 7 Sept. 1784 and J. Watt to J. Watt jun., 17 Oct. 1784.
78 JWP, Letter Book 1, J. Watt to G. Hamilton, 17 Oct. 1784.
79 Ibid, J. Watt to J.A. De Luc, 14 Nov. 1784.
80 Dickinson & Jenkins, (1927), p. 43, quoting from Boulton Papers, 5 July 1775. See also Smiles, (1878), p. 164.
81 Dickinson & Jenkins, (1927), p. 43.
82 Boulton 348.52, M. Boulton to J. Watt, 1775.
83 Boulton 348.59, J. Watt to M. Boulton, 3 July 1776.
84 Boulton 348.60, J. Watt to M. Boulton, 8 July 1776.
85 Muirhead, (1854), Vol. II, p. 98, Letter 157, M. Boulton to J. Watt, July 1776; Boulton 348.62.
86 Ibid.
87 Dickinson & Jenkins, (1927), p. 45. See also JWP, C3/4, 'Journal', 18 and 20 April 1779.

5
Fair Prospects in Cornwall

The Mining Background

Back in 1771, Small had drawn the attention of Watt to four or five copper mines in Cornwall being about to close owing to the high price of coals. At that time, Cornwall was the world's major source for copper. He recognised that there might be a demand for Watt's more economical steam engine in the Duchy which contained one of the highest concentrations of steam pumping engines at that time in the world. A very approximate number of Newcomen engines in 1775 has been given as Newcastle district 60; other coal mines 10; Cornwall 30; London waterworks 10; others 20 making a total of one hundred and thirty. [1] These figures show the scale of the Cornish prize for Boulton and Watt. Waterpower was scarce in the region and most of the mines were situated in the higher, central part of the peninsula where there were few streams. Man or animal power was insufficient to drain the deeper mines. William Pryce, author of *Mineralogia Cornubiensis*, wrote in 1778:

> ... that, happily for us and our posterity, Mr. Newcomen's invention of the steam fire engine, even in the weakness of its infancy, promised that future excellence to which it is since arrived, whereby we are enabled to sink our Mines to twice the depth we could formerly do by any other machinery. [2]

Newcomen may have erected an engine at Wheal Vor tin mine in Breage where it worked from 1710 to 1714 but the first one of which there is a positive record in Cornwall was erected at Wheal Fortune, Ludgvan, near Penzance, in about 1719. By the end of 1733 when Savery's master patent expired, four or five Newcomen engines had been installed in the county. [3]

The use of steampower was restricted through the high cost of importing coal on which a tax was levied. In 1730, the Cornish mine captains petitioned the Government to remit the duty of 5s.5d. on a chaldron but this plea was not heeded until 1741 when Pryce says there was only one fire engine at work in the county. Then numbers rose dramatically for Pryce continued, 'since which time above three score have been erected, and more than half of them have been rebuilt, or enlarged in the diameter of their cylindrical dimensions'. [4] Jonathan Hornblower, one of the leading Cornish engineers, wrote to Watt about Wheal Virgin where he 'began with a waterwheel afterwards I put up a 46 inch cylinder, since that have encreased to four Engines with 60 inch cylinders each'. [5] But the situation was not as rosy as these figures suggest, for Boulton told Watt in September 1775, that a Mr. Papps of Truro, Cornwall:

> Hath given me a list of all ye Engines in that County with their respective names which I have deposited amongst the archives of our Engine.
> The whole number in ye County of Cornwall are exactly 40 but there are only 18 of em in work on acct. of ye high price of Coals. [6]

Pryce gave a figure of £3,000 worth of coal consumed every year for a 'fire engine of magnitude'. [7]

The introduction of a rebate on the coal tax had stimulated copper mining. One high point had been reached earlier in 1732 with production at 7,290 tons, which declined before rising to 11,000 tons in 1739. However production fell to only 5,000 tons in 1740, possibly the cause of rescinding the tax. But, for the next thirty years, figures show a fluctuating although rising rate of production until in 1770 a high of 30,776 tons was attained. The following trough bottomed at 27,654 tons in 1772. [8] This slump, taken together with a gradual decline in price per ton of copper of over a pound from £7.15s.10d. in 1726 to £6.14s.6d. by 1775 caused the mine captains to search for ways of reducing their costs.

Yet the experience in Cornwall of inventors proffering improvements to steam engines had been disastrous. Pryce pointed out how 'Many trials of mechanical skill have been made by our engineers, to very little purpose, for the total application of heat and the saving of fuel'. [9] Blakey had erected one of his improved Savery engines with a form of water-tube boiler but it had exploded when it was being demonstrated down a mine. [10] Pryce commented merely that these engines had 'been lately improved by Mr. Blakey, though not to a degree of power sufficient to unwater a deep mine'. [11] Hornblower warned Watt:

> Here I shall make a remark by the by to you (viz) tis generally attended with some difficulty & delays when any new system is introduced among any people when the principle is not understood but by few of them; more so here [Cornwall] as Mr. Blakey, Hunt, and sundrie others have failed; I say these things makes them Diffident & Cautious in their proceedings. [12]

The Cornishman's innate suspicion of outsiders caused grave problems for Boulton and Watt later. Also there was hostility from some of the established engineers who saw their livelihoods threatened. John Budge, a respected engineer at Wheal Union, was one who refused to undertake the erection of a proposed new engine there if Boulton and Watt received the order. [13] Pryce, while respecting Watt's 'skill in pneumaticks, mechanicks, and hydraulicks', as well as the 'regular set of experiments' he had made upon steam, still doubted the great claims for the economy and other features of this new engine:

> The working of these engines [Watt's] is more regular and steady than the common ones, and from what has been said, their other advantages are apparently very considerable; but to say exactly how much they excel common engines, is difficult, as common engines differ very much among themselves - I am told, that the savings amount at least to two-thirds of the fuel, which is a very considerable object where coals are as expensive as they are in Cornwall. [14]

Here we may note as an aside that Boulton, Watt, Fothergill and 'Derwin' were subscribers to Pryce's book.

Cornishmen's doubts may have been increased through Smeaton's failure with one of his improved Newcomen engines at the Chacewater mine. After the poor performance of his engine for the New River Company in 1769, Smeaton had made over 130 tests on an experimental engine which he had constructed at his home at Austhorpe, near Leeds. [15] (See Vol. I, p. 323) Based on these results, he designed a replacement pumping engine for Long Benton Colliery, Northumberland, in 1772. Tests in May 1774 on his new engine, which had the same cylinder diameter of 52 ins. as the old one, showed an increase from 30 to 40 h.p. as well as an impressive increase in 'duty' from 4.6 to 9.1 million. This duty figure, which showed the pounds weight of water raised one foot high with a bushel of coal, represented an increase of 25% over the most efficient engine previously built, and an improvement of 40% on the typical performance of the average Newcomen engine at that time.

In 1773, Fenton & Co. took over Chacewater mine and asked Smeaton to install an engine incorporating all his improvements. [16] This mine was so deep and so much water had to be pumped out that two large Newcomen engines were needed. One lifted the water from the bottom of the mine halfway up the shaft from where the second lifted it to adit. The Chacewater engines were situated at the top of the hill above the village at a height of about 330 feet [17] but an adit or tunnel had been driven from the valley below into the mine so that the engines did not pump the water up to the top but only to the level of this adit, something

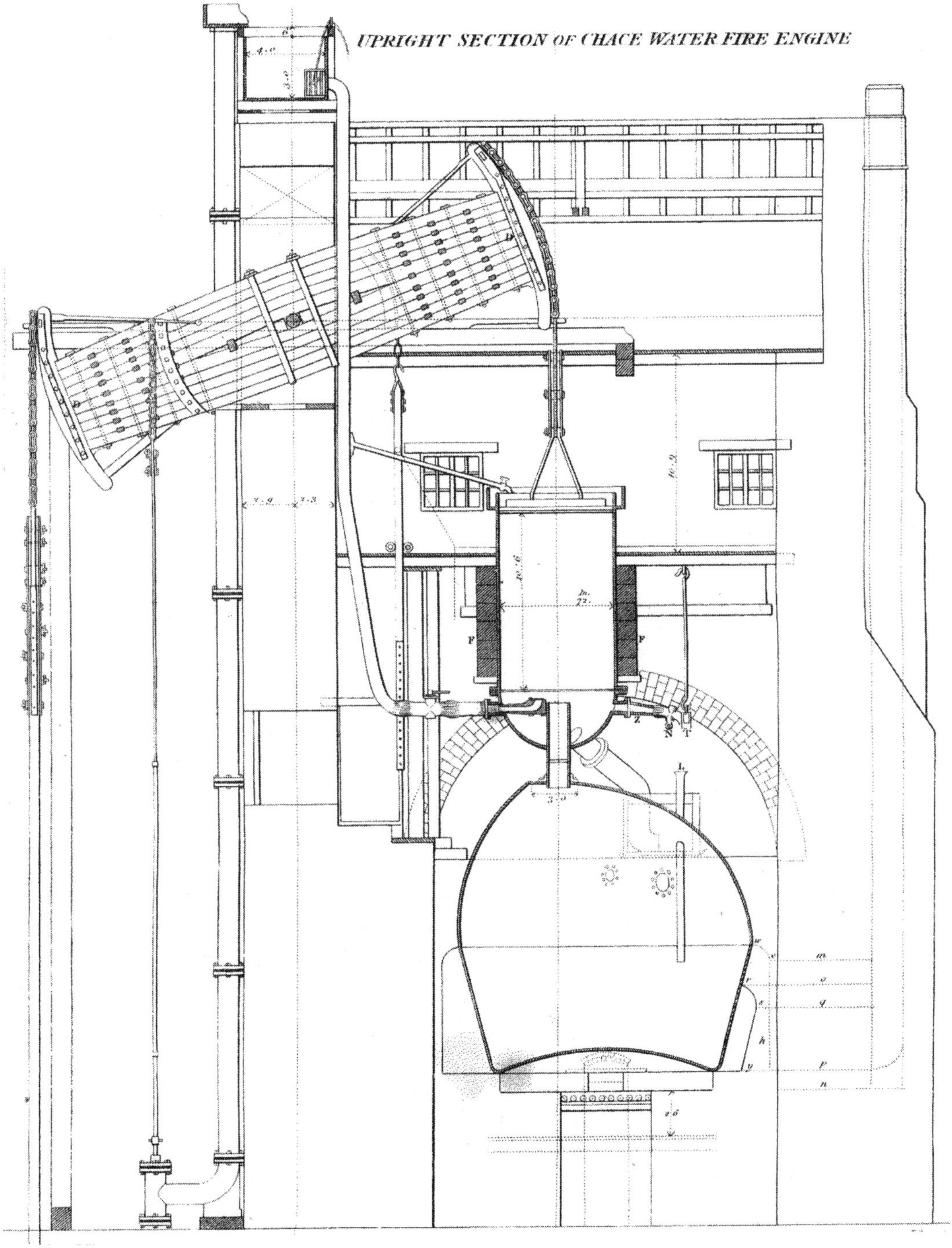

Smeaton's atmospheric pumping engine for Chacewater mine, 1776.

common at other Cornish mines. These engines had cylinder diameters of 64 and 66 ins. and together developed 60 h.p. with a duty of 6.9 million. The engine Smeaton designed had a cylinder diameter 72 ins., 9 ft. stroke, rated at 72 h.p. It was expected to have a duty of 10 million. He wrote in his report:

> The performance therefore of the new engine, in point of effect, will be greater than the two former actually were in the proportion of four to five nearly, but the effect produced by the same quantity of coals will be as 16 $^1/_2$ to 9 $^1/_2$, that is, nearly as seven to four...
>
> N.B. This engine, though not the largest that has been built, will be of considerably greater power than any I have seen, and when worked at its full extent, will work with a power of 150 horses acting together, to keep which power throughout the twenty-four hours would require at least 450 horses to be maintained. [18]

The engine was set to work in August 1776:

> ... but after working several months at a considerable expence it was found ineffectual for the purpose wanted having never been able to sink the water more than 2/3ds of the depth required to work the Mine to effect, & the Engine was of necessity obliged to stop till a Level which had been begun at near a mile distance from the mine in 1773 could be finished. [19]

This level was to be linked to the Great County Adit which was not cut through to Chacewater until January 1780. [20] Smeaton had probably improved the Newcomen engine as far as was practicable but his work was already being overshadowed by that of Watt.

Perhaps hearing of the success of the Bloomfield and New Willey engines, a deputation from Cornwall, led by Thomas Ennis of Redruth, visited Birmingham probably in the summer of 1776 while Watt was in Scotland to find out all they could about this new type of engine. But, after they had left the Soho Manufactory, it was discovered that a drawing of an engine was missing. Boulton wrote to Ennis, 'We do not keep a school to teach fire-engine making, but profess the making of them ourselves'. [21] The missing drawing was returned quickly, for it had been taken by Richard Trevithick senior under a misapprehension, as he claimed. In 1776, Trevithick senior occupied the position of manager at Roskear (Wheal Chance) and in the following year was, in addition, manager of Dolcoath, Wheal Treasury and Eastern Stray Park mines. He had the reputation of 'being the best informed and most skilful Captain in all Western Mines'. [22] Here was another potential opponent to Watt.

The First Boulton and Watt Engines in Cornwall

The background to the Boulton and Watt enterprises in Cornwall was the growing scale of the war in the American colonies and with it the rising demand for copper. On 16 December 1773, ships had been boarded in Boston harbour and cases of tea were thrown overboard; the Boston Tea Party. The real war may be considered to have begun in 1776 with the Battle of Bunker Hill followed by that of Long Island and the demand for independence. France entered the war against England at the beginning of 1778, a war which continued into 1783 when the Treaty of Paris was signed on 3 September. [23]

The first three Boulton and Watt engines in Cornwall were at Wheal Bussy, Ting Tang and Chacewater mines. Wheal Bussy was ordered in January 1777. Watt paid his first visit to Cornwall from early August to the end of September 1777 when he supervised the erection of this engine and saw it start up in that September. During this visit, he lodged with Thomas Wilson but, when he returned in May 1778, he rented a house at Plain-an-guary Green on the outskirts of Redruth, which was closer to Ting Tang and was more convenient for the other northern mines than Chacewater. This time, he saw the Ting Tang engine set to work. It had been ordered in November 1776, but started only at the end of July 1778 on the same day as that for Chacewater which had been ordered in January 1778. He left Cornwall shortly after Christmas 1778. He made two further visits to this county. In 1779, he was there from the latter part of September until the end of February 1780. He did not set out again until the middle of June 1781 when he stayed for almost a year until towards the end of April 1782. By this time he had appointed people to look after the interests and engines of Boulton and Watt in that important region so that he did not return again for possibly ten years. [24]

The visit of the Cornish deputation may not have been the first time that Boulton and Watt had been contacted about an engine for that area. In Watt's memorandum book for 1776 there are two entries, the first for 17 April, 'Sent of[f] drawing Hornblower's condenser' and for 3 May, 'Wrote Mr. Wilkinson with drawing of Hornblower's & Bow Pump'. [25] However it was not until that November when Hornblower placed a firm order for an engine at Ting Tang mine near Redruth, the first one for a Boulton and Watt engine in a Cornish mine. Hornblower's father, Joseph, moved to Cornwall from Staffordshire in 1725 to erect a Newcomen engine and settled at Chacewater in 1748. He must have trained his son in the engine business for Jonathan, after erecting Newcomen engines in Derbyshire, Shropshire and Wales, returned to Cornwall in 1745 where he was to remain for the rest of his life. In the 1740s, he erected engines for William Lemon, using cylinders supplied from Coalbrookdale through Thomas Goldney of Bristol. [26] Between 1744 and 1768, Goldney recorded details of forty-nine engines which passed through his hands. By 1758, Hornblower may have erected twenty engines and became engineer to many Cornish mines such as Wheals Virgin, Maid, Sparnon, Carharrack as well as Ting Tang. He was highly regarded in his profession and, by 1776, was one of the leading steam engineers in Cornwall. His assistance greatly eased the problems Watt faced when erecting the early Boulton and Watt engines in Cornwall.

Hornblower wrote to Boulton on 10 October 1776 asking if he would send the Adventurers of Ting Tang mine proposals for erecting one of Boulton and Watt's improved steam engines. Foundations had already been dug for a Newcomen engine with a cylinder 52 ins. diameter. He thought 'this place a very good one for introducing your engine among the mines in Cornwall having very good water for the Boiler, which is otherwise in many places among the copper mines'. [27] A further letter followed a fortnight later with more details. [28] Watt suggested that Boulton and Watt would erect an engine at their own expense and remove it if it failed to answer but, if it did succeed, then the Adventurers would pay for the whole of it. [29] At last there was a prospect of Boulton and Watt breaking into the Cornish market.

Hornblower wrote on 23 November 1776 saying that 'Ting Tang Copper Mine wants a 50 or 52 inch cylinder'. [30] Boulton transmitted the good news through their banker, Matthews, to Watt who was in London:

> We have a positive order for an engine for Ting-Tang mine, and from what I heard this day from Mr. Glover, we may soon expect other orders from Cornwall. Our plot begins to thicken apace, and if Mr. Wilkinson don't bustle a little as well as ourselves, we shall not gather our harvest before sunset... I perceive we shall be hard pushed in engine work, but I have no fears of being distanced, when once the exact course or best track is determined upon'. [31]

This was the tenth order, including the larger engine installed at Soho, which had to be designed that year. Rumours of many inquiries had reached Hamilton in Scotland who wrote, 'I rejoice to learn there is such an appearance of demand for them, & hope a few years will give you so much of the ready as to allow you to indulge your favourite laziness for the rest of your days'. [32] There was little chance for laziness during the next ten years.

Hornblower sent further details of the Newcomen engine which they were proposing to erect. Watt required changes because, on his engine, the boiler was placed at one side of the cylinder and not underneath. Hornblower pointed out that oak 30 ins. square for the main beam was not available so it would have to be made from smaller pieces. He agreed to Watt's suggestion for cast iron links in the main chains but the mine manager, Capt. Paul, refused. Paul wanted the outer steam casing omitted which Watt refused. Hornblower pointed out that there were competent smiths in Cornwall who could carry out this type of work. [33] Watt started preparing 'our orders' for this engine on 18 January 1777 and Hornblower acknowledged that he had received drawings at the end of the month. [34] The drawing for the beam is dated February.

Wheal Bussy

But at this point it is convenient to leave the Ting Tang engine and turn to the one which was the first Boulton and Watt engine set to work in Cornwall, Wheal Bussy or Wheal Spirit, at Chacewater. We have seen that Smeaton erected an engine on the main shaft of this mine which failed to clear the water. (See p. 96) £12,000 or £14,000 had already been spent but it became obvious that the mine could not be worked until the Great County Adit was completed which would reduce the height to which the water had to be raised from 62 to 48 fathoms (a fathom was 6 ft.). In the meantime, the Adventurers agreed in January 1777 to build a small engine to expedite work on extending the Adit by draining the intermediate construction shafts for the tunnel. Thomas Wilson, the mine manager, recalled what happened next:

> At this time, Thos. Dudley an engineer, was at Soho from Cornwall & had examined 2 or more engines built near that place on Mr. Watt's construction, he wrote to his friends such flattering accounts of those Engines superiority over the Engines in common use, that the Managers of Chasewater were induced to apply to B. & W. to erect a small Engine on the above level; which engine was set to work under the inspection of Mr Watt in September the same year. [35]

The cylinders and other castings for both Ting Tang and Wheal Bussy were ready at Bersham in May 1777 [36] but, when it came to shipping them, it was found that the Ting Tang cylinder would not pass through the hatches. The Wheal Bussy parts were sent off first, to Watt's annoyance. Dudley, who had been sent to Birmingham from Cornwall to be trained as an engine erector, started its erection when the parts arrived so that Watt found 'Wheal Bussy is in considerable forwardness... All the world is agog to see its performance' when he arrived in Cornwall on 9 August. [37] He commented about it a little later, 'Our engine Carpentry and stone work at Wl Bussy is very rough but seems to be sufficient & well enough done for an engine to stand only a year after which it is to sleep another and then go to Wheal Chance a new mine in Chacewater sett'. [38]

It must have been at work a few days before 13 September when Watt told Boulton, 'It has forked the water in the engine shaft'. [39] It worked at a disadvantage for it pumped water by a series of flat rods out of two shafts along the line of the new level. The engine itself was typical early Boulton and Watt practice, with steam jacketed cylinder open to the boiler all the time. The condenser lay beyond the beam wall with single air pump and hot water pump above it. The beam was a single baulk of timber, not trussed, but with unequal centres. Pryce described the layout of the pumps:

> The small engine at Huel-Bussy Mine... has a cylinder thirty inches diameter, that works a pump of six inches and a half diameter in two shafts by flat rods with great friction, three hundred feet distant from each other, forty five fathoms high in each shaft, equal in all to ninety fathoms, and can make fourteen strokes of eight feet long in a minute, with a consumption of coal less than twenty bushels in twenty-four hours. [40]

Watt proudly wrote they had had many spectators, 'several of which are become converts. I understand all the West-Country captains are to be here tomorrow to see the protégé'. [41] One of them was Budge, about whom it was said to Boulton that 'almost one half of those [engines] in this country are under his care, and were built by his direction, and he is esteemed one of the best practical engineers in it'. [42] Watt told Boulton that 'Bouge has been to wait upon me & promises to read his recantation so soon as he is convinced & never to touch a common engine again'. [43] Budge later helped erect some of the Watt engines.

Another spectator was Wilson himself who was so impressed that not only did he order an engine for Chacewater mine itself but became agent for Boulton and Watt in Cornwall. Because the Wheal Bussy engine was only draining the working shafts during the construction of this part of the Great County Level, it finished this task in June 1778 when it was removed to Scorrier where it worked until September 1782. It was moved at least a couple more times and finished at Poldory as an auxiliary in the 1790s. [44]

Watt's Impressions of Cornwall

Watt had set out for Cornwall with Annie on 4 August 1777 with 24½ guineas (£25.14s.6d.) in his pocket. [45] Their journey was quite arduous but not untypical of travel in those days:

> Tuesday 5 o'clock left Birmingham; by 9 o'clock at night arrived at Bristol very much in want of sleep having got little the night before - Found all the places in coaches & diligences taken to Exeter for next day & day after - so took chaise at 6 o'clock in the morning & kept company with the diligence to Bridgewater, where places becoming empty, we went in Dilly to Exeter. Found the town full of people and all the stages to Plymouth full, so went on in the Chaise way... Thursday, a few hours at Exeter & a few at Plymouth. Friday at Plymouth Dock [Devonport] - crossed over to Saltash (which I advise you not to do), lay at Lostwithiel; Saturday Truro. [46]

They had spent four days on the journey before reaching Chacewater where they lodged with Wilson. Boulton and Watt were charged £15.4s.6d., showing how expensive travel was in those days. [47] We should note Watt's economy in not using the expensive post chaise where possible.

Watt was not impressed with either the Cornish people or the standard of engineering there. 'The people here... have the most ungracious manners of any people I ever saw', [48] while the engines were 'clumsy and nasty, the houses crackt & everything dripping with water from their house cisterns'. [49] Three years later, Boulton reported to Watt, 'You would be pleased and even astonished to see the neatness, exactness and quietness of the two engines at ye United Mines and at Tresavean. We have at last raised a spirit of emulation amongst the

Cornish enginemen, each to exceed his neighbour in neatness and good order'. [50] Annie was not greatly impressed with Cornwall either, for she told Mrs. Boulton:

> I scarcely know what to say to you of the country. The spot we are at is the most disagreeable in the whole county. The face of the earth is broken up in ten thousand heaps of rubbish, and there is scarce a tree to be seen. But don't think that all Cornwall is like Chacewater. I have been at some places that are very pleasant, nay beautiful. The sea-coast to me is charming, but not easy to be got at. In some cases my poor husband has been obliged to mount me behind him to go to some of the places we have been at. I assure you I was not a little perplexed at first to be set on a great tall horse with high pillion [Annie was well advanced in pregnancy!]... One thing I must tell you of is, to take care Mr. Boulton's principles are well fixed before you trust him here. Poor Mr. Watt is turned Anabaptist, and duly attends their meeting; he is indeed, and goes to chapel most devoutly. [51]

Hornblower had built the chapel at Chacewater, which may have been the reason for Watt's diplomatic presence there.

While Hornblower wrote to Watt in November 1776, 'I should be glad to see you in Cornwall', [52] Watt was much more suspicious. Just after his arrival in Cornwall, he told Boulton, 'I have been at Chacewater & have seen Messrs Wilson Dudley and Hornblower, the latter seems a very pleasant sort of old presbyterian', [53] but a few days later he said, 'Hornblower seems a good sort of man and carries himself very fair but is I hear an unbelieving Thomas'. [54] This suspicion also extended to others involved in Cornish mining for Watt was annoyed that they spread all sorts of prejudicial rumours and reported to Boulton:

> I chuse to keep out of their company, as every word spoken by me would be bandied about and misrepresented. I have already been accused of making several speeches at Wheal Virgin where, to the best of my memory, I have only talked about eating, drinking, and the weather. The greater part of the adventurers at Wheal Virgin are a mean dirty pack, preying upon one another, and striving who shall impose most upon the mine. [55]

Here we see the seeds of the troubles that would beset Watt later when dealing with the adventurers. Yet at the same time he recognised that the whole future success of Boulton and Watt in Cornwall lay in the operation of Wheal Bussy engine, 'which if as successful as I hope I shall make it my business to get acquainted with the principal miners'. [56]

He expected that 'all the West Country [mine] Captains are to be here tomorrow to see the prodigy'. [57] Orders for many engines depended upon the performance of Wheal Bussy for Watt understood that if that engine gave 'satisfaction we shall have several customers for them' [58] and he had 'heard of several small engines wanted for the tin mines which cannot afford to use the common kind'. [59] He left Cornwall on 28 September knowing that 'The voice of the country seems to be at present in our favour and I hope will be much more so when the engine [Wheal Bussy] getts on its whole load'. [60] But he left unfinished business because the cylinder for the Ting Tang engine had not yet arrived. The delays in the erection of this engine caused a strain in the relationship with Hornblower. However, Watt had good cause to be satisfied with the results of his first visit to Cornwall.

Constructing the Engines and Ting Tang Engine

On his return to Birmingham, as well as dealing with enquiries,Watt was faced with the task of the erection of the other engines which had been ordered outside Cornwall. The total numbers of orders received were six in 1776 and seven in 1777. Out of nine in 1778, three were from Cornwall. In May 1778, Boulton and Watt were refusing to accept orders for engines to be begun before 1779. [61] Just before that, Watt had complained to Boulton, 'I sett off for Salmons tomorrow and have been so occupied with Reynold's drawings I have not got out to Soho since came home... Tingtang people want me too among others. I fancy I must be cut in pieces and a portion sent to every tribe in Israel'. [62]

The initial work of determining the size of the engine and preparing drawings fell on Watt's shoulders. While he soon developed a standard basic type and layout, the design of each engine had to be adjusted to suit both local requirements and technical modifications. Boulton and Watt undertook to supply drawings and recommend an erector but it is not clear who ordered the necessary parts from various suppliers since the final responsibility for the erection of the engine lay with the person ordering it, which was not always understood by the purchaser. In 1779, use was made of a printed list of parts, with hand-written entries against appropriate ones. [63]

The Ting Tang engine provides a good example of the problems Watt had to overcome in order to complete the erection of an engine. Watt sent Hornblower 'a list of all the parts of the engine, distinguishing where they are to be made':

> Bersham. Inside and outside cylinders, bottoms, piston, lid and stuffing box. The condenser cast iron work complete.
> Bradley. The nozzles and short pipe. The gudgeon and bed, the plummer block and brasses. Cast iron chains for cyl^{r} end.
> Soho. The regulators, brass and ironwork. The F and Y [valve gear levers], the plug frame chains, two chains for condenser pumps, gearing the buckets and clacks of condenser, two round piston rods for do., all the screws & burrs for the two cylinders, condenser and boiler steam pipes; screwed ends for holding down screws, the iron work for the great piston rod top, the piston rod itself, 3 two inch dia^{r} adjusting screws and burrs for top of piston chains, 3 do. for pump chains, if you chuse them. 4 screwed ends for the stirrup to hang the beam in, 2½ screws, screwed ends to shoot to martingale bolts.
> Tingtang. The boiler, firedoors and grates. The pumps and ye appurtenences. The pithead iron work, the poise beam, the pump chains, the screw bolts and arch plates for beam, the catch pins, the two glands which keep down the gudgeon, the stirrups for do., excepting the screwed ends which will be brought ready to shoot to. The martingales. The house and all the wood work. [64]

The piston rod was made in London by Coulson who supplied others to Boulton and Watt. [65] It was sent to Falmouth in July 1777 where it was found to be slightly rusty. [66]

After his arrival in Cornwall in 1777, Watt realised that manufacturing costs for smith's work was cheaper there and told Boulton, 'We must give up the manufacturing of any parts of the large engines excepting the nozzles, our prices so much exceed what they are done for here that we shall never get our bills pleasantly paid and to be paid with a grudge and to lose by the business at the same time will never suit'. [67] This would push Boulton and Watt further towards a consultancy business at this period and meant that they had to provide adequate drawings and erection instructions. It still leaves the question unanswered of who was responsible for checking progress on the manufacture of the parts.

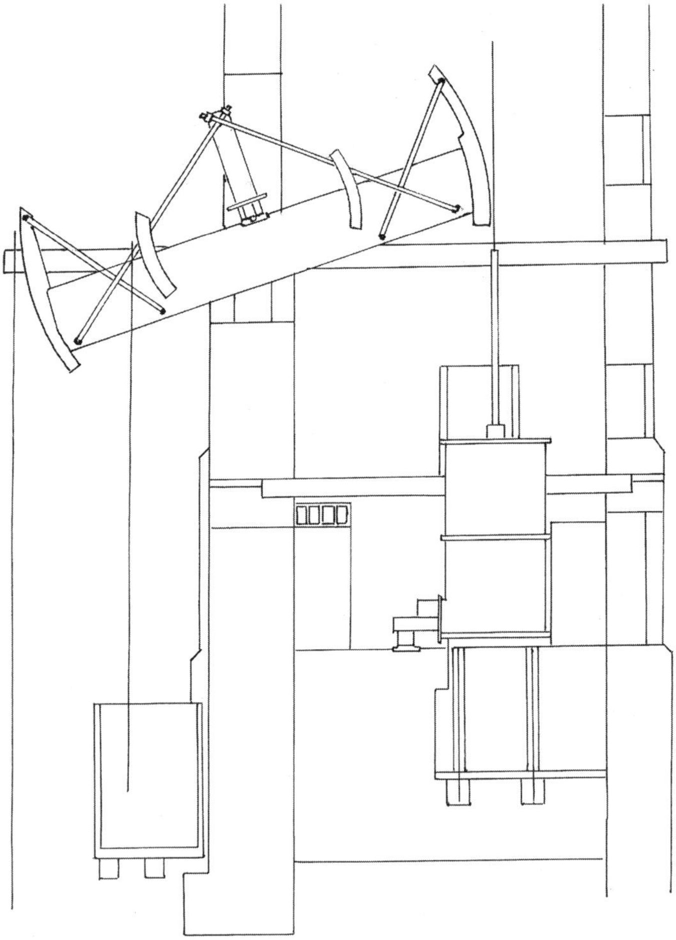

Watt's single-acting engine for Ting Tang mine, 1778.

Since placing the order for the Ting Tang engine in November 1776, Hornblower had been discussing its design with Watt and making preparations for building the engine house. That and water reservoirs were to be begun in early 1777 as soon as the snow had gone although they were delayed through the state of the roads. [68] Some parts came during the summer so that Watt found on his arrival in August that 'there is a very good engine house built and a good Beam and the boiler stands by' [69] but no cylinder. Watt was afraid that, if the cylinder failed to arrive immediately, 'it [the engine] would be cast upon our hands as they are now fairly put out by water & the work at stand'. [70] Watt had to leave the engine unfinished to go back to Birmingham.

In the middle of December, Hornblower wrote forlornly, 'Having had at sundry times fine northerly winds gave us daily expectation of an acct. from Falmouth of the arrival of Ting Tang cylinders but after all can't hear anything of them - The water is let in and all quiet at Ting Tang at present'. [71] The Adventurers were in desperate need of their engine. Luckily the vessel arrived soon afterwards. [72] There is no mention of how this heavy cylinder, or for that matter any of the later ones, was moved from Falmouth over the poor roads and up the hill to the site of the mine.

Watt returned to Cornwall on 28 May 1778, coming via London. The following morning he visited Ting Tang where he found the cylinder and steam case in place and the condenser and eduction pipe screwed together. Work continued during June and July with Watt paying regular visits until at last on 28 July he wrote in his Journal, 'the engine all ready to sett to work excepting ye injection gear and the Buckets of ye pit pumps'. The engine was finally started on 30 July '& seems to do very well'. [73] Its design was still the early type, with steam jacket open to the cylinder top, communicating directly with the boiler. There was a fire below the cylinder to keep it hot. The beam was made from four baulks of timber. Valves were some sort of circular type but Watt modified the design of the operating mechanism. The condenser had twin air pumps. The engine house was Watt's standard pattern. Richard Michell built a model of this engine which so impressed Watt that he recommended Michell as an erector. The model may have been taken to London in the 1790s as an exhibit in the trials concerning the validity of Watt's patent where it disappeared. [74] After some teething troubles, the Ting Tang engine performed satisfactorily.

The Chacewater Engine

In the meantime, Boulton and Watt had received an order for an engine to replace Smeaton's on Chacewater Mine. This was the result of the impressive performance of the Wheal Bussy engine. Wilson wrote:

> The engine [Wheal Bussy] so built was evidently to all unprejudiced persons so much superior to any Engine then in use, both with respect to power & saving of fuel that the writer was convinced an Engine of the same construction of 63 Inches would do all the business that was intended to have been done by the engine before built by Mr. Smeaton which was of 72 inches Diam that he immediately wrote to his Co & had their consent to order such engine as it would save them considerable time & money, to know whether the Mine on which they had already expended such large sums was worth prosecuting without being under the necessity of waiting from that time as it afterwards proved more than 3 years before the Level could be completed till when the Mine could not have been set to work by the Engines then in use. The engine built by Smeaton was converted into one of Boulton & Watts and was set to work in Augst 1788 [sic] & tho it certainly was not so perfect as if it had been entirely new, yet it got out all the water to the depth of 62 fathoms & also enabled the Co to sink to the Depth of 72 fathoms before the Level was completed which was not done till Jan 1780. This Engine was immediately the means of convincing the most prejudicial against this improvement in Steam Engines, for it was known to most Adv[enturer]s in mines as well as to the Men particularly working in that Mine that the 2 Engines which had been employed previous to 1770... had with great difficulty drawn the water to the same depth particularly in winter & that they consumed on the Average of the year 6 Weys of Coals in 24 Hours; wheras [sic] this Engine of B. & W. kept out the water effectually & with ease during the winter & consumed for some months not as much as 2 Weys... This was a proof convincing to the meanest capacity, the facts were plain & needed not either argument or calculation to convince all of the great utility of Mr. Watts improvement. The consequence was that applications were made to B. & W. to build by most of the Mines in working & several Mines which were idle, which could not be wrought by the Old Engines on acct. of the great expence attending them. [75]

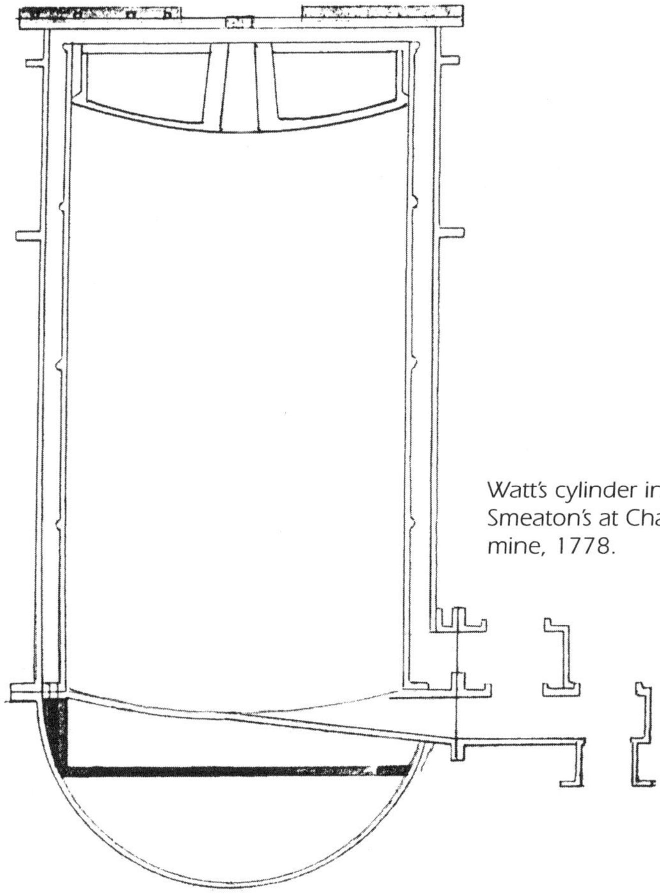

Watt's cylinder inside Smeaton's at Chacewater mine, 1778.

Watt started the drawings in January 1778. The boiler underneath Smeaton's cylinder was taken out and replaced by a masonry pillar which supported Watt's new, smaller steam cylinder. Smeaton's larger cylinder was retained as the outer casing of the steam jacket with Watt's new cylinder ingeniously placed inside. [76] After three attempts at casting the cover in Cornwall, Wilkinson may have been asked to supply this. [77] Watt found the work going on briskly when he arrived in Cornwall in May 1778 although there were fears that the mine might stop owing to the low price of copper. The boiler was filled on 29 July and the fire lit. Next day, 'The engine sett agoing very well but was heavy to hand on account of ye great weight necessary to open ye Exhaustion regulator'. [78] This day, 30 July 1778, must have been a very happy one for Watt with both this and the Ting Tang engines being started.

As at Ting Tang, modifications and adjustments had to be carried out and Watt commented, 'The Connoissieurs say we shall never fork ye water, or if we do we may fork anything as ye water is reckoned as the heaviest in ye whole county'. [79] Watt realised that 'Chacewater is our capital card for should it succeed in forking the mine all doubts will be removed'. [80] The weight of the three lifts of pumps and the rods was reckoned at nine tons. The engine worked at eleven strokes per minute. He entered in his Journal for 18 September the joyful news, 'Chacewater going in fork'. [81] In December he wrote to Black, claiming that it had taken three months to unwater the mine. This letter made no mention of Smeaton's engine:

> The water of this mine formerly baffled 2 Engines, one a 66 and the other a 64 but though this is the rainiest season and the water the most plenty below ground we keep it very well under hitherto – but that you may know what a job it was, we were three months going at the above rate in forking or unwatering the mine. The whole country declared it impossible some on account of the known great qu[anti]ty of water and other from a belief that the engine could not work the pump to that depth. [82]

With that triumph under his hat, Watt left Cornwall after packing up all his furniture during the morning of 31 December 1778. He took the evening coach and recorded in his Journal, 'Set out for Bodmin - stormy night horses broke the traces – left 2 hours in the midst of Bodmin Moor got to Bodmin 1.30'. [83]

Organisation of Boulton and Watt

At the end of August 1778, the usually pessimistic Watt had told Boulton, 'By attending to the business of this county alone, we may at least live comfortably for I cannot suppose less than 12 engines will be wanted in two or 3 years but after that very few more as there will be sufficient to get one enough but you cannot reckon the average profits to us above £200 pr. Engine.' [84] In fact, orders were placed by Cornish mine owners for around a further five engines in 1779 and nine in 1780. It became evident that Boulton and Watt needed a permanent presence there to negotiate new orders, to supervise erection and maintenance of the engines, to collect the premiums due on their patent, as well as attend to any other business such as meeting the mine owners. But running the business in Cornwall was expensive. During his visit in 1778, he spent £109.19s.5½d. on furniture, £71.10s.11d. on 'housekeeping', £12.7s.3d. on 'Horsehire & expences at mines', £3.7s.9d. on postages and £16.0s.4d. on the journey home. [85] Annie was with him at least some of the time.

Both Boulton and Watt recognised the importance of the Cornish market because the product of the metalliferous mines had a high value and so were worth exploiting but coal was expensive. Hence there were rosy prospects for sales of their more economical steam engines. It fell on Watt's shoulders to convince the Cornish mine owners that Boulton and Watt's engines would not only fulfil their promise of more economical performance but also would function reliably. So Watt had to ensure that the first engines were erected properly. We can follow his movements because the journal he had been keeping up to his arrival at Birmingham in May 1774 was recommenced in February 1778 - at least that is the date when the next surviving one starts. [86] For a person who had been sickly as a youth as well as suffering later from severe headaches, his activities in Cornwall show that in fact he had a remarkably tough constitution.

Before his 1778 visit to Cornwall, Watt had been in London with Boulton and Henderson assessing the performance of the Newcomen engine at Chelsea which they were going to replace. He left London at 6 o'clock in the morning of Monday 25 May, reaching Salisbury at 7 in the evening. A short day followed, arriving at Exeter at 3 o'clock in the afternoon. But on 27 May, 'came to Truro about 11 at night'. He went on to Redruth on the Thursday when a rest might have been expected but 'in evening walked to Chacewater, Both cylinders in place & condenser screwed together'. That was a good three miles each way. On Friday morning he went to Ting Tang, only about a couple of miles from Redruth, and then on to Chacewater in the afternoon. On Saturday, he was at Chacewater where he 'sett cylinders upright'. Sunday 1 June at Ting Tang, 'noted geological formations'; Monday, 'Wheel Union' was visited; Tuesday, Chacewater again and Wheal Spirit. At Wheal Spirit he noted 'water from a red Killas to be highly vitriolic and moreover it appears that some of ye acid is raised with the steam... brass & copper not affected'. On Friday 5 June he returned from Ting Tang in rain which continued on Saturday. He spent Whit Monday writing letters. Wednesday 10 June must have been taken as a day off at Falmouth where he 'saw the Plymouth fleet pass by 14 in number'. One result of all this activity was 'a very violent headach that night'. As well as supervising and assisting his engine erectors, Watt also met influential people in the area. On Tuesday 30 June, in spite of being 'ill with Headach, dined at ticketing at Redruth with Messrs Edwards, Ennis, Richards, Trevithick, etc.' A ticketing was when sales of copper and ore were made and the settling of mine profits. His headaches are mentioned frequently in subsequent entries and correspondence. Watt had taken the opportunity of a holiday on Midsummer Day (24 June) to finish some drawings. [87] Such activity was typical during all his visits to Cornwall.

Watt's letters to Boulton reveal the strain he was under. On 27 June, he reported, 'I am quite tired with riding yesterday and walking today'. [88] A few days later he said that he was arriving home tired between three and four o'clock and then had to write. In August, he sent some drawings to Birmingham and said that he 'must press dispatch in ye copying and forwarding these drawings which would have done myself but have no time except when I am too tired to do anything correct'. [89]

Boulton arrived in the middle of September 1778 which may have lessened the load. For example, Boulton went to Chacewater where he assisted with packing grease in the piston to cure that engine's bad running as well as trying the boiler there. [90] He met the Wheal Union Adventurers and other mine captains before leaving in early December. A trip to Lands End in October for both of them with their wives may have been a rare holiday. [91] However Watt did not accompany Boulton and both their wives to Falmouth in early November, a day he spent at Chacewater and Scorrier engines. [92] In an undated letter written after Boulton had returned to Birmingham, Watt complained, 'I lead a very solitary life here, excepting Mr. Wilson no person deigns to visit me as I suppose they do as they would be done by I visit as few; nor indeed have I much time to waste upon them having besides the daily visit to the Engines many occasional drawings to make'. [93] He must have missed the Lunar Society meetings.

By late 1778, it had become obvious that Boulton and Watt needed some regular organisation in Cornwall to look after their interests since neither Watt nor Boulton could remain there permanently and also that it was necessary to establish some regular technical staff who could keep an eye on the performance of the engines. The house at Redruth was retained when Watt returned in the autumn and winter of 1779 to 1780 while Annie took the children to Scotland but then he and Boulton decided to buy one at Cosgarne in the Gwennap valley on the southern edge of the mining district. Here the Watt family made a prolonged stay from June to the end of December 1781. Watt remained in Cornwall. The house was in a more agreeable situation than either Chacewater or Redruth for he told Hamilton:

> This is a most delightful place a neat roomy house with sash windows double breadth the front to the south covered with vines loaded with young grapes – a walled garden with excellent peaches & plums – plenty of currants – two orchards – a lawn before the door – the whole surrounded with tall elm trees – it stands in a valley safe from the north & south winds but the east & west winds can get at us – everything has its drawback – the Blackbirds have eaten all the currants of which there was an amazing crop – the snails eat the strawberry the house is infested with Rats & mice - and there is scarce any apples in the orchards all through this county this year. [94]

This house still survives in its delightful rural setting through the narrow Cornish lanes.

Cosgarne House, Cornwall.

Personnel in Cornwall

The person they chose to become their agent handling the business affairs was none other than Thomas Wilson. During his lodging with Wilson in 1777, Watt had been helped in many ways and became impressed with him, telling Boulton, 'I go upon Wednesday to Wl Union wch is about 20 miles distant Mr Wilson accompanies me as he did to Falmouth last week he is most exceedingly obliging in a plain unaffected manner'. [95] Wilson was equally impressed by the performance of the Wheal Bussy engine. He had been sent from Yorkshire by a 'Company of Gentlemen', Fenton & Co., who had started to work the Chacewater mine a couple of years previously. In November 1778, Wilson became involved again with Boulton and Watt through giving them details of the performance of the engines at Wheal Virgin and telling them that all was going well at Chacewater. In December, Watt gave Wilson some articles to be copied and engrossed as well as discussing terms for the Chacewater engines. Watt visited him on Christmas Day. [96] It could be soon after this that Wilson became their agent for during 1779 he was corresponding with Watt about a variety of problems over engines and erectors. To him fell the task of collecting the monthly premiums due when engines were running. He received a commission of 2 1/2% on these which amounted to £3,845 between 1781 to 1800, or an average of £174 per annum, not a vast reward for the hostility he had to face in later years. [97]

While Wilson superintended the commercial side of the business in Cornwall, someone was needed to supervise the engine erectors. Hornblower took on this role, at least for replacements of the Newcomen engines over which he had been manager. He had been consulted about the design of the Ting Tang engine and then in December 1778 he and Capt. Paul called on Watt about a new engine for Wheal Maid. [98] During 1779 while Watt was in Birmingham, there was constant correspondence between the pair. This covered engines at

Chacewater, Tresavean, Poldice, Ale and Cakes and Wheal Virgin. [99] When Watt was in Cornwall during the autumn of 1779, they consulted about Wheal Virgin. But Hornblower became ill around this time suffering from a 'fit of the stone & gravel, found it tedious not from the Gravel so much as the piles'. [100] From this point in time, he may have withdrawn from some of his work through further illness so in April 1780 he wrote to Watt:

> Mr Henderson will give you a more particular acct. of matters in general than I can pretend to occasioned by confinement. Poldice acct. last was the first of my going abroad [I] was poorly several days after and one day last week after returning from Poldice was seized with the stranguary again & altho the weather is this day become much better & warmer am afraid to venture out. I am very glad to hear you are better. [101]

This is the final letter from him in the James Watt Papers. He may not have recovered sufficiently to have assisted Boulton and Watt again for he died of stones on 7 December 1780, aged 63. [102] Watt must have felt the loss of one who had been such a help with his first engines in Cornwall.

Through Hornblower's death, Watt needed to find a replacement and the choice fell on Henderson who was certainly in Cornwall by April 1780. (See above) It had always been the intention of both Boulton and Watt to appoint Henderson as their technical assistant and relieve Watt of the supervision of actual erection of engines. After Torryburn, he 'set Byker to work prosperously' at the beginning of October 1778. [103] In the following February, he was to visit Wrens Nest near Dudley, Malehurst and, in Shropshire, Ketley and Donnington engines. [104] On the other hand, Watt had earlier complained to Boulton:

> Henderson has confused the Chacewater Eduction pipe for Wheal Union... I beg of you to take care that it be immediately got done as our reputation depends upon it and such repeated disappointments will undoubtedly ruin our credit in this country [Cornwall]... The sooner you can return to Birmingham the better it will be for our affairs there for I am sure H. will do nothing but multiply mistakes to our ruin. [105]

It seems that Watt found Henderson bad-tempered and sullen. Certainly after his arrival in Cornwall there was a good deal of friction between him and Wilson. When in Birmingham earlier, he had stayed at Boulton's house and may have upset him too through his manner of leaving. [106] Henderson returned to Birmingham in February 1782 when Watt became uneasy about not hearing of his safe arrival. [107] But Watt must have been dissatisfied with Henderson's performance in Cornwall, for, although we do not know the terms on which Henderson had been engaged, he must have asked for some form of regular payment and Watt pointed out to Boulton, 'Henderson's salary must be properly agreed... if he is disposed to go on as he did some part of the time he was here, I must vote for parting with him as a person we cannot afford to keep'. [108] Perhaps Watt, the rather dour Scottish Presbyterian, was upset by Henderson's philandering and setting up house with a 'Miss Peggy'. He reported to Boulton in April 1781, 'Henderson says Falmouth is threatened by an attack from the French & will have to help defend the place. He is also keeping a woman who is plundering him'. [109] However terms were agreed, only for Henderson to leave suddenly in February 1783. [110] Yet, for all his failings, Henderson suggested several important improvements to the engines. The first was the addition of the cataract at Torryburn to control the timing of the strokes. Another was recommending that, when first starting from cold, steam should be let into the eduction pipe and passed out through a 'blowing valve' into the condenser tank to remove the air. [111] In addition, he showed Boulton a pendulum counter for recording the number of strokes of the engine. He suggested alterations to some of the valves, but which are not known. Watt wrote to him, 'I shall be as glad to see your improved valves answer as you will my new way of writing which however notwithstanding many expts I have not brought to its ultimate perfection'. [112] Watt was referring to his method of copying letters. (See p. 190)

The answer to Boulton and Watt's need for a technical supervisor in Cornwall was already to hand in the person of William Murdock. He was twenty-three when he called at Soho in 1777, seeking work. Boulton interviewed him and, according to Smiles, was impressed with Murdock's hat. Murdock had made it himself on a lathe, which would have involved some oval turning, needing considerable expertise. [113] He quickly proved his competence since he had a natural affinity for understanding steam engines. In August 1778, Boulton and Watt decided to give him his first independent commission to erect the engine for Gilbert Meason at the Wanlock Head lead mines. [114] Here he took it upon himself to alter the mechanism operating the valves so that the steam valve was driven from the exhaust shaft. Jabez Hornblower, who was erecting an engine later at Donnington Wood, asked permission to modify that engine in the same way which Watt refused. [115] Watt had spent many years developing his 'perfect' engine so we should not be surprised to find him taking umbrage at the suggestions of a very raw engine erector. That he was willing to accept improvements from others we have seen in the case of Henderson's suggestions. It is probable that linking the steam and exhaust valves was not advantageous because both these engines had three valves and it would mean that the inlet valve remained open too long. Boulton visited the Donnington Wood engine towards the beginning of August 1779 and found that the top inlet valve did not shut until the piston reached the bottom of the cylinder although the

William Murdock,
1754 – 1839.

engine was only lightly loaded. He directed that the valve gear should close the steam inlet valve when the piston had made two-thirds or three-fourths of its stroke. [116]

Murdock was recalled to Birmingham in August 1779, leaving Meason's engine in good order. [117] He was sent to Donnington Wood to put that engine right. He followed this by examining the steam regulator at Ketley. Then he was despatched to Cornwall in the company of James Law with letters to Wilson [118] where he had arrived by the end of September 1779, probably never thinking that this would be his home for nearly twenty years. Here he was quickly accepted by the Cornish men and married a Cornish woman, before leaving in 1798 for the Soho Foundry.

Much has been written about the ill-treatment of Murdock by Boulton and Watt and their suppression of his inventions. Yet Boulton described him as 'the most active man and best engine erector I ever saw'. [119] It has been claimed that Watt treated him meanly by refusing to grant an increase in pay in 1780 after Murdock had proved his worth following a year in Cornwall It had been agreed earlier in February 1780 that Murdock, who had started at 15s. per week should be given 26s. when Law was receiving only 20s. [120] Therefore he was already earning considerably more than other erectors although it has been claimed recently that he was 'earning only a shilling a week more than the run-of-the-mill mechanics who assisted him'. [121] Remembering the dire plight of Boulton and Watt's finances at this time, we should not be surprised to find Watt protesting at a further pay increase. He wrote to Boulton:

> As to Wm. I agree with you in his merit but am sorry to find that he has got such notions of self importance which will very soon render him perfectly unserviceable to me, for wherever a man once gives way to that sort of ambition, he is never satisfied but wishes always for something impossible. As to giving such wages as two Guineas per week I think an example of that kind would ruin us by stimulating every other man we had to similar demands. – I therefore look upon Wm. as a lost man to us. [122]

Murdock had not yet become virtually indispensable to Boulton and Watt in Cornwall and a balance in the rates of pay among their employees had to be maintained. Yet there was a streak of parsimony in Watt who tried in other cases such as Rennie to pay as little as possible. In Murdock's instance, Boulton persuaded the mine owners to make him a gift of £20. [123]

A closer investigation of the various Boulton and Watt archives shows that Watt was willing to accept Murdock's ideas when he thought them practicable. Unfortunately we do not know what the mechanism was over which Watt made a mistake and was corrected by Murdock in June 1780. Watt wrote to Wilson, the person we would expect to find in charge of Boulton and Watt's Cornish operations:

> I approve of the method W^{m} proposes to the reverse motion which indeed is the one I first thought of but laid it aside thinking the other would be less work – I do not see how I came to be so stupid as not to see that 3 wheels could not reverse the motion. Two wheels can but not 3...
>
> At any rate in any case when a thing is found not to answer, if the matter presses, Willm is certainly right to go on in such way as will do. A man who only speculates in his closet, cannot have always such clear ideas of objects as the man who has them before him. I am glad to hear that the engine is so far advanced & I hope it will answer. [124]

Here we see that Watt was willing to allow some latitude to the man on the spot.

Watt was certainly willing to agree to Murdock's proposals when he thought them practicable. When some small portable engines were being discussed in 1783, Watt told

Wilson, 'I shall be glad to see Williams drawing and will readily agree to try anything he proposes to which I do not see such objections as would hinder him from wishing it'. [125] This was a time when Murdock was suffering from 'ague' and Watt was very concerned about him. He obtained various prescriptions from Withering which he sent to Wilson asking that:

> ... his apothecary should immediately make up for him [Murdock] and administer according to circumstances which we cannot judge of at this distance; but if after all he continues dangerously ill or is not soon relieved by these medicines, we desire that you would call in Dr. Gould or Dr Woolcat, and if judged necessary have William removed to Truro or Falmouth to be near his Doctor, and while he feels himself weak he ought not to undergo any fatigue. [126]

Watt certainly tried to look after his valued employees. Later Watt implemented some of Murdock's schemes but not always successfully.

If Murdock was a success, some other engine erectors certainly were not for Watt's letters are full of complaints. We hear mostly of the problems and not about those who carried out their tasks efficiently. These men were trained and recommended by Boulton and Watt and on them rested Boulton and Watt's reputation. If there was no engine available for erection, they may have been kept on at Soho. For example, Watt asked Boulton whether Henry Williams could be employed at Chelsea, 'his wages are running on here needlessly and he is no workman and can be of little service to us here' at Soho. [127] In Cornwall, Boulton and Watt employed full time a small team to assist with erection and maintenance. James Lawson was there for five years from 1780 to 1784, starting at 7s. rising to 9s. per week. [128] The career of Thomas Dudley is fairly typical of those who upset Watt. We have seen that he had been to Birmingham from where he had written enthusiastically to Wilson about the new Watt engines. At first all went well and Watt was pleased with progress at Wheal Bussy on which Dudley was working during August 1777, finding him 'exceeding manageable' so that he was put in charge of this engine when it was finished. [129] But then a year later, Watt reported to Boulton that Dudley was doing 'bad work'. [130] He was probably erecting an engine at Wheal Union where, at the beginning of September 1778, Watt listed a catalogue of problems:

> On Friday I went to Wl Union where I found them in ye dumps On Thursday they had attempted to sett ye Engine to work before they had got it ready. I had told them that they had not cold [water] for above 7 or 8 strokes pr. minute. However as there was a great number of spectators Dudley thought he would show them somewhat and accordingly sett off at the rate of 24 str[okes] pr. minute; he soon got all his water boiling hot and then they seemed to be at a loss why the Engine would not go, but as they found by experiment that cold [water] improved it they set about putting down a cold [water] pump to add it, wch was finished by Saturday morning. D. complained that ye Barometer had choaked soon after they began but I found it was fairly soldered up & had never gone. I also found many air leaks wch convinced me he had never sought after them, he had neglected to put oakum about the necks of ye regulator Spindles and had used no means to prevent over opening of exhaustion regulator by wch they had such rapps as to break one of ye spring beams off all these things he had been fully warned on Monday & before - before they set to work at all, the fire would not burn wch was caused by a parcel of Bricks left in ye flues near ye damper. I staid to 9 o'clock on Friday night and saw most things cured. I was at them again by 7 o'clock on Saturday morning and got the engine to work by 10 o'clock quietly & peaceably 18 strokes pr. minute 5 $^1/_2$ feet long only. [131]

This well illustrates the responsibilities undertaken by these erectors as well as the knowledge and skills necessary. It also shows Watt's exceptional ability in general engineering practice to have been able to see such projects through from start to finish.

But alas Dudley seems not to have mended his ways. In May 1779, he was found to have committed a series of blunders on the engine for Hallamannin in Cornwall so that Watt exclaimed, 'Dudley will ruin us'. [132] Watt told Henderson, 'Dudley has lett the engines in his charge gett most damnably out of order and has not life or brains to mend them so Joseph [Harrison] is to be sent down to rectify them'. [133] Watt wondered if his employment should be continued and in the same letter also congratulated Henderson on the good work being done by Law, 'You have great credit in him he seems much improved under your care'. One of the problems with Dudley, as with other erectors, was heavy drinking. Hornblower commented, 'I am inclined to believe that T.D. has been induced by such company much to his hurt and will endanger a good constitution if he is so unhappy as to persevere'. [134] Watt also had to tell Boulton, 'Joseph has pursued his old practices of drinking in this neighbourhood in a scandalous manner untill the very enginemen turned him into ridicule... there was several days when he was never sober'. [135]

At least Dudley went to apologize to Watt for his mistakes at Hallamannin, possibly realising that his future career with Boulton and Watt was at stake but Watt remained sceptical:

> T.D.s carelessness and bungling has been much more than we were informed of, he was here last night and pleaded very hard. I told him very calmly that as the case stood, he could not expect we would blacken our character to save his and that we would not run any further risks by employing him untill some proof was exhibited of his abilities. [136]

Here at last we see Watt facing up to an employee.

One outcome of all these labour problems was that Watt decided to examine the way in which these men received their instructions. He was thinking about this in June 1778 when the workload was increasing with more orders:

> I am very sensible that in ye mult[itude] of things I have to think of & the vexations I sometimes meet with I am not so accurate in giving out orders as I might be but in ye present case I did my part & the blunder lies among Mr. H[enderson] Playfair & Hall - and at any rate I see a determined resolution in him Hall to understand nothing without as much a mere drawing or explanation as would serve a workman who had never seen one of our Engines - In so much that I should save myself trouble by giving the orders to the Engineers upon ye spot instead of him and it is really impossible for me to comply with all that accuracy in giving orders to our workmen who ought to understand these minutiae better than me otherwise are not fit for their places. [137]

We have a clue to the way Watt's mind was working in a letter a few days later, 'I have no time to make drawings or write studied directions it being as much as I can do to keep the people here right'. [138] During his civil engineering practice, he had written out instructions for the contractors and so he determined to produce something similar for his engine erectors and made a start with two and a half sheets later that summer. [139]

After his return to Birmingham in the New Year of 1779, Watt wrote his *Directions for Erecting and Working the Newly-Invented Steam Engines.* [140] It contained four sections: General Directions for Building the Engine House; Directions for Putting the Engine Together;

Directions for Working the Engine; Additional Directions and six drawings or plates. In early June, he spent one Sunday 'writing directions for putting engines together & for them to be printed'. [141] Soon he was correcting proofs and sending more text. On 12 June, he went to the printers and 'ordered 50 copies on copy paper and 50 on thin post paper'. [142] These were ready before the end of the month because a copy was despatched to Hornblower. [143] It was a typical Watt work, very detailed, for the text as reprinted by Dickinson and Jenkins covers nearly twenty pages. [144] It has been described as the earliest text on steam engine construction and played an important role in standardising the procedures of Boulton and Watt.

Premiums

There still remained the problem of settling the premiums to be paid for the use of Watt's patent or privilege. His 1769 patent was not for a steam engine but for 'A method of lessening the consumption of steam and fuel in fire engines'. Watt could neither own his 'perfect' engine nor exploit it at mines or mills as Arkwright could with his water-frame for spinning cotton. Arkwright charged a fixed, single sum when others wished to use his invention. Later Joseph Bramah stated that he made a single charge when he sold one of his inventions. [145] Watt had decided that he should receive a third of the savings of the cost of fuel in his engine when compared with the consumption of the common Newcomen type, doing the same work. This was a good advertising and publicity concept because not only did it show the superior performance of his own engine but it was linked to the profits which the purchasers would obtain since they retained the other two-thirds. Yet, while this sounds delightfully simple, Watt was soon forced to admit how problematic it was in reality:

> The principal objection that lies against [his conditions] is the difficulty of ascertaining what is the real consumption of fuel by common engines in performing a certain quantity of work; though this may at first sight appear an easy matter to determine, yet it is in reality more intricate than it appears, for it is necessary to ascertain the quantity of coals consumed in some given space of time, the number of strokes made by the engine during that time, the length of these strokes, the weight of the column of water raised by the engine, besides many other less material circumstances, which do in some degree affect the result. [146]

Watt was aiming to compare the actual performance of engines in terms of the weight of water raised with the amount of coal consumed, showing his delight in mathematical calculations.

But there were few instances where direct comparisons could be made. It was obviously impossible where a Boulton and Watt engine was being installed in a new situation where there was no earlier engine or, as in the case of Wilkinson's New Willey ironworks, where the engine was being applied to a new task such as blowing the furnaces which had never been done before directly by steam power. At many mines, the owners might want to be able to sink the shaft lower, draining deeper, and so needed a more powerful engine or, as sometimes happened in Cornwall, two Newcomen engines might be replaced by a single Boulton and Watt engine. Then until the first Boulton and Watt engines had been running for some time, it was impossible either to determine their performance or to convince the sceptics of their greater economy. Therefore Boulton and Watt had to erect their first engines and set them to work without obtaining any agreements over premiums to be paid. And of course once an engine had been set to work, there was less incentive for the owners to make any settlement.

Wilkinson was granted special terms and had to pay only one shilling a year, possibly because not only he had received one of the first prototype engines but also it was only he who could bore cylinders with sufficient accuracy. [147] (See p. 37) The Bloomfield engine may also have been supplied under special terms since Watt had agreed to install his engine for the same price as the Newcomen one it replaced. [148] This engine had to be extensively rebuilt to bring it up to the standard of later ones. It took a while to settle the premiums for other early engines. Agreement for Hawkesbury Colliery, Bedworth, which was erected at the beginning of 1777, was not reached until two years had passed and after it also had been fitted with improved drop valves. A trial was held in March 1779 in the presence of an arbitrator who awarded Boulton and Watt £217 per annum as opposed to the £30 which the owners thought sufficient. [149] Watt took Keir for moral support to a meeting at Bedworth to reach an agreement and said, 'I shall act with all the temper and boldness I can'. [150] This was the occasion of Watt's joyful letter to Boulton in June 1779:

> Hallelujah! Hallelyjee!
> We have concluded with Hawkesbury
> £217 - pr annum - from Ladyday last
> £275.5 for time past, £117 our account
> We make them a present of £100 guineas
> Peace and good fellowship on earth -
> Perrins and Evans to be dismissed -
> 3 more Engines wanted in Cornwall -
> Dudley repentant and amendant
> Yours rejoicing
> James Watt
> Please burn this nonsense. [151]

Watt's relief at this settlement was the result not only of the long struggle to achieve it but also at the prospect of an end to the financial crises which Boulton and Watt were facing through not receiving their patent dues. Back in August 1778, Watt wrote sternly to Boulton, wanting a trial at Bedworth:

> At any rate send some steady person there to superintend the working for a week or two that a true average of the fuel may be ascertained upon oath - their tardiness & others to settle should make us obstinate in not taking in hand any Engine untill a positive Legal agreement be signed, and for my part you must excuse me when I tell you that I will not put pen to paper on a new subject untill that is done. Untill we have ordered the engine our power is greater than that of the Lord Chancellor as I believe even he cannot compell us to do it unless we chuse. Lett our terms be moderate & if possible consolidated into money a priori and it is certain we shall gett some money, enough to keep us out of jail, in continual apprehension of which I live at present. [152]

This letter was followed by another a few days later when Watt tried to put his foot down even more firmly because no agreements had been reached over the engine for Shadwell either. Negotiations there had been dragging on since at least the beginning of December 1776 when Watt had been arguing with them:

> We came to no terms further than what we wrote them before which I confirmed and offered moreover to keep the engine in order for one year; when they modestly

> insisted we should do so for the whole 25 years which I firmly refused - They seemed to doubt the reality of the performance of the Bow engine... I should not have been so sharp with them had they not been bullying me selon la mode de Londres. [153]

No settlement had been reached by May 1780. [154]

Watt wanted an agreement to be reached at Chelsea before any work started there:

> Every article must be settled & sealed with Cornish adventurers before we begin... therefore let not Chelsea be begun untill signed & sealed. I hope you will not take amiss my writing so positively on this subject of agreements but realy my faith in mankind will carry me no further and if I can't get money I am resolved to save my bacon and to live in hunger & ease. As it is, we don't even gett such a share of reputation as our works deserve for every man who cheats us defames in order to justify himself. [155]

Meason at Wanlock Head would not settle until agreement had been reached at Chelsea; obviously everybody was waiting to see what settlements were reached by others. [156]

The situation was similar in Cornwall. In January 1778, Watt had valued Ting Tang at £100 per annum but it took nearly a year before the agreement was finally settled with the owners there. [157] Even at Chacewater, where Wilson was on such friendly terms with Watt, it took a long time for agreement to be reached after starting the large engine at the end of July 1778. Wilson and Watt met in early December 'in evening to settle about engines our demand £700 for the large engine and to pay nothing for the small engine so long as they use the large one'. [158] Even so another seven weeks passed before the draft agreement was received from Wilson. [159] That had to be sent to Yorkshire for confirmation by the proprietors where it was not ratified until Boulton was there in March 1780. Watt commented to Annie, 'They use us vilely'. [160]

It was evident that Boulton and Watt needed to determine standard terms for their premiums which would be applicable to any engine based on the performance of a standard common Newcomen engine and adjusted for their own. The opportunity came when an engine was wanted for Poldice mine. This mine was a large one with three 60 ins. and two 66 ins. diameter Newcomen engines. [161] In September 1778, it was agreed to replace the two Eastern engines of 66 ins. and 60 ins. diameter with a single Boulton and Watt engine of 63 ins. diameter. [162] Watt reported to Boulton, 'Poldice people are now welding hot & must not be suffered to cool - They are exceeding important as they loose £150 p. month untill our engine is going'. [163] Accordingly Watt was very busy preparing drawings and ordering materials in October and early November. The cylinder was set up a year later on 1 October 1779 and the engine started on 30 December 1779. It worked so well that a second engine was ordered in early 1780. [164]

But the Poldice engine is important not so much for the performance of a Boulton and Watt engine as the agreement of the mine owners to allow records to be kept of their two original engines on the Eastern shaft from which a standard could be made to determine the saving of coals with the new type of engine. Records were kept of the coals consumed, rate of working, length of stroke, etc., to calculate the amount of water raised during the months of August and September 1778. Watt said, 'The old engines were both in good order, one of them was lately built, and the other had undergone a thorough repair'. [165] The loads on their pistons were 5.7 and 5.2 lbs. This was below the average for the sixty-four engines listed in Watt's 'Engine Book'. [166] It is noteworthy that he recorded those engines erected by Smeaton as achieving a pressure of 7 lb. or over. Most of the other engines in Cornwall were over 6 lbs. pressure so it is evident that these Poldice engines were not overloaded. It was agreed by other mine owners and captains that the performance of these engines represented a fair and reasonable average. Coal consumption would have had to be measured in any event to obtain the drawback on the excise duty and so could not have been falsified.

Watt therefore drew up ten initial proposals for the Poldice adventurers which he also forwarded to Boulton.

1. They to build Engine according to plans & directions wch are to be strictly adhered to & will be [erected] by Hornblower if I may judge by Tingtang.
2. The strokes pr. minute & Coals now used to be fixed by the Books, wch are regularly sworn to monthly the same to be done with the new Engine.
3. We to receive 1/3 savings monthly - while ye Engine is used at Poldice.
4. Our premium to increase as ye strokes do, & to encrease 1/12 for every 5 fathom of depth additional beyond 60 fathom.
5. No extra premium during time of forking pool at beginning.
6. Books always liable to our inspection, and us to appoint at our charge supervisor or cheque of Coal measure.
7. Absolute power of displacing engine men & of rejecting bad coals.
8. Power of inspecting and directing repairs of Engine.
9. Liberty of changing boiler if not equal to Chacewater.
10. New agreement if sold off premises. [167]

Watt obviously wanted to ensure that his engines continued to function properly and not let down his reputation.

One necessity if this scheme were to operate satisfactorily was to have a device which would record the number of strokes accurately. A ratchet and pawl counter, which Watt had arranged for the early trials at Soho, was not considered to be sufficiently tamper-proof. When testing the Chelsea and Shadwell engines, Boulton had an apparatus actuated by a pendulum which had originated as a pedometer made by Wyke and Green of Liverpool. It had been brought to the notice of Boulton and Watt by Henderson. [168] This was fixed to the beam and it counted each stroke as the beam oscillated. Those made subsequently at Soho were not as reliable as the ones from Liverpool so the Wyke instrument became standard on later Boulton and Watt pumping engines. [169]

Watt recognised that the performance of their engine at Poldice would be very important because it would set the standard for future premiums. Accordingly he instructed that the Soho Manufactory was to 'Lett the utmost care be taken that these goods [parts for Poldice] be executed in ye very best manner for the Engine is likely to be of more consequence to us than any we have made yet'. [170] Murdock was to be recalled to make the patterns. Watt also set about preparing 'Proposals' which would form a standard agreement for other engines. He was working on them during August and received the printer's proofs in the middle of September 1779. He told Boulton in October, 'I am calculating the effects of all the engines in the county and reducing them to a 60 [ins. diameter] five & half feet stroke so that a general average may be drawn from them. but I mean not to include any engines of whose effects we are uncertain of vizt Dolcoath & Wheal Prosper'. [171]

Watt drew up instructions for calculating the savings of fuel which would be applicable to any engine based on the figures obtained at Poldice. For those engines which were not the same size, he showed how the comparison could be made by first reducing the Poldice figures to a cylinder diameter of 10 ins., with a 1 ft. stroke loaded to 1 lb. These results could then be adjusted for the actual dimensions of the appropriate engine. He was trying to be absolutely fair and to devise a foolproof system but, as will be seen in the draft proposal printed in Appendix 3, his system became excessively complex so that it is difficult to imagine how the ordinary mine captain or clerk of those days could understand it. Added to which were the suspicions of the Cornish people that Boulton and Watt were trying to defraud them as well as suspicions about the reliability of the counters. Then it was impossible to know in advance what the premiums might be since these varied according to the amount of water pumped and depended on seasonal fluctuations in rainfall as well as the price of coal.

Meanwhile Watt's Journal for the latter part of 1778 shows how these tests at Poldice became the basis for negotiating the premiums. He met the Wheal Union adventurers and could not agree with their figures but 'after some juggling agree to settle a loaded engine by the average of the 3 at Poldice' and so he duly calculated their premiums based on these experiments. Within a couple of weeks there was another meeting with 'Messrs Phillips & Edwards with Capt. Grundy and Hodge', another with Capt. Simmons over sizes of engines for Poldory and Ale and Cakes and yet another with Capt. Paul and Hornblower about Wheal Maid. Tables were drawn up for all these engines as well as Ting Tang. Then on 14 December, Watt received an order from Wheal Chance for beginning their engine but ominously there was a 'dispute over their table'. [172]

While these terms for premiums were appropriate at places where coal was not readily available, they could not be applied easily at coal mines where fuel was available at often no cost. Nicholas Walton of the Long Benton Colliery told Smeaton in 1779 that 'The quantity of coals we use at the Fire Engine are now hewed for nothing'. [173] This is why Boulton and Watt supplied so few engines to collieries. Watt realised that it was not worthwhile trying to break into this market and told Smeaton, 'We have no intention to combat you neither on the coal hill nor anywhere else, and I think neither you nor us have a chance of much employment in Newcastle, where we have never sought after it'. [174]

But the scene in Cornwall was totally different where orders for engines were being placed one after the other now that those of Boulton and Watt were seen to be so superior. Watt wrote to Black in 1778:

> An universal confidence of the whole county in the abilities of the engine is now fully established, and we have executed agreements for several others, one of which will pay us better still, and is also to do the work of two other engines larger than itself. Several mines, formerly abandoned, are likely to go to work again through virtue of our engines; we have five engines of various sizes actually going here now in this county, and have eight more in contemplation, so that our affairs wear a most smiling aspect to human eyes... Our affairs in other parts of England go on very well; but no part can or will pay us so well as Cornwall, and we have luckily come among them when they were almost at their wit's end how to go deeper with their mines. [175]

A year later at the end of October 1779, Watt wrote excitedly from Cornwall that Boulton must 'come down directly on account of Wheal Virgin engines for which mine 5 engines are wanted forthwith'. [176] Negotiations took a while so it was not until October 1780 that Watt told Hamilton, 'We have at last concluded with Wheal Virgin for a good round annual sum for the use of 5 engines which will take us about a year to execute and untill they come to pay we shall remain poor people after that better things is expected'. [177] This was the largest single order Boulton and Watt ever received. Prospects in Cornwall looked fair to bring both fame and fortune to Boulton and Watt.

Notes

1 Lord, (1923), p. 151.
2 Pryce, (1778), p. 308.
3 Rolt & Allen, (1977), p. 44.
4 Pryce, (1778), p. xiv.
5 JWP, 4/36, J. Hornblower to J. Watt, 26 Oct. 1776.
6 Boulton 348.52, M. Boulton to J. Watt, Aug. or Sept. 1775.
7 Pryce, (1778), p. 308.
8 Ibid, p. xv.
9 Ibid, p. 308.
10 Hills, (1989), p. 39.
11 Pryce, (1778), p. 153.
12 JWP, 4/20, J. Hornblower to J. Watt, 2 Nov. 1776.
13 Smiles, (1878), p. 177.
14 Pryce, (1778), p. 312.
15 Skempton, (1981), p. 182.
16 Wilson Papers DDX 318/11, T. Wilson, 'An Account of the State of the Principal Mines in the County of Cornwall at the Time of the Introduction of Mr. Watts improved Steam Engines into the County', no date.
17 Buckley, (2000), p. 47.
18 Smeaton, (1812), Vol. II, 'Chase Water Engine', pp. 348 – 49.
19 Wilson Papers DDX/11, 'An Account...'
20 Ibid.
21 Quoted by Dickinson & Jenkins, (1927), p. 129, no reference.
22 Dickinson & Titley, (1934), p. 12.
23 Clark, (1983), various pages.
24 See Robinson & McKie, (1970), p. 188, Letter 134, A. Watt to J. Black, 6 Nov. 1792.
25 JWP, C3/1, 'Memorandum', 17 April & 3 May 1776.
26 Rogers, (1976), p. 10.
27 JWP, 4/36, J. Hornblower to M. Boulton, 10 Oct. 1776.
28 JWP, 4/36, J. Hornblower to Boulton & Watt, 26 Oct. 1776.
29 Ibid, 2 Nov. 1776.
30 Ibid, 23 Nov. 1776.
31 Muirhead, (1854), Vol. II, p. 103, Letter 161, M. Boulton to J. Watt, 31 Nov. 1776; Boulton 348.66.
32 JWP, 4/20.32, G. Hamilton to J. Watt, 9 Oct. 1776.
33 See JWP, 4/36, J. Hornblower to J. Watt, Letters 7, 23 Dec. 1776, 6 & 25 Jan. 1777.
34 JWP, 4/36, J. Hornblower to J. Watt, 30 Jan. 1777.
35 Wilson Papers DDX 318/11, 'An Account...'
36 Dickinson & Jenkins, (1927), p. 129, J. Watt to M. Boulton, 20 May 1777.
37 Boulton 348.75, J. Watt to M. Boulton, 9 Aug. 1777.
38 Boulton 348.76, J. Watt to M. Boulton, 14 Aug. 1777.
39 Boulton 348.78, J. Watt to M. Boulton, 13 Sept. 1777.
40 Pryce, (1778), p. 313.
41 Boulton 348.78, J. Watt to M. Boulton, 13 Sept. 1777.
42 Dickinson & Jenkins, (1927), p. 308, Edwards to M. Boulton; B. & W. Col.
43 Boulton 348.79, J. Watt to M. Boulton, 20 Sept. 1777.
44 Wilson Papers DDX 318/11, 'An Account...' and JWP, C3/3, 'Journal', 3 June 1778.
45 JWP, C3/2, 'Journal', 4 Aug. 1777.
46 Dickinson, (1935), p. 101, J. Watt to M. Boulton, 9 Aug. 1777.
47 B. & W. Col., MI, 2/9, 'Account Book', 10 Aug. 1777.
48 Boulton 348.74, J. Watt to M. Boulton, 4 Aug. [Sept.] 1777.
49 Boulton 348.76, J. Watt to M. Boulton, 14 Aug. 1777.
50 Dickinson & Jenkins, (1927), p. 135, M. Boulton to J. Watt, 2 Sept. 1780.
51 Smiles, (1878), p. 179, A. Watt to Mrs. Boulton, 1 Sept. 1777.
52 JWP, 4/36, J. Hornblower to J. Watt, 2 Nov. 1776.
53 Boulton 348.75, J. Watt to M. Boulton, 9 Aug. 1777.

54 Boulton 348.76, J. Watt to M. Boulton, 14 Aug. 1777.
55 Boulton 348.77, J. Watt to M. Boulton, 25 Aug. 1777.
56 Ibid.
57 Boulton 348.78, J. Watt to M. Boulton, 13 Sept. 1777.
58 Boulton 348.76, J. Watt to M. Boulton, 14 Aug. 1777.
59 Boulton 348.79, J. Watt to M. Boulton, 20 Sept. 1777.
60 Ibid.
61 Lord, (1923), p. 153, M. Boulton to Hodgkinson, New Burlington Street, 8 May 1778; Tangye Mss.
62 Boulton 348.84, J. Watt to M. Boulton, 2 Feb. 1778.
63 B. & W. Col., Portfolio 1070, engine for Périer, Paris, 1779.
64 Dickinson & Jenkins, (1927), p. 131, no reference.
65 Boulton 348.84, J. Watt to M. Boulton, 2 Feb. 1778.
66 JWP, 4/36, J. Hornblower to J. Watt, 21 July 1777.
67 Boulton 348.78, J. Watt to M. Boulton, 1 Sept. 1777.
68 JWP, 4/36, J. Hornblower to J. Watt, 6 Jan. & 22 March 1777.
69 Boulton 348.76, J. Watt to M. Boulton, 14 Aug. 1777.
70 Boulton 348.74, J. Watt to M. Boulton, 4 Aug. [Sept. in pencil] 1777.
71 JWP, 4/36, J. Hornblower to J. Watt, 17 Dec. 1777.
72 Ibid, J. Hornblower to J. Watt, 26 Jan. 1778.
73 JWP, C3/3, 'Journal', 28 & 30 July 1778.
74 Michell, (1984), p. 20.
75 Wilson Papers DDX 318/11, 'An Account...'
76 B. & W. Col., Portfolio 1315.
77 Boulton 348.89, J. Watt to M. Boulton, 27 June 1778.
78 JWP, C3/3, 'Journal', 29 & 30 July 1778.
79 Dickinson & Jenkins, (1927), p. 133.
80 Boulton 348.107, J. Watt to M. Boulton, 29 Aug. 1778.
81 JWP, C3/3, 'Journal', 18 Sept. 1778.
82 Robinson & McKie, (1970), p. 42, Letter 36, J. Watt to J. Black, 12 Dec. 1778.
83 JWP, C3/3, 'Journal', 31 Dec. 1778.
84 Boulton 348.107, J. Watt to M. Boulton, 29 Aug. 1778.
85 B. & W. Col., MI, 2/9, 'Account Book', 31 Dec. 1778.
86 JWP, C3/3, 'Journal', 3 Feb – 21 Dec. 1778.
87 Ibid, see various entries.
88 Boulton 348.89, J. Watt to M. Boulton, 27 June 1778.
89 Boulton 348.92, J. Watt to M. Boulton, 6 July 1778 and Boulton 348.105, J. Watt to M. Boulton, 22 Aug. 1778.
90 JWP, C3/3, 'Journal', 10 Oct. – 9 Dec. 1778.
91 Ibid, 17 Oct. 1778.
92 Ibid, 6 Nov. 1778.
93 Boulton 348.120, J. Watt to M. Boulton, no date.
94 JWP, 3/8, J. Watt to G. Hamilton, 23 July 1781.
95 Boulton 348.77, J. Watt to M. Boulton, 25 Aug. 1777.
96 JWP, C3/3, 'Journal', 7, 17 Nov., 1, 3, 25 Dec. 1778.
97 Dickinson & Jenkins, (1927), p. 290.
98 JWP, C3/3, 'Journal', 4 Dec. 1778.
99 JWP, C3/4 and C3/5, 'Journal', various dates.
100 JWP, 4/36, J. Hornblower to J. Watt, 25 Sept. 1779.
101 Ibid, J. Hornblower to J. Watt, 24 April 1780.
102 Anon, 'The Hornblower Family', *Journal of the Trevithick Society*, No. 4, 1976, p. 18.
103 JWP, C3/3, 'Journal', 1 Oct. 1778.
104 Ibid, 23 Feb. 1779.
105 Boulton 348.86, J. Watt to M. Boulton, 8 June 1778.
106 Tann, (1981), p. 46, Doc. 11, M. Boulton to L. Henderson, 13 Aug. 1779. See also Ibid, pp. 52 – 3, Doc. 13, M. Boulton to J. Watt, 13 Jan. 1781.
107 Muirhead, (1854), Vol. II, J. Watt to M. Boulton, 9 Feb. 1782; JWP, 3/18.
108 JWP, 3/18, J. Watt to M. Boulton, 9 March 1782.

109 JWP, 3/8., J. Watt to M. Boulton, 7 April 1781.
110 JWP, C3/7, 'Diary', 25 Feb. 1783.
111 JWP, C3/4, 'Journal', 5 March 1779.
112 JWP, W5, J. Watt to L. Henderson, 19 May 1779.
113 Smiles, (1890), p. 127.
114 Boulton 348.106, J. Watt to M. Murdock, 27 Aug. 1778. See Dickinson & Jenkins, (1927), p. 138 for Murdock going there in March 1779 through delays in getting the materials to the mine.
115 Dickinson & Jenkins, (1927), pp. 181 – 82.
116 Ibid, p. 122, M. Boulton to J. Watt, 12 Aug. 1779.
117 JWP, C3/5, 'Journal', 26 Aug. 1779.
118 Ibid, 22 Sept. 1779.
119 Dickinson, (1935), p. 115.
120 JWP, C3/5, 'Journal', 13 Feb. 1780.
121 Griffiths, (1992), p. 103.
122 Ibid, J. Watt to M. Boulton, 28 Sept. 1780.
123 Roll, (1930), p. 64.
124 Wilson Papers DDX 318/1, J. Watt to T. Wilson, 5 June 1780.
125 Ibid, J. Watt to T. Wilson, 18 Sept. 1783.
126 Ibid, J. Watt to T. Wilson, 16 Aug. 1783.
127 Boulton 348.131, J. Watt to M. Boulton, 28 Jan. 1779.
128 Wilson Papers DDX 318/1, 'Lawson's Account'.
129 Boulton 348.76, J. Watt to M. Boulton, 14 Aug. 1777 and Boulton 349.79, J. Watt to M. Boulton, 20 Sept. 1777.
130 Boulton 348.98, J. Watt to M. Boulton, 2 Aug. 1778.
131 Boulton 348.108, J. Watt to M. Boulton, 6 Sept. 1778.
132 Boulton 348.137, J. Watt to M. Boulton, 11 May 1779.
133 JWP, W5, J. Watt to L. Henderson, 19 May 1779.
134 JWP, 4/36, J. Hornblower to J. Watt, 22 July 1779.
135 Boulton 348.153, J. Watt to M. Boulton, 4 Oct. 1779.
136 Boulton 348.154, J. Watt to M. Boulton, 6 Oct. 1779.
137 Boulton 348.87, J. Watt to M. Boulton, 19 June 1778.
138 Boulton 348.88, J. Watt to M. Boulton, 25 June 1778.
139 JWP, C3/3, 'Journal', 23 Aug. 1778.
140 Dickinson & Jenkins, (1927), pp. 377 – 98 and Musson & Robinson, (1969), pp. 126 – 54.
141 JWP, C3/4, 'Journal', 6 June 1779.
142 Ibid, 12 June 1779.
143 Ibid, 29 June 1779.
144 Dickinson & Jenkins, (1927), pp. 377 – 98.
145 Fitton, (1989), p. 93 and Bramah, (1797), p. 66.
146 JWP, 4/69, 'Proposals to Adventurers'.
147 Lord, (1923), p. 158, J. Watt to J. Wilkinson, 17 July 1777.
148 Dickinson & Jenkins, (1927), p. 113, J. Watt to Bentley & Co., 15 June 1775.
149 Ibid, p. 117.
150 Boulton 348.142, J. Watt to M. Boulton, 30 June 1779.
151 Boulton 348.143, J. Watt to M. Boulton, 31 [sic] June 1779
152 Boulton 348.107, J. Watt to M. Boulton, 29 Aug. 1778.
153 Boulton 348.68, J. Watt to M. Boulton, 3 Dec. 1776.
154 JWP, 3/8, J. Watt to M. Boulton, 1 May 1780.
155 Boulton 348.108, J. Watt to M. Boulton, 6 Sept. 1778.
156 Boulton 348.139, J. Watt to M. Boulton, 26 June 1778.
157 Boulton 348.82, J. Watt to M. Boulton, 13 Jan. 1778 and JWP, C3/5, 'Journal', loose notes, 29 Dec. 1778.
158 JWP, C3/3, 'Journal, 3 Dec. 1778.
159 Boulton 348.126, J. Watt to M. Boulton, 23 Jan. 1779.
160 Tann, (1981), p. 206, Doc. 84, M. Boulton to C. Vere, April 1779 and JWP, W5, J. Watt to A. Watt, 4 March 1780.
161 B. & W. Col., MI, 2/10, 'Engine Book'.
162 Boulton 348.110, J. Watt to M. Boulton, 12 Sept. 1778 and JWP, C3/3, 'Journal', various entries for Sept. 1778.

163 Boulton 348.113, J. Watt to M. Boulton, 24 Sept. 1778
164 See JWP, C3/5, 'Journal', for various entries.
165 JWP, 4/69, 'Proposals to Adventurers' and Tann, (1981), pp. 281 – 83 for Watt's figures.
166 B. & W. Col., MI, 2/10, 'Engine Book'.
167 Boulton 348.111, J. Watt to M. Boulton, 16 Sept. 1778.
168 Dickinson & Jenkins, (1927), p. 227.
169 Boulton 348.161, J. Watt to M. Boulton, 28 Oct. 1779.
170 Boulton 348.114, J. Watt to M. Boulton, 2 Oct. 1778.
171 Boulton 348.160, J. Watt to M. Boulton, 21 Oct. 1779.
172 See JWP, C3/3, 'Journal', entries between 19 Nov. and 14 Dec. 1778.
173 N. Walton to J. Smeaton, 28 May 1779. I wish to thank A.P. Woolrich for drawing my attention to this letter.
174 J. Watt to J. Smeaton, 12 July 1779. I wish to thank A.P. Woolrich for drawing my attention to this letter.
175 Muirhead, (1854), Vol. II, p. 112, Letter 170. J. Watt to J. Black, 12 Dec. 1778.
176 JWP, C3/5, 'Journal', 28 Oct. 1779.
177 JWP, W5, J. Watt to G. Hamilton, 28 Oct. 1780.

6
Development of the Single-Acting Pumping Engines and Foreign Patents

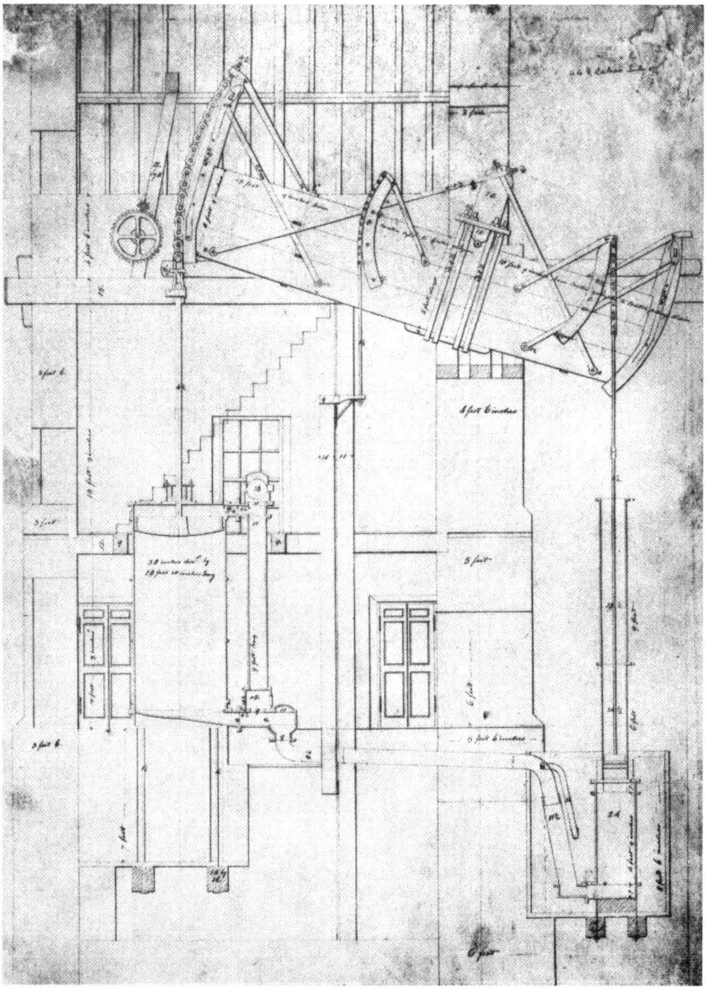

Watt's single-acting pumping engine for Ale and Cakes mine showing stepped construction of engine house and beam made from six baulks of timber with short king-post.

Improvements to Engines

By the end of 1778, the Boulton and Watt engine had proved to be far more economical than any previously known. Changes to its valve gear and other parts increased its reliability so that Watt had a design which led the world and far outstripped any rivals. Earlier he had expressed a desire to lead a life of leisure, so we might have expected that he would rest on his laurels and not continue to improve his engine. After all, the Newcomen engine, on which he had based his own, had scarcely changed from the original one of 1712. It is true that Smeaton had increased the efficiency, but this was through systematic alterations rather than radical changes in design or mode of operation. Watt had obtained the Act of Parliament in 1775 to extend his patent partly through pointing out that he had invented the perfect engine, implying that it could not be improved. This concept was quietly dropped as he continued to develop it both through modifications to detail parts based on experience as well as radical alterations based on theory. These improvements show the remarkable versatility of Watt's mind which could abandon old ideas and introduce new ones which kept him ahead of rivals.

While we are concentrating on Watt as the subject of this biography, it is important to remember that this was a partnership of Boulton and Watt with both contributing to the organisation of the firm as well as to the technical development. Boulton was modifying the engine at Soho while Watt was in London during the early part of 1775. While Watt spent those many months in Cornwall, Boulton had to supervise the engine business at Birmingham. In May 1778 when Watt went to Cornwall, he left Boulton and Henderson testing the Newcomen engine at Chelsea which they were going to alter. When in Cornwall during the autumn, Boulton visited their engines and gave assistance. Boulton may never have given up hopes of establishing a full steam engine manufactory at Soho for in 1778 he told Smeaton:

> We are systematizing the business of engine making, as we have done before in the button manufactory; we are training up workmen, and making tools and machines to form the different parts of Mr. Watt's engines, with more accuracy, and at a cheaper rate than can possibly be done by the ordinary methods of working. Our workshop and apparatus will be of sufficient extent to execute all the engines which are likely to be soon wanted in this country; and it will not be worth the expense for any other engineers to erect similar works, for that would be like building a mill to grind a bushel of corn. [1]

Watt was no stranger either to organising production for he had set up his scientific instrument business in Glasgow along these lines. (See Vol. I, pp. 84-5) Even if the original partners did not establish a manufactory for complete engines at this time, we can see that Watt had Boulton's active co-operation, practical assistance and suggestions in forwarding their enterprise.

We have examined how it had been necessary to change the type of valves controlling the steam on its passage through the engine from an oscillating sector type to the lifting poppet form with conical seating. (See p. 67) Boulton had commented in 1776 that, 'The Regulators in an Engine are certainly of as much Consequence as ye left & right Ventricles of ye heart in Animals & I should be happy to see a better valve made for I have never thought it a bad one as those in our Engine seem to be as perfect as one can expect'. [2] Both the valves and their operating mechanisms received attention to improve them since it was crucial to have an engine which was not only economical but also operated smoothly. These objectives depended upon the operating mechanisms. The steam valve on the Newcomen engine was opened by one leg of a pivoted upside down 'Y' lever. A peg on the moving plug-tree rotated the lever

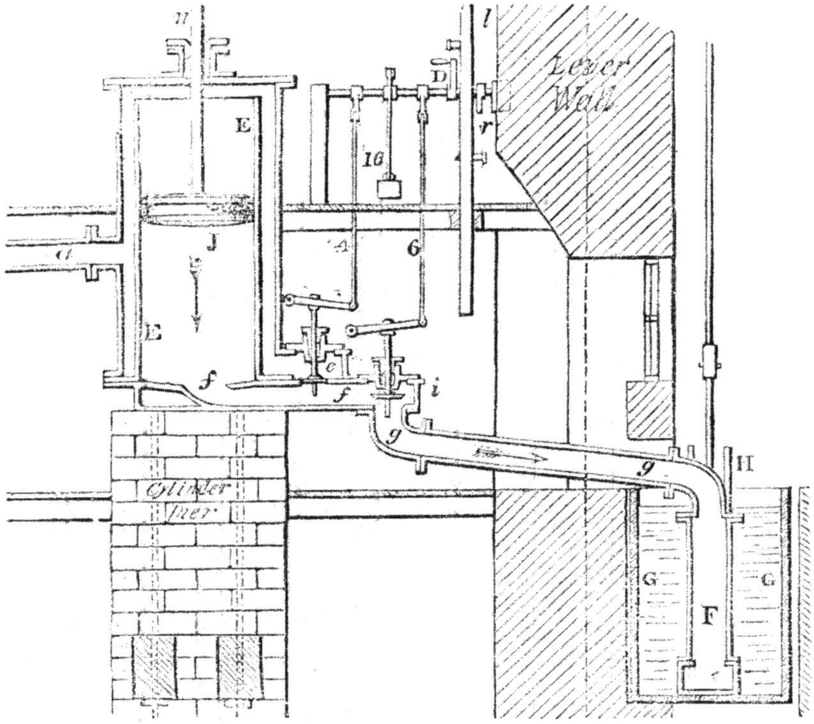

Early layout of two conical valves, one to transfer the steam from the top to the bottom of the cylinder and the other opening to the condenser.

during the ascent of the piston until a weight on the tail of the 'Y' overbalanced and the other leg pushed the valve shut. This had to be changed because, instead of only two valves, the steam and the water injection, Watt needed three, one to control the steam passing from above to below the piston, which he called the steam regulator and later the equilibrium valve, one to let the steam pass to the condenser, the exhaustion regulator, and a third to open the injection of cold water.

To the operating mechanisms he added better catches and release levers which also enabled the valves to be opened and shut by hand for starting and stopping as well as more convenient linkages for weights to open the valves. Slightly different systems of levers and linkages were fitted on the Wheal Bussy, Chacewater and Ting Tang engines. Soon after starting Wheal Bussy, Watt told Boulton, 'We are now obliged to bridle the regulators so as they may not open above half an inch otherwise it would tear the house down for the vacuum is perfect whenever we give it injection'. [3] Watt drew out on paper curved operating levers for Ting Tang and when it was set to work found that it '... seems to do very well. The Elliptical curve answers very well and requires 104 pound at 6 inch radius to open it $2^1/_2$ high. The steam regulator is thrown open with great violence without any great effort on ye part of ye Y shaft which is owing to very light Rods ye beam has to pull out'. [4] Here we see not only Watt's debt to the earlier Newcomen engine but also the scale and weight of the engineering. The valve operating weights were a problem too at Chacewater where Watt commented, 'The engine sett agoing very well but was heavy to hand on account of ye great weight necessary to open ye Exhaustion regulator'. [5]

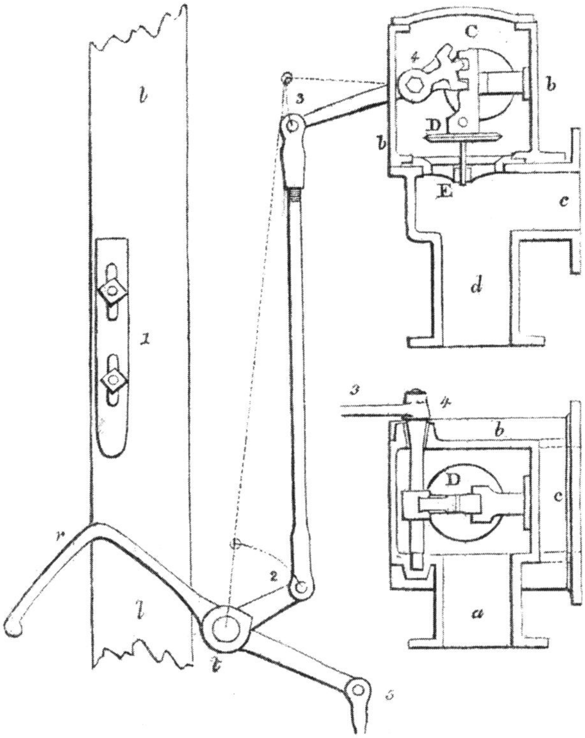

Rack mechanism for opening a valve.

On the larger engines, considerable force was needed initially to lift the large exhaustion regulators off their seats against the pressure difference of the steam on one side and the vacuum on the other as well as the large weight used to close them. A new type of valve was fitted to the Poldice engine in 1779 which had a cylindrical box, open at the bottom to the eduction pipe. Its rim fitted into the conical seat when shut. The closed top of the box contained a small valve which was raised first to allow a little steam to pass through and reduce the difference in pressures. Then the whole box could be lifted more easily off the conical seating, opening the major steam passage. [6] Slightly modified forms continued to be fitted on pumping engines until at least 1797. [7]

One of Watt's note books shows that by July 1783 he had calculated the optimum sizes for his valves, based on his experiments with steam. He knew approximately that:

> ... the quantities of different fluids which will go through the same hole under the same pressure being inversely as the sq[uar]e roots of their specific gravities, it follows that 42½ times the q[uanti]ty of steam, would go through any given hole in any given time, under any given pressure as would of water in the same circumstances. [8]

He realised that the passage of steam through the exhaust valve was different since that opened into a vacuum. Then there had to be sufficient space around the valve to give free passage to the steam. Allowance had to be made for a round valve in a square box so these boxes were made larger at three times the area of the valve. The openings were determined

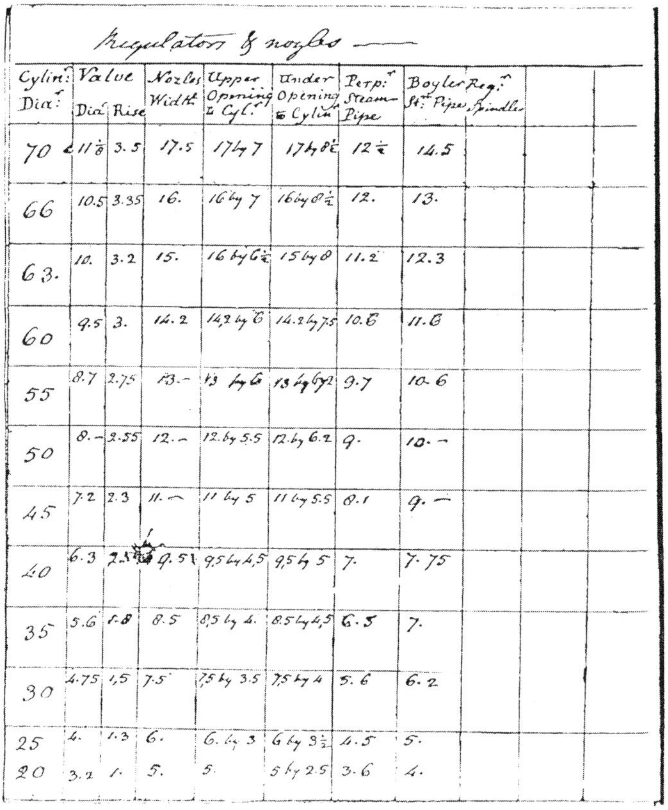

Regulators & nozles —

Cylin: Dia:	Value Dia:	Value Rise	Nozles Width	Upper Opening to Cyl.	Under Opening to Cylin	Perp: Steam Pipe	Boyler St. Pipe	Reg: Spindle
70	11 ⅛	3.5	17.5	17 by 7	17 by 8½	12½	14.5	
66	10.5	3.35	16.	16 by 7	16 by 8½	12.	13.	
63.	10.	3.2	15.	16 by 6½	15 by 8	11.2	12.3	
60	9.5	3.	14.2	14.2 by 6	14.2 by 7.5	10.6	11.6	
55	8.7	2.75	13.–	13 by 6	13 by 6.2	9.7	10.6	
50	8.–	2.55	12.–	12. by 5.5	12. by 6.2	9.	10.–	
45	7.2	2.3	11.–	11 by 5	11 by 5.5	8.1	9.–	
40	6.3	2.5	9.5	9.5 by 4.5	9.5 by 5	7.	7.75	
35	5.6	1.8	8.5	8.5 by 4.	8.5 by 4.5	6.5	7.	
30	4.75	1.5	7.5	7.5 by 3.5	7.5 by 4	5.6	6.2	
25	4.	1.3	6.	6. by 3	6 by 3½	4.5	5.	
20	3.2	1.	5.	5.	5 by 2.5	3.6	4.	

Page from Watt's note book showing sizes of valves and other adjacent parts.

at one and a half the area of the valve and the height the valve had to be raised was based on the cylindrical area. The size of the passages of the steam pipes was also based on one and a half times the area of the valves. All of these sizes were calculated for an engine with a 63 ins. diameter cylinder of 9 ft. stroke and then worked out in a table for engines of other sizes. Watt probably had tables of sizes for other parts of his engines which have not survived but would have assisted with standardisation of his designs. This was certainly the case with his contemporary John Curr, coal viewer at Sheffield, who published his tables in 1797 which were 'the result upwards of twenty years study and practice... They were originally composed for my private use and that of my assistants, in order to facilitate and dispatch business'. [9] Many similar tables were published later by John Farey and John Bourne.

When first starting from cold, the cylinder and valve chests had not only to be brought up to the temperature of the incoming steam to prevent condensation but all these parts together with the eduction pipe and condenser had to be cleared of air to achieve a high vacuum. The steam and exhaustion regulators were opened to allow steam to pass through from the casing, into the cylinder and down to the condenser. Farey commented in 1827:

> The steam must be sufficiently accumulated in the boiler, to cause it to rush with rapidity through these passages, so as to blow forcibly through the condenser, in order to expel the contained air therefrom through the valves of the air-pump; and this action which is termed blowing through must be continued till all the air is expelled from all parts of the engine. [10]

Front elevation of Penrhyn Du engine with lever for raising the bucket in the later single cylinder air pump with hot water pump above. For side elevation see p. 53 and plan p. 135.

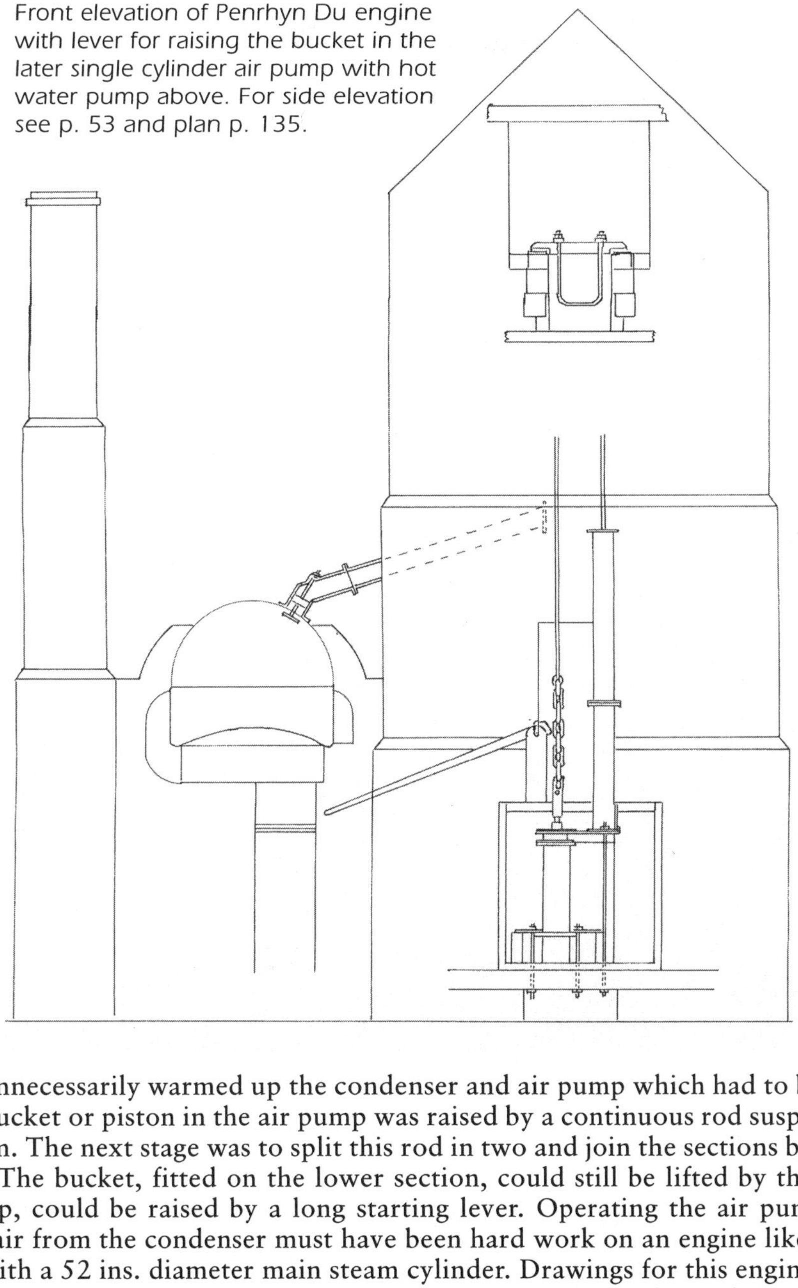

But this unnecessarily warmed up the condenser and air pump which had to be kept cold. At first the bucket or piston in the air pump was raised by a continuous rod suspended from the main beam. The next stage was to split this rod in two and join the sections by a short length of chain. The bucket, fitted on the lower section, could still be lifted by the beam but, on starting up, could be raised by a long starting lever. Operating the air pump manually to clear the air from the condenser must have been hard work on an engine like that for Byker colliery with a 52 ins. diameter main steam cylinder. Drawings for this engine in 1777 show that this new system was being installed by then and it continued on others until after 1780.

Then a much simpler method was introduced, possibly copied from the Newcomen engine. On these, a litte snifting valve was fitted to the eduction pipe near the bottom of the cylinder. When the main steam valve was opened, air and steam were blown through the snifting valve which was sealed by being immersed in a small tank of water. On Boulton and Watt engines, a similar valve was connected to the body of the condenser and drawings for J.M. Jary's engine in 1779 show that a pipe led from the condenser, out of the side of the condenser tank to a small box on the side in which this 'blowing' valve was submerged. [11] On some engines, such as that for Wheal Virgin, Watt played safe by fitting both the blowing valve and the manual air pump lever. This blowing valve is most likely to have been suggested by Henderson for in March 1779, Watt wrote to Jabez Hornblower 'with Mr. Hendersons scheme of blowing'. [12] This blow-through valve was retained on new beam engines until well into the 1840s.

The steam cylinder also received attention. The tin cylinder used at Kinneil and then at Soho had been surrounded with a wooden casing with a gap which was filled with steam to keep the whole of the working cylinder hot. The first cast iron cylinder for the Soho engine also had a wooden steam casing. Wood was of course a good insulator so prevented heat loss to the outside atmosphere. To prevent internal condensation, Watt recognised that it was necessary to keep the cylinder up to the temperature of the incoming steam. He realised that, if water were formed on the cylinder walls as it was in the Newcomen engine, it would evaporate again when the vacuum was formed so partially vitiating the vacuum from the steam thus produced. In practice however, the condenser would be quite effective in drawing off this vapour. [13]

The New Willey engine seems not to have had a casing [14] but other early engines were fitted with ones cast by Wilkinson. The inner working cylinders stood inside their casings without any connection except at the bottom. They were open at the top to the steam coming directly from the boiler. The top cover was bolted to the outer casing so it may have been difficult to align the stuffing box for the piston rod with the working cylinder. There were complaints from customers because Wilkinson charged the same rate for the unbored outer casing as for the bored inner cylinder. [15] Also it was difficult to gain access to the joint at the bottom where the working cylinder had to be secured. This was especially so where the bottom was made from sheet copper as at Bloomfield. Watt referred to it as 'that plagued joining of the copper bottom to the nozzle pipe which we could not get at & always went wrong'. [16]

The second engine at Soho had no steam casing. It was ordered in February and was running by August 1777 before Watt left for Cornwall. [17] Doubtless the cylinder would have been covered with some insulating material but Boulton later told Smeaton, 'We have also executed some engines without outside cylinders altogether; but have had no reason to applaud our economy, as the consumption of fuel was considerably greater'. [18] On the Soho engine, the cylinder cover had to be secured to the working cylinder itself so turning it into a main structural part of the engine. Since this would also have made erection and alignment easier, the advantages were not lost on Boulton and Watt and it became the standard design for later engines. It was also simpler to add an upper steam admission valve to the cylinder.

But this left the difficulty of how to gain the advantages of the steam casing. Again Boulton told Smeaton:

> We have now laid aside the clumsy cast-iron outside cylinder, and have substituted a case of wrought iron plates, which fits within an inch and a half of the cylinder, and extends from within 6 inches of the upper flange, to within 3 inches of the lower flange of the cylinder. This seems to answer as well as the outside cylinder. [19]

These jackets received a supply of steam separate from that to the working cylinder and would have been insulated on the outside. Later there was a reversion to cast iron, this time made into segments often supplied by the Eagle Foundry, Birmingham, which was cheaper than wrought iron. [20] Sometimes a false lid was fitted on top of the cylinder cover and the gap also filled with steam. Condensate from the casing was run to waste through a siphon which prevented escape of steam. [21]

The Expansive Engine

The second engine at Soho was used to experiment with 'cutting off' or stopping the entry of steam part way through the stroke and allowing the steam to expand within the cylinder during the rest of the stroke. This was controlled by a steam inlet valve. This engine also suffered an unfortunate fire during the night of 10 – 11 July 1778 [22] which consumed the roof of the engine house, part of the framing and burnt the main beam so badly that it had to be replaced. While the engine was running again within about three weeks, Watt lost a lot of models which had been stored there. He made a list, perhaps for an insurance claim, so that we are able to see what interesting historical items were destroyed.

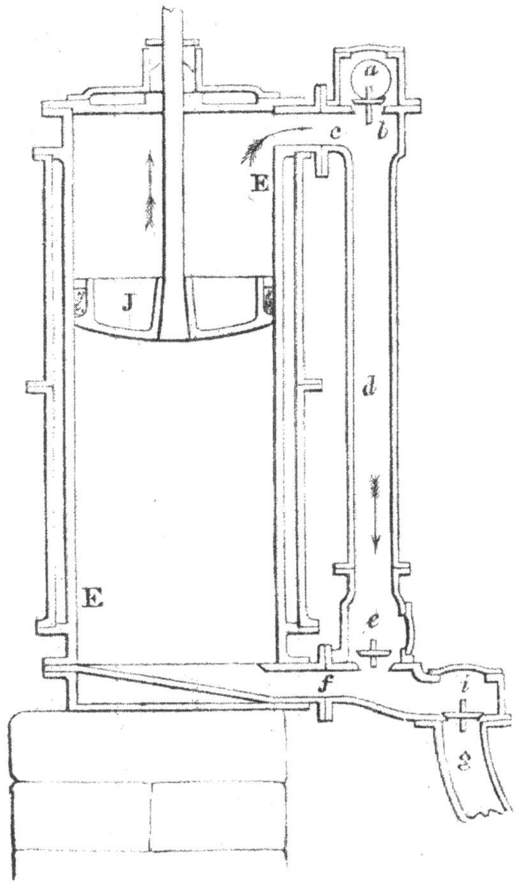

Right: Cylinder with three valves, inlet at top and transfer and exhaust valves at bottom. The cylinder now forms the main structural unit with subsidiary steam jacket.

Items Destroyed in Soho Fire.

Model of a Fire Engine brought from Scotland cost above	£30
Model of a wire mill	£ 5.5
a Block tin Condenser	£ 5.5
Another do.	£ 1.-
Tools for making pipe condensers	£ 2
clock dividing machine [crossed out]	£ 3.3
Rolled metal for pipe condensers & pipes	£ 2.-
A large brass syringe	£ 2.2
Brass and copper work belonging to another small engine	£ 6.-
A great variety of models parts of Engines and other things which I cannot pretend to give particulars of but could not rate under	£ 5. [23]

This list gives us a glimpse of the costs of Watt's experiments on his steam engine in Glasgow.

But to return to the expansive engine. He had tumbled on this phenomenon through his experiments but never understood the true principles through lack of knowledge of thermodynamics and reluctance to use high pressure steam. We find that he investigated four different types, showing how comprehensive were his ideas. Our problem in understanding what he was doing is that he used the term expansive in a variety of ways as well as 'elasticity' probably with the same meaning. He was well aware of the expansive force of steam through his very early experiments with high pressure steam from Denis Papin's digester, a sort of boiler, pushing against a small piston in a syringe to raise a weight. (See Vol. I., p. 301) That steam was released directly into the atmosphere. He never developed this type of high pressure non-condensing engine, perhaps because it did not need a separate condenser. He referred to it in his 1769 patent by including the paragraph:

> FOURTHLY, I intend in many cases to employ the expansive force of steam to press on the pistons, or whatever may be used instead of them, in the same manner as the pressure of the atmosphere is now employed in common fire engines: in cases where cold water cannot be had in plenty, the engines may be wrought by this force of steam only, by discharging the steam into the open air after it has done its office. [24]

This paragraph was of course repeated in the 1775 Act of Parliament.

After his trials with the University of Glasgow model Newcomen engine, Watt decided to investigate the properties of steam and was puzzled by the phenomenon of what Black called latent heat. Both realised that a great deal of heat was essential to convert water into steam which did not change the temperature, hence the term latent. Watt had also corrected earlier notions about the volume of the expansion of steam when water was turned into steam at atmospheric pressure. Then he drew up tables showing the relationship between temperature and the pressure of steam, so he experimented to see if an engine could be built using expansion and contraction of steam in a closed system at temperatures higher than boiling point to avoid crossing the latent heat boundary. This was his second type of expansive engine. Once again, this appears in the 1769 patent; SIXTHLY, I intend, in some cases, to apply a degree of cold not capable of reducing the steam to water, but of contracting it considerably, so that the engines shall be worked by the alternate expansion and contraction of the steam'. [25] Watt envisaged that this principle could be applied to both his reciprocating and 'circulating' engines. In a letter to Small, Watt mentions a temperature of 224°F, or 4 p.s.i. which he expected would raise water 6 ft. high with a 1 ins. diameter piston. [26] A sketch shows a very complex sort of vertical boiler with a small horizontal cylinder protruding at one side. The cooling medium proposed was boiling water so the steam in the cylinder would not be condensed. [27] Watt must have realised quickly that this type of engine was impractical although this was also included in various pleas before Parliament in 1775. [28] Boulton later suggested that one should be made. [29]

The fourth type of expansive engine was Watt's double cylinder type which will be described in the discussion on the 1782 patent. (See Vol. III) This leaves the third which became the true expansive engine in which steam is stopped from entering the cylinder part way through the stroke so that it expands during the remainder. Watt had discovered the advantages of this early in his steam engine experiments for he wrote to Small in May 1769:

> I mentioned to you a method of still doubling the effect of the steam, and that tolerably easy, by using the power of steam rushing into a vacuum at present lost. This would do little more than double the effect, but it would too much enlarge the vessels to use it all; it is peculiarly applicable to wheel-engines, and may supply the

> want of a condenser where force of steam only is used; for, open one of the steam-valves and admit steam until one-fourth of the distance between it and the next valve is filled with steam; shut the valve, the steam will continue to expand, and to press round the wheel with a diminishing power ending in one-fourth of its first exertion. The sum of this series you will find greater than one-half, though only one-fourth steam was used. The power will, indeed, be unequal, but this can be remedied by a fly, or several other ways. [30]

This has been underlined at sometime in the original. It appears that Watt had been experimenting with his steam-wheel, which would have acted as a sort of flywheel and so continued to revolve through its initial momentum as the steam pressure fell. Perhaps this was the most important contribution from all his efforts with these rotary machines. Watt realised that, from a quarter of the steam used, he could obtain half the power. The disadvantage was that, to obtain the same power, the engine had to be bigger.

Watt must have been puzzled by this phenomenon because he returned to examine it several times over the next few years. He noted in his 'Journal' for 20 February 1774, 'broke the machine for trying expansion of steam'. [31] Then, at the end of 1776, he drew a diagram showing the curve for falling steam pressure after the steam had been cut off part way through the stroke. [32] This is the earliest representation to have been found of what later became the steam engine indicator diagram. This may have been in preparation for the expansive engine at Soho for in April 1777, Boulton in London wrote to Watt in Birmingham, 'I wish our great Expansive Engine was done as the power of that I hope will be sufficient to depress Smeaton's spirits & I think the putting the Cylinder into the boyler hath many good conveniencies'. [33] Apparently there was a scheme for mounting the cylinder within the boiler, most likely to keep it as hot as the steam but also to stop spies finding out how this engine worked. [34]

The Soho engine was fitted with a plug valve at the top for controlling steam admission. A drawing shows a sort of rotary valve at the bottom of the transfer pipe for acting as both the equilibrium and exhaust valves so possibly a vacuum was maintained in the lower part of the cylinder all the time. [35] Watt held out great hopes for the expansive engine and began to publicize it. Pryce recorded:

> I am informed that Mr. Watt's improvements do not rest here; for he means soon to exhibit to the world engines upon the same principles, though differing somewhat in construction, which will use much less fuel than those described, and will also be more convenient for the purposes of mining, than any kind of engine yet used. [36]

Boulton was also enthusiastic, because he recognised that, by varying the admission of steam through a fourth valve, the top regulator, the power output of the engine could be varied too. He pointed out to Smeaton:

> Mr. Watt's engine has a very great advantage in mines, which are continually working deeper; suppose for instance, that a mine is 50 fathoms deep, you may have an engine which will be equal to draining the water, when the mine is worked, to 100 fathoms deep, and yet you can constantly adapt the engine to its load, whether it be 50 or 100 fathoms, or any intermediate depth; and the consumption of coals will be less in proportion, when working at the lesser, than at the greater depths; supposing it works, as our engines generally do, at 11 lbs. per square inch, when the mine becomes 100 fathoms deep. [37]

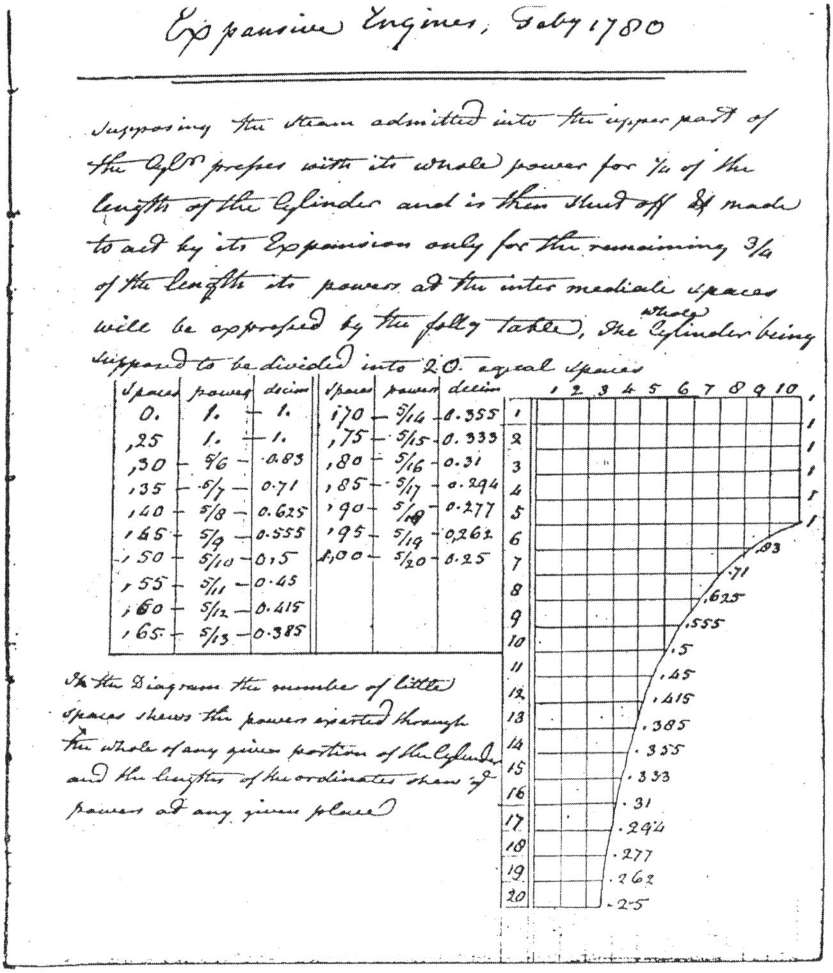

Expansive Engines, Feby 1780

Supposing the steam admitted into the upper part of the Cylr presses with its whole power for 1/4 of the length of the Cylinder and is then shut off & made to act by its Expansion only for the remaining 3/4 of the length its powers at the intermediate spaces will be expressed by the follg table, the whole Cylinder being supposed to be divided into 20 equal spaces

Spaces	power	decim	Spaces	power	decim
0.	1.	1.	,70	5/14	0.355
,25	1.	1.	,75	5/15	0.333
,30	5/6	0.83	,80	5/16	0.31
,35	5/7	0.71	,85	5/17	0.294
,40	5/8	0.625	,90	5/18	0.277
,45	5/9	0.555	,95	5/19	0.262
,50	5/10	0.5	1,00	5/20	0.25
,55	5/11	0.45			
,60	5/12	0.415			
,65	5/13	0.385			

In the Diagram the number of little spaces shews the power exerted through the whole of any given portion of the Cylinder and the lengths of the ordinates shew the powers at any given place

Page from Watt's note book showing calculations and diagram for fall in steam pressure in the cylinder after early cut-off, 1780.

This gave a totally unexpected improvement because, by regulating the power of the engine with the throttle valve and not the injection valve, the engine became 'much more manageable and is more easily adapted to every variation in its task'. [38]

But the expansive engine at Soho quickly acquired a reputation for being a devil to control and was nicknamed Beelzebub. Watt wrote to Boulton in picturesque language soon after his journey to Cornwall in August 1777:

> I send drawing of the best scheme I can at present devise for equallizing the power of Belzebub [sic] and so obliging him to save part of his youthful strength to help him forward in his old age – It appears at first view to be a rather clumsy contrivance but it is attended with very little additional friction and will produce the effect for at the beginning the power of the cylinder will be diminished 6/10 and at last will be nearly doubled. [39]

Another letter was sent three weeks later, with the suggestion that a steam casing be added. [40] The expansive engine had more power at the beginning of the stroke with full steam pressure in the cylinder and then the power diminished as the steam expanded and the pressure dropped. Boulton thought that Beelzebub was too lightly loaded at only 5.7 p.s.i and would not be so fierce if loaded at $7^1/_2$ or 8 p.s.i. Although more modifications were tried, Watt agreed with Boulton in 1781 that it was one of the worst engines they had built. [41]

A clue to why Beelzebub performed so badly may be found in another engine fitted with expansive steam cut off, that for the Shadwell waterworks in 1778. Farey noted that the pegs on the plug-tree to operate the inlet valve were:

> ... regulated so as to shut the upper steam valve, when the piston had descended two-thirds of the stroke, so that the remaining third of the motion was performed by the expansion of that two-thirds charge of steam. With this moderate degree of expansion, no machinery was required for equalizing the decreasing force, the impetus of the great lever, piston, pump-rods, and other moving parts, being sufficient for the purpose. [42]

Both the Soho and Shadwell engines had quite short pump rods for low lifts when compared with mine pumping engines. At Shadwell, the rods and other moving engine parts contained sufficient mass so that the momentum was sufficient to complete the stroke with a very late cut-off. The lesser mass of the Soho engine would have made control more difficult. Cornish engines developed after 1800 had higher pressure steam cut off much earlier but the acceleration given to the much heavier mass of their pumps and balance bobs was enough to finish the stroke.

The expansive principle may have been tried on one of the engines for the Birmingham Canal, again a low lift pump when compared with a mine. In June 1779, Watt reported to Boulton, 'I was at the Navigation engine yesterday. It uses by Thermometrical experiment about 50 cube inches per stroke when the upper working gear is used and 60 without it as near as I can guess, however the difference is sufficiently remarkable'. [43] He had been continuing to devise methods of equalizing the power of expansive engines and trying experiments [44] but for the time being reverted to regulating the power by using the inlet valve to throttle the steam and wire-drawing it. Watt's Directions for Erecting and Working the Newly-Invented Steam Engines contain the following instructions:

> There is a third regulating valve, called the TOP REGULATOR, placed in the cross pipe at the upper end of the perpendicular steam pipe, which serves to proportion the Quantity of steam to be admitted from the boiler, to the load of the Engine; so that when the load is less than ten pounds and a half on the inch, the steam in the upper part of the cylinder, which presses on the piston, may be less dense or weaker than what is required when the engine is fully loaded. This regulation may be effected in two ways; either by opening the top regulator fully, at the beginning of the stroke, and shutting it before the piston arrives at the bottom; or by opening it so far as just to give the piston a sufficient velocity, and keeping it open until the end of the stroke... (the degree of this opening must be determined by experience). [45]

Watt had recognised that, for the best efficiency, the engine should be worked on the first principle and have an early cut-off with greater expansion but with an irregular power output. His knowledge of steam was insufficient to show why this was preferable to using lower pressure steam without cut-off. He followed the phlogiston theory current at that time which suggested that heat was a sort of elastic fluid, working through pressure. Now if this is compared with waterpower, the same power can be achieved by having a little water falling through a greater height or more water falling through less height. With the boilers of Watt's day, it was safer to use more steam at lower pressures.

In the second method of working, by reducing the opening of the inlet valve and retaining it at that setting for the whole of the stroke, steam pressure would at first build up in the cylinder until there was sufficient pressure to overcome the weight of the pumps, the water to be raised and the friction so that the engine started moving. Once moving, friction decreased so that the piston would accelerate until steam pressure fell since the flow was restricted, with the result that the engine would slow down allowing steam pressure to be built up again. While this was liable to give a jerky motion, once the degree of opening had been determined by experience, the engine would be almost self-regulating. Therefore the Boulton and Watt engines erected after 1779 were this simple three valve type, regulated by wire-drawing the steam rather than earlier cut-off. The water injection valve was not included in this number of valves.

Watt may have preferred this because it gave a higher terminal pressure and so a higher temperature of the steam. He was obsessed with preventing condensation or waste of steam. We have seen his insistence on keeping the steam cylinder as hot as the steam entering it with steam casings and even a fire below the cylinder. While this was excellent for the start of the stroke, it was incorrect for the end. Watt knew very well that, as the steam pressure inside the cylinder fell with expansion, so did its temperature, and then some of the incoming steam would condense as it reheated the cylinder walls if there was no steam jacket. He must have been aware of this when he experimented at Chelsea and possibly at a few other engines where he made the piston rise in a vacuum.

Watt had recognised that power was lost when, in his early engines, the steam in the cylinder below the piston rushed 'by its elasticity or expansive quality' into the condenser so that the cylinder was 'left empty without being cooled'. [46] By taking the steam directly from the upper side of the piston to the condenser so that there was then a vacuum above and below the piston, he may have hoped to reduce this loss. He also hoped to improve the vacuum through condensation occurring for the duration of the ascent of the piston. But it was found in practice that the Chelsea engine did not come up to expectation. More air was drawn into the vacuum through leaks and the stuffing glands. The narrowness and length of the eduction pipes slowed the creation of the vacuum so that the engine needed to be counterbalanced more heavily. In addition, Farey's comments concerning the temperature in the cylinder are illuminating:

> When the cylinder is continually exhausted, the rare steam which always remains in it cannot be above 102 degrees of temperature, and about one pound per square inch elasticity. The metal of the cylinder, being kept at 212 deg. by the steam surrounding it in the steam-case, must continually communicate heat to the rare steam; this is a complete waste, and also tends to render that rare steam more elastic, so as to offer a greater resistance to the motion of the piston, than if it had not been heated. Hence the steam-case would be a source of waste; and such engines would probably perform better, without keeping the cylinder hot. [47]

But not keeping the cylinder hot would have been anathema to Watt. So, regarding the engine as a pressure engine and not a heat engine, he opted to abandon the vacuum engine and retain a high terminal pressure at the end of the stroke, which meant less condensation for the next charge of incoming steam.

However the prospect of greater savings in fuel with the expansive engine, and so greater premiums for Boulton and Watt, induced Watt to keep on experimenting to find ways of evening out the force being developed. His methods were covered in a patent taken out in 1782. (See Vol. III) In the meantime, engines were still fitted with the top regulator valve to control the steam inlet, but this was pre-set so that it operated 'by opening it so far as just to give the piston a sufficient velocity, and keeping it open to the end of the stroke'. [48] With all these various improvements, the Boulton and Watt single-acting pumping engine had reached the first stage of its development around the end of 1779.

Standardisation of Engines

Watt had evolved a standard layout for a mine pumping engine together with the house which could be modified to suit the particular circumstances of situation and size. Its origins can be traced back to the design for the Torryburn engine and later drawings show that it was continued until the late 1790s at least. The stone structure of one still survives at the Penrhyn Du mine, Llangenan, near Abersoch in North Wales. [49] It shows how the design was put into practice since drawings of this engine have been preserved as well. For these mine pumping engines, the mine shaft was not covered by the engine house. The main wall to support the rocking beam fronted onto the mine shaft. The condensing apparatus was fitted into a recess in this wall at the top of the shaft. A doorway gave access to the condenser from the interior but otherwise this wall was solid. This front wall was thicker than the others, giving greater strength, but also recognising that the others could be thinner for more economical construction. Another economical feature was the reduction in wall thickness by stepping in the outsides at the successive floor levels. This seems to have been characteristic of Watt's designs.

Just behind the beam wall at ground level on either side was a doorway to give access to the boilers. Larger engines required two boilers, so there would be one on each side, which necessitated two chimneys. Small engines, such as Penrhyn Du, had only one which might be on either side of the engine house. The design could be easily adapted to suit the location. The main entrance to the engine house was through the rear wall which gave access to the bottom of the cylinder. There were usually no windows at this level. Stairs led up to staging at cylinder top level. Here the side walls each had a window through which the steam pipe passed from the boiler to the upper valve chest at the top of the cylinder. There was a third window in the rear wall. All these windows were rectangular with wooden lintels, another Watt feature. A second flight of stairs led up to the beam floor. One window in the rear wall was all that admitted light here since the roof came down too low to admit windows. A winch might be fitted in the gable above the cylinder.

The whole engine house would be lower than a comparable Newcomen type because steam was taken to the top of the cylinder and not the bottom. Watt secured his cylinder with tie rods to a stone pedestal near the middle of the engine house and removed the boiler to the outside. The valves and valve gear were placed between the

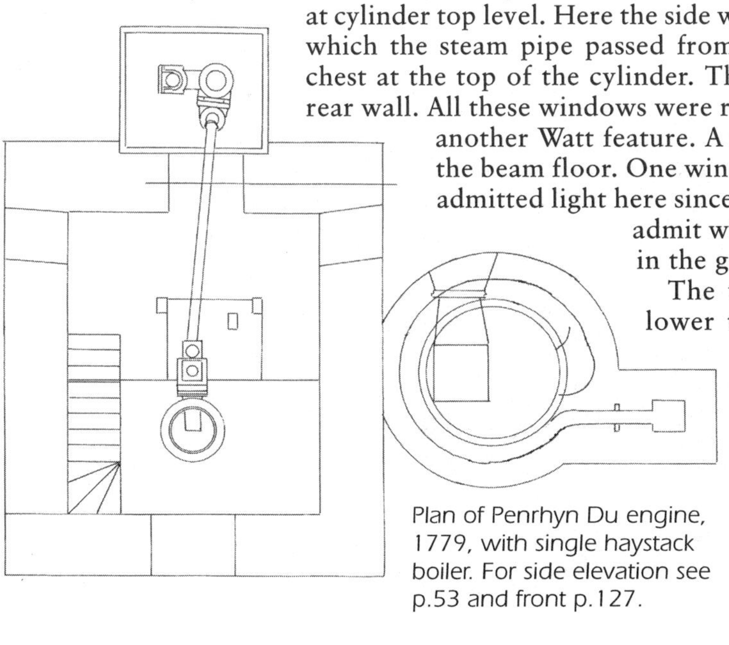

Plan of Penrhyn Du engine, 1779, with single haystack boiler. For side elevation see p.53 and front p.127.

cylinder and the beam wall. Below them, the eduction pipe took the steam to the condensing cistern at the top of the mine shaft. Water for the condensing jet was taken out of this cistern, a weakness in Watt's design because this water would have become warm. Inside the cistern was the air pump to remove condensate and air. This lifted the water to the hot pump which raised it high enough up a stand-pipe so it could run by gravity into the boiler against the low pressure steam. Both pumps were driven from an arch-head near the outer end of the beam.

The steam piston at one end and the pump rods and pistons at the other end of the beam were suspended from chains fitted round arch-heads so the rods were made to run truly vertical and central. Beams for larger engines had to be assembled from multiple baulks of timber but on small engines, such as Penrhyn Du, this could be a single piece. The addition of a king post or a double truss system above the beam took some of the strain, reducing the size of the beam. This was not fitted at Penrhyn Du. The beam pivot might be off-set from the centre, giving the longer stroke, and hence the advantage, to the steam cylinder. On each side of the main beam were placed long timbers which stretched from back to front of the engine house and out over the mine shaft. Not only did they form the support for flooring right round the beam and over the shaft, but additional spring beams were fitted on them against which blocks on the main beam could strike in the event of overstroking to lessen the shock.

Watt took great care in proportioning all his engine parts to give maximum performance and economy. Surviving drawings suggest that he extended this to the engine houses as well, standardising as much as possible. This helped the engine erectors to build the engines on plans they knew well with the minimum supervision from Watt himself. There were also his written instructions. Drawings were prepared, parts ordered and it was left to Jabez Hornblower to go and erect the Penrhyn Du engine. Even when he encountered difficulties, he was left to sort them out himself. However, in this case, he was dismissed and another erector was sent to finish the job. Watt never went to see this engine being erected or how it performed, which in the end was satisfactory after initial teething problems. The organisation of Boulton and Watt for designing and erecting engines had been developed successfully so it operated with the minimum of supervision. It lessened the need for a layer of middle management between Watt and his erectors. At the beginning of 1779, Watt was able to write to Black with some satisfaction:

> The anxiety which I have suffered since I began this [the engine] business has been exceedingly great, as the difficulties have also been. It now begins to assume a more pleasing appearance; but it has unavoidably incurred a very considerable debt, we cannot with propriety call the profits our own until that debt be discharged, which I hope will soon be the case. Another campaign in Cornwall will, I expect, do very great things; but to my sorrow I find, that as business increases, my cares increase in the same ratio, which in a very considerable degree prevents the enjoyment of life. [50]

Foreign Patents

News of Watt's improved steam engine spread quickly overseas. As early as February 1769 a Mr. Enslie had enquired about draining a lake near Rotterdam. Watt told Roebuck that 'the quantity of water raised by the water-mills is immense' which, together with the high costs of dykes, meant that the scheme was not pursued further. [51] While the Act of 1775 extended the area covered to Scotland, it could give no protection overseas. Erection of their engines abroad presented Boulton and Watt with the difficulties of first preventing others from copying their design without authorization and second of receiving any dues should an engine be erected. One answer should have been to secure their rights through obtaining patents in foreign countries while another was through preventing foreigners learning about the secrets of their designs.

Draining the low lands in the Netherlands by steam power appeared to present an outlet for their steam engines so, in July 1775, Watt sent van Liender at Rotterdam an outline of his improvements and the conditions for using his engine. He continued, 'Out of Britain I would not chuse to erect my engines unless the state where they are to be erected will first grant an exclusive privilege for a reasonable number of years and no person here can make them for exportation without my consent'. [52] Shortly afterwards, Boulton sought advice from Sir Joseph Yorke, the British Minister at The Hague, about taking out a patent for the province of Holland or for all the United Provinces. [53] At this time, van Liender was helping Jabez Hornblower to erect a Newcomen engine at Rotterdam. [54] When it was finished, Jabez returned to Cornwall during 1778 where he assisted his father to erect the Boulton and Watt engines, so learning about their construction. Although a good engineer, Jabez had the unfortunate propensity of upsetting his employers which presented Watt with a dilemma. Should Boulton and Watt stop employing him, Watt was afraid 'he will go to Holland & can hurt us. I must therefore try and get him bread here'. [55] He was transferred from Cornwall to work in the Midlands where he caused further upsets, resulting in his being offered the erection of the Penrhyn Du engine, North Wales, in August 1779 since Watt still feared 'to turn him desperate untill we secure Holland'. [56] Watt's anxiety concerning the Dutch discovering their secrets was well founded because that autumn he was advised by Henderson how the Batavian Society had sent people to Newcastle-upon-Tyne to 'get drawings of every part of the Byker engine - I think no time should be lost in applying for the Dutch patent'. [57] While this did not happen for six years, it shows how open to industrial espionage were Watt's steam engines.

Another method of industrial espionage was to visit the Soho Manufactory and inspect the engine there. Boulton was always pleased to show people round but some of them abused Boulton and Watt's hospitality. An example was that of some Germans who wanted to discover the secrets of the engine which Watt tried to prevent:

> All my care in concealing matters from the Germans was in vain it seems they came to the warehouse Newhall under pretence of taking leave of Mr. Fothergill and thence made some errand to Soho whither they went under the conduct of Collins and went directly to the engine where they found only Henry the Cornishman who knowing no better showed and explained to them all he could - The only chance is in their stupidity for they are neither scholars nor mechanics but may give such account of things that somebody else may be able to make them. Please acquaint Magellan with their behaviour which has been extremely ungenteel after the civility with which we received them. [58]

In practice, such people do not seem to have gained much benefit.

Watt told van Liender that 'no person here can make them [his steam engines] without my consent'. It was only Wilkinson who could bore cast iron cylinders accurately enough and who of course knew the secrets of Watt's designs through possessing their engines. We have no idea of how many people approached Wilkinson to supply them with a Boulton and Watt type of engine, knowing that the patent was not valid overseas. At first Wilkinson chose not to do this and turned down enquiries from two Frenchmen, J.M. Jary and J.C. Périer. Périer had the concession to supply Paris with water. He came to England in the spring of 1777 to order pipes and a Newcomen engine from Wilkinson but then heard about the superiority of Watt's and tried to order one from Wilkinson. Boulton's comments to Watt are very instructive not only about his suspicions of both Périer and Wilkinson but also about his attitude to Boulton and Watt's strict control of their business:

> The Frenchmen are not yet returned to London but am now fully convinced they only meant to pump & not employ us. I have been with Hadley [at the Bow engine] who told me they had sollicited him to make for them one of our Engines but he says he told them he w'd do no such thing nor do I see how they can accomplish their wishes without the assistance of Mr Wilkinson (as a founder) & I must say that I think it inconsistent with that mutual interest & friendship that at present subsists between us & W: for him to cast other cylinders that will answer for any other person. I am sure we are likely to be infinitely better customers to him than ye frenchmen can be but I suspect if we had given W. a power for Shropshire he wd have stretched it into France & therefore let us keep the Manufacture within ourselves by wch alone we shall preserve our powers & our reputation. I should not think it impolitick with Wilkinson & to intimate that we had rather he did not cast cylinders for them but to let them pass through our hands wch will be more to his advantage in the end. [59]

Watt replied, 'You may think I have been too open with Wilkinson. I did it with a view of leading him by his own interest as the surest motive and if he should not behave as becometh I assure you I will not be his friend'. [60] Unfortunately neither Boulton nor Watt followed up their suspicions of Wilkinson and Périer by whom they were tricked later.

The interest shown by people in France made the acquisition of a patent there essential. Watt was able to announce to Black in December 1778, 'Our affairs wear a most smiling aspect to human eyes Le Roi de France dans son Conseil nous a Accorde un Privilege exclusif pour faire et vendre ces machines en france and we have now an engine actualy under hand for that kingdom - but this entre nous'. [61] This privilege had been granted on 14 April 1778. [62] It had been obtained for them through the good offices of the Comte d'Heronville who was hoping to drain the marshland around Dunkirk with a Boulton and Watt engine. Magellan was acting as intermediary. But the patent would not become valid until an engine had been erected to prove the claims made. Boulton and Watt included their steam-wheel in their application as well as the reciprocating type. [63] In spite of protracted negotiations, no engine was ever erected for d'Heronville although the conditions for obtaining the patent were that an engine had to be constructed either at Paris or in the low lying 'Moeres' of Dunkirk. [64]

In the meantime, Jary had consulted Boulton and Watt about an engine to drain a coal mine at Nantes in Brittany. Therefore they tried to have the trial switched from Dunkirk to Nantes. Other advantages were that Jary already had a Newcomen engine for comparison and that he was 'a very ingenious man possest of the necessary \previous/ knowledge [and] has undertaken the care of the erection himself all to the finishing which will require the attendance of some person practically acquainted with the putting our Engines together'. [65] Jary had visited Cornwall in October 1778 with Boulton where he would have seen the engines at work. [66] Jary ordered an engine so that Watt prepared drawings during the early part of 1779. [67] Complex arrangements with special permission had to be organised about shipping the parts because England was at war with France and Jary wanted to obtain copper as well. The goods were to be sent via Rotterdam. [68] This engine became the first Boulton and Watt type erected outside Britain. Jary may have been offered the concession for engines in France but Boulton and Watt reneged on this so that they received neither payment nor return of any other kind from him. [69]

The reason for dropping Jary was almost certainly because an alliance with Périer seemed to offer better prospects. At the end of August 1778, Périer had been able to set up the powerful organization of the Compagnie des Eaux with the backing of the Orleans interest. [70] By late October, the Company had empowered him to contract with Boulton for a new steam engine. Périer was in Birmingham during January 1779, negotiating terms for two engines. Watt wanted his usual agreement of a third of the savings of fuel but instead Périer offered twenty-one shares which was later raised to twenty-five. Negotiations were broken off and then reopened with a tempting cash offer of £2,000. [71] Périer was able to drive a hard bargain so that Boulton and Watt 'found ourselves obliged to grant him on very moderate terms because our Arret not having yet the force of a patent we durst not risk any opposition'. [72] Lack of any engine erected so far had delayed granting the patent. Périer stuck so close to Watt during his stay that Watt barely had an opportunity to talk with Boulton.

Drawings were being prepared during February and parts ordered. Once again passports had to be obtained, not only for England but also France and America. [73] *The Mary* was able to avoid privateers and reached France where the engine erector was refused permission to land although the engine parts were forwarded to Paris. Périer, an outstanding engineer, was determined to learn the secrets of the Watt engine so he could construct them himself. He had also asked Boulton and Watt to supply various metals including 2,000 lbs. of steel 'and particularly of the sort which is most successful for turning metals'. [74] He established a fine engineering works in Paris. The first engine for the Paris waterworks was successfully operated on 8 August 1781 amid celebrations when little mention was made of Boulton and Watt. [75] There were delays in payment because the water company lacked the resources. Also no progress was made concerning the confirmation of the French patent in spite of the success of the Paris engine. While this might have been anticipated due to the war, other impediments were raised, such as the need to change the wording covering the steam-wheel. Périer was in an influential position to conduct delaying tactics.

Watt had recognised that it was necessary to have somebody in France to keep an eye on their business and told Boulton, 'If we mean to keep this our kingdom of France in proper subjection, it will be necessary that one of ourselves go over there soon; and if our subjects do not declare independence, leave some person in power to govern them, suppose Le Camus as Heronville's friend'. [76] In fact trade in steam engines with France never prospered. Failure to secure confirmation of their patent was one of the reasons why Boulton and Watt visited Paris over Christmas 1786.

Watt felt that he had been very badly treated and tricked by the French so that he was very cautious when van Liender approached him in 1785 about Boulton and Watt supplying an experimental engine to the Batavian Society for draining the polder of Blydorp and Kool near Rotterdam. [77] He told van Liender that not only would the Dutch have to meet the costs of erection and other expenses but also 'that you will oblige yourselves to procure us an exclusive privilege for the use of our invention'. [78] But there were difficulties in procuring the patent because Watt was not a resident in the Netherlands. This made him even more cautious and he pointed out that one reason was owing to his ill-treatment by the French:

> We feel ourselves very much hurt by the treatment we met with in France, where they granted us an arret de conseil for the exclusive use of our engines, and as soon as they got us to erect some for them told us the Arret was of no validity & even insisted on our giving up the names of the ministers who granted it & the friends who solicited it, that the granting it might be brought as a crime against them. They have gone farther they have refused to pay us one part of the sum which was [stipulated] as a premium on the engines we did make for them. This however we still hope of obtaining. We do not pretend to be so unjust as to say that the States of Holland would imitate the arbitrary Government of France, but it certainly should be a lesson to us to act with all honourable caution in similar actions. [79]

Watt wanted to know whether any privilege would be granted for all the Dutch provinces or only that of Holland. Van Liender tried to reassure him that 'It is morally impossible that a treatment as you have met with in France can ever befallen you in this Country'. [80]

It was agreed that the privilege be taken out in the names of the Directors of the Batavian Society and van Liender himself. [81] Accordingly Watt submitted a text of great length, covering progress with his steam engines up to the beginning of 1786. [82] The text was to be engrossed on parchment during the summer and it was probably officially accepted by the Dutch authorities in November. [83] Later attempts were instigated to secure patents in America and Spain [84] as well as Austria, Belgium, Prussia and possibly Italy. [85] But overseas sales were sparse and certainly failed to match expectations. Between 1775 and 1800, Spain received the most engines at five, followed by France at four. The Netherlands had three, Germany and Russia two each and Italy one, making a total of seventeen during the duration of the 1775 Act. [86]

Notes

1 Farey, (1827), Vol. I, p. 330, M. Boulton to J. Smeaton.
2 Boulton 348.67, M. Boulton to J. Watt, 3 Dec. 1776.
3 Boulton 348.78, J. Watt to M. Boulton, 13 Sept. 1777.
4 JWP, C3/3, 'Journal', 12 June & 30 July 1778.
5 Ibid, 20 July 1778.
6 B. & W. Col., PF 1286 for Poldice Eastern mine which contains a description.
7 Dickinson & Jenkins, (1927), Plate XLVI.
8 I am undebted to the late Norman Bertenshaw for sending me a copy of this notebook.
9 Curr, (1797), p. 5.
10 Farey, (1827), Vol. I, pp. 323 – 24.
11 B. & W. Col., PF 648. See also PF 1292 for possibly earlier engine.
12 JWP, C3/4, 'Journal', 3 March 1779.
13 Farey, (1827), Vol. I, p. 380.
14 Dickinson, (1914), p. 24 and Dickinson & Jenkins, (1927), Plate XVIII.
15 Dickinson & Jenkins, (1927), p. 193.
16 Ibid, p. 193, J. Watt to J. Wilkinson, 16 July 1779; B. & W. Col., Letter Books.
17 Ibid, p. 120.
18 Farey, (1827), Vol. I, p. 329, M. Boulton to J. Smeaton, 1778.
19 Ibid.
20 JWP, C3/5, 'Journal', 20 Sept. 1779.
21 Farey, (1827), Vol. I, p. 381.
22 JWP, C3/3, 'Journal', 18 July 1778.
23 Boulton 348.119, no date.
24 Muirhead, (1854), Vol. III, pp. 13 – 4, Patent 913, 5 Jan. 1769.
25 Ibid, p. 14.
26 Boulton 348,21, J. Watt to W. Small, 20 Oct. 1770.
27 B. & W. Col., MI, 1/16, 'Journal', 20 Oct. 1770.
28 JWP, 4/69, 'Characteristics of J. Watt's Steam Engines', 1775.
29 Tann, (1981), p. 57, Doc. 15, M. Boulton to J. Watt, 30 July 1781.
30 Muirhead, (1854), Vol. I, p. 62, Letter 58, J. Watt to W. Small, 28 May 1769; Boulton 348.9.
31 B. & W. Col., MI, 1/20, 'Journal', 20 Feb. 1774.
32 Ibid, MI, 2/8, 'Notebook', no date but after 9 Dec. 1776.
33 Boulton 238.71, M. Boulton to J. Watt, 20 April 1777.
34 Dickinson & Jenkins, (1927), p. 121.
35 B. & W. Col. PF 626 and Andrews, (1988), p. 19.
36 Pryce, (1778), p. 312.
37 Farey, (1827), Vol. I, p. 330, M. Boulton to J. Smeaton, 1778.

38 Robison, (1822), pp. 126 - 30.
39 Boulton 348.74, J. Watt to M. Boulton, 4 Aug. 1777.
40 Boulton 348.77, J. Watt to M. Boulton, 25 Aug. 1777.
41 Dickinson & Jenkins, (1927), J. Watt to M. Boulton 15 Aug. 1781. See also Tann, (1981), p. 47, Doc. 11, M. Boulton to L. Henderson, 13 Aug. 1777 and p. 48, Doc. 12, M. Boulton to J. Watt, 18 Aug. 1777, 'Belzibub improving'.
42 Farey, (1827), Vol. I, p. 341.
43 Boulton 348.139, J. Watt to M. Boulton, 26 June 1779.
44 JWP, C3/4, 'Journal', 6 & 25 March 1779.
45 Dickinson & Jenkins, (1927), p. 390.
46 JWP, 4/69, 'Plea before Parliament', 1775.
47 Farey, (1827), Vol. I, p. 336.
48 Dickinson & Jenkins, (1927), p. 390.
49 See R.L. Hills & D. Gwyn, 'Three Engines at Penrhyn Du, 1760 – 1780', Vol. 75, No.1, 2005.
50 Muirhead, (1854), Vol. II, p. 114, Letter 171, J. Watt to J. Black, 13 Jan. 1779; JWP, W5.
51 JWP. 4/32.20, J. Watt to J. Roebuck, 10 Feb. 1769.
52 B. & W. Col., Letter Book 1, J. Watt to J.D.H. van Liender, 18 July 1775.
53 A.o.S., MB3147/3/554/23, M. Boulton to Sir Joseph Yorke, [1776].
54 See van der Pols & Verbruggen, (1996), pp. 31 – 7.
55 Boulton 348.108, J. Watt to M. Boulton, 6 Sept. 1778.
56 Boulton 348.50, J. Watt to M. Boulton, 13 Aug. 1779.
57 Boulton 348.158, J. Watt to M. Boulton, 16 Oct. 1779.
58 Boulton 348.130, J. Watt to M. Boulton, 27 Jan. 1779.
59 Boulton 348.71, M. Boulton to J. Watt, 20 April 1777.
60 Boulton 348.72, J. Watt to M. Boulton, 2 May 1777.
61 Robinson & McKie, (1970, p. 42, Letter 36, J. Watt to J. Black, 12 Dec. 1778.
62 Dickinson, (1935), p. 100.
63 See Harris, (1998), pp. 297 – 313 for a detailed account of these intricate negotiations and dealings.
64 Dickinson & Jenkins, (1927), p. 48.
65 Robinson & McKie, (1970), p. 45, Letter 38, J. Watt to J. Black, 13 Jan. 1779.
66 JWP, C3/3, 'Journal', 10 to 22 Oct. 1778
67 See JWP, C3/3 and C3/4, 'Journals'.
68 JWP, C3/5, 'Journal', 24 Jan. 1780.
69 Dickinson & Jenkins, (1927), p. 48.
70 See Barker, (1992), pp. 57 – 76 for a brief history of the Paris Waterworks.
71 See Lord (1923), p. 211 and Boulton 348.127, J. Watt to M. Boulton, 24 Jan. 1779.
72 Robinson & McKie, (1970), p. 49, Letter 40, J. Watt to J. Black, 13 Jan. 1779.
73 JWP, C3/5, 'Journal', 2 Dec. 1779.
74 Harris, (1998), p. 312, see B. & W. Col., Box 3/9/14, French Business 1777 to 1786, Périer to Boulton, Paris, 3 May 1779.
75 Lord, (1923), p. 211.
76 Muirhead, (1854), Vol. II, p. 121, J. Watt to M. Boulton, 4 May 1780; JWP 3/8. See also Boulton 348.129, J. Watt to M. Boulton, 26 Jan. 1779.
77 Hills, (2003), p. 65 – 6.
78 B. & W. Col., Letter Book 9, J. Watt to J.D.H. van Liender, 29 March 1785.
79 Ibid, 4 Sept. 1785.
80 B. & W. Col., Box 36, J.D.H. van Liender to J. Watt, 4 Oct. 1785.
81 Ibid, 29 Nov. 1785.
82 A.o.S., MS 3147/2/28.
83 See B. & W. Col., Box 36, J.D.H. van Liender to J. Watt, 17 Nov. 1786 and B. & W. Col., Letter Book 13, J. Southern to J.D.H. van Liender, 27 Nov. 1786.
84 Tann (1981), p. 96, Doc. 35, Boulton & Watt to R. Barclay, 22 July 1788 and ibid, p. 100, Doc. 38, from Don Garcia Gaslon & Rowe Co., Cadiz, 22 Feb. 1791.
85 Ibid, p. 4.
86 Ibid, p. 7.

7
The Financial Struggle

While the period between 1775 and 1779 saw intense activity on the part of Watt developing his steam engine, the subsequent years up to 1785 were to be even more strenuous, covering a much wider range of interests. They were 'the period during which he seems to have combined the greatest maturity with the greatest activity of intellect; and the time of his life which they comprehended, was from his fortieth to his fiftieth year'. [1] There was the engine business to be administered which necessitated preparing the plans and drawings for the engines as well as ensuring that the erection was carried out. There was the need continually to improve the engines in the light of knowledge gained through experience. Watt faced competition from other inventors such as Matthew Wasborough and Jonathan Hornblower junior so that, in order to keep ahead of them, he introduced his expansive and rotative types of engines. Such new designs had to be tested first at Soho. His ideas for these were covered in three long patents in 1781, 1782 and 1784. Boilers were not forgotten either in his final patent of 1785. People from overseas tried to copy his engines without permission so that attempts were made to obtain foreign patents.

Then there were Watt's other interests in the philosophical field. One outcome of his interest in chemistry was patenting his method of copying letters in 1780. His continuing experiments on steam resulted in his paper to the Royal Society of London on the composition of water which helped towards his election in 1785. All this was carried out against a background of financial crises, caused partly through failure to obtain the expected patent dues. One reason was that the price of copper fell through exploitation of new sources on Parys Mountain, Anglesey. This dramatically affected Cornish mines with many forced to close, so rousing hostility to Boulton and Watt there. Also the debts of Boulton and Fothergill were increasing so that the engine business had to shore up that concern. The death of Fothergill in 1782 caused further complications. Therefore we should not be surprised to find Watt's letters full of complaints about the state of his health. Watt junior well remembered 'during the greater part of the time... seeing him suffer under most acute sick headaches, sitting by the fireside for hours together, with his head leaning on his elbow, and scarcely able to give utterance to his thoughts'. [2] Yet these headaches often preceded some of the most inventive periods of his life.

There is no doubt that this was the most critical period for Boulton and Watt. How close they came to failure we shall never know. In the end it was the payments from the Cornish mines which proved their salvation. Watt junior said about his father, 'It was unquestionably the busiest as well as the most anxious period of his life, and fraught with the most important results'. [3] Not only was the organisation of Boulton and Watt in Birmingham put on a proper basis which lasted until 1800 but some of the new ideas contained in the patents proved to be the foundations for the further development of the steam engine, particularly the rotative type, which carried steam power into new industries.

Indebtedness

At the beginning of their partnership, it had been agreed that Boulton would handle the finances. He would meet the expenses of the 1775 Act. He had also agreed to finance the experiments on the engines which would have included both the reciprocating and rotary types. This had to cover the manufacture of parts and wages of men like Harrison. Watt himself received a salary of either £300 or £330 per annum until 1785. [4] Then, once orders for engines had been received, further expenses were incurred, such as setting up a place for manufacturing selected parts, financing their manufacture, training engine erectors, paying travelling expenses, establishing an office for handling correspondence and preparing drawings. But the way Watt had decided to charge for the use of his patent meant that no income was received until after the engines had been set to work and their savings in coal calculated. Therefore considerable indebtedness accumulated during the first years of the partnership. Smiles says that 'It was estimated that upwards of £40,000 were invested in the engine business before it began to yield profits; all this money was found by Boulton'. [5] An earlier estimate was for nearly £50,000. [6] It has not been possible to check either figure. In November 1774, Roebuck had pointed out to Watt how fortunate he was, 'You have now effectually established the justness on which your machine is constructed and the generous spirited gentleman you are connected with will never suffer it to fail for want of exertion to carry it into execution'. [7]

But even Boulton's generosity had its limits, especially since there were financial difficulties with Boulton and Fothergill at the Soho Manufactory. This was an ambitious venture built on the fortunes of Boulton's wives and probably was always undercapitalised and unprofitable. Even though they did not lose a great deal through the bankruptcy of the London bankers, Neale, James, Fordyce and Down in 1772, which had a serious effect on trade, they had sustained losses during the previous four years amounting to nearly £7,000 on top of which they were faced with repaying a Dutch loan of £8,000 as well as other debts. Losses continued so that by 1777 the Bill Account had risen to nearly £25,000. [8] This had been financed by the bankers Lowe, Vere, Williams and Jennings but Lowe and Vere could give little help in the summer of 1778 because the junior partner, Jennings, had accepted bills to the amount of £180,000 which had not been entered in the books and which became due. He absconded to France with £500. [9] This coincided with the need for more money to finance not only the engine business but also the Manufactory.

One of the problems was the character of Boulton himself who entered into too many projects and could not keep an adequate check on them. He was described in 1776 as a:

> ... sensible, ingenious and enterprising man, who plans and executes with equal expedition, but like many other great men he has his hobby horse. He is scheming and changeable, ever some new matter on the anvil to divert his attention from a steady pursuit of some grand object. He is always inventing, and by the time he has brought his scheme to perfection, some new affair offers itself. He deserts the old, follows the new, of which he is weary by the time he has arrived at it. This volatility prevents him from becoming very rich. [10]

At this period he entered into partnership with Francis Eginton over a novel method for copying pictures by mechanical means which was losing about £500 per annum. [11] Then in 1778 the production of the highly expensive ormolu ornaments was palpably outstripping demand. Possibly due to the severe financial pressures or a desire to move stock, a sale was arranged hastily at Christies in May 1778 which was a failure, raising only £200. [12] The engine business was occupying more and more of Boulton's time, possibly because he found this was more interesting. He had told Watt in 1776, 'I assure you that of all the Toys & trinkets which we manufacture at Soho none shall take the place of fire Engines in respect to my attention'. [13] In fact it would be the steam engine business which came to the rescue of Boulton's many other ventures, even the famous Soho Manufactory itself.

> Boulton was a superb promoter with considerable industrial vision, yet he does not seem to have been able to manage a business. It was his association with Watt that saved his reputation because the steam engine gave him a product which was badly needed and for which Boulton and Watt held a monopoly of the best type. At the same time, the association with Boulton saved Watt because it was Boulton's vision that recognised the possibilities of the Watt engine which Watt himself did only with reluctance and it was Boulton's energies that succeeded in pushing forward the development of the engine. The premiums from the steam engines enabled Boulton to weather his financial storms. [14]

Watt must have known that the steam engine business was proving costly for Boulton but we do not know when he became aware of the extent of the debts. In May 1778, he 'signed a bond with Mr. Boulton to Messrs Vere and Co for cash account to be opened with them'. [15] This and other bonds caused much anguish later. Boulton must have revealed something in July 1778, perhaps the parlous state of Lowe and Vere. Watt tried to encourage Boulton from far away Cornwall, 'I beg of you to employ some part of your spirits to support the rest & as far as you can turn your mind from ye distressing object... I shall not allow myself to despond in this hour of affliction & anxiety'. [16] But he could hold out little hope of obtaining any payment from Ting Tang because that engine was not yet running while there had been problems at Wheal Union through Dudley as well as delays in arrival of parts. A further letter a couple of days later was just as depressing. Nobody was willing to believe the claims made by Boulton and Watt for the performance of their engines until one was actually running. Neither the Cornish banks nor anyone else would lend money on a mortgage except Wilson who had none. Prospects for mining were bad because ore prices had slumped by a third. Watt proposed to Boulton:

> If you can by any means push away the evil hour for a month or six weeks untill these large engines are in prosperous working the minds of men will alter & the people here will be as obliging & cringing to us as they are now distant and supercilious & then I doubt not you may have the desired credit from ye Cornish bank. [17]

We shall see that Watt seriously misjudged the character of Cornishmen by hoping that they would be grateful for the improved steam engines.

Watt suggested a proposal to bring in more finance by expanding the partnership to include Wilkinson, perhaps by Wilkinson buying out Fothergill:

> You know I am a bad ways & means man, but however the following thought may merit your consideration. You know W[ilkinson] has several times hinted a wish to be concerned in this scheme, to wch we have material objections but rather than founder at sea we had better run on shore - suppose he was to purchase F[othergill']s and besides to furnish some stipulated sum of money to B. & F. on ye engine security, say ye ammount of ye Engine debts to M.B. - one of the first things to be enquired is his ability to perform this. [18]

If Wilkinson had been drawn into Boulton and Watt through becoming a partner, later events might have turned out very different. As it was, Watt hoped that Boulton was 'applying the pruning knife to every branch that needs it' [19] for he was being as frugal as possible. The immediate crisis passed because, by the beginning of August, Boulton had 'weathered the storm' and Watt would be receiving £386 from Ting Tang in the following week to pay for some engine parts. [20] Boulton was successful in obtaining £2,000 from the Truro bankers, Elliott and Praed. [21] The commitment to Lowe and Vere was standing at £8,400. [22]

Lowe and Vere

The relief however was short-lived since by the beginning of January 1779 more money was needed. The debit account with Lowe and Vere had been increasing. Boulton blamed Fothergill for 'having been too free in paying off our Xmas Ballances'. [23] Also £6,000 which they had hoped to receive from someone had failed to materialise. This may have been a loan of £7,000 from another merchant banker, a Dutchman James Wiss, which was to be charged against money due on the patent premiums. Wiss was anxious not to lock up money in annuities unless there was a distinct prospect of a good, secure return. He wanted the advance allocated against certain engines and that the proprietors would pay their dues directly to him and not through Boulton and Watt. Also should an engine cease to work, he asked to be compensated by revenue from the others. [24]

All this put Watt in a state of great consternation. There must have been discussions before Boulton left for London. Watt had been reflecting on the terms and wanted to see a copy of the mortgage:

> ... for my mind was too uneasy at that time to remember anything of it... I cannot rest in bed untill they [these affairs] have some determinate form... I beg you will pardon my importunity but I cannot bear the uneasiness of my own mind and that they are as much your interest as mine to have settled. [25]

Not only did he think Wiss's suggestions impossible, [26] but he went on to stress his own concerns in a further letter to Boulton:

> I need scarce inculcate any more that I wish you to understand that I cannot execute any mortgage to Mr. Wiss or any other unless it be applied in relief so far of the mortgage to L. V. & Co. for the thought of the debt there lye too heavy on my mind to leave me the proper employment of my faculties in the prosecution of our business and besides common honesty will prevent from loading the scheme with Debts which might be more than it could pay. [27]

It must be remembered that Watt was still encumbered with debts incurred in Scotland by borrowing from relatives and friends such as Black to support his ventures with the Delftfield Pottery and his merchanting business. He probably felt that his chances of escaping from these burdens were rapidly disappearing, just when prospects should have been brightening through the success of his engine. On the other hand, Boulton was much more optimistic. He pointed out to Lowe and Vere how the engine business was flourishing and that during the course of that summer as more engines were set to work the amount received through premiums would double. He held out prospects of being able to 'discharge ye Greatest part of the sum which your house has been pleased to grant us... I therefore hope that neither you nor Mr. Low nor Mr Willms will be so severe with us as to bring on any [interruption] in our affairs which otherwise will in the course of this summer be in comfortable train'. He pointed out the premiums which he expected they would be receiving soon.

Premiums Expected in 1779

Chasewater	7000	
Chelsea & Shad[well]	2500	
J. Scott	1220	
Reynold of Ketley	1000	
Hawkesbury	1500	wch they offer but we contend for more
Birmg Navig[atio]n	210	
Ld Gower	150	
Chapman Newcastle	770	
Hull	300	
Lady dumfries in Scotld	1200	
Wrens nest forge	500	
	16350	[28]

Therefore the outlook should have been very promising, but this fails to take into account Boulton's character for whom money was there to be spent and not conserved for a rainy day. In addition, we have seen that agreements still had to be reached with some of the proprietors. Watt pointed out, 'In relation to undertaking engines, I would not undertake any without it was either for a fixt sum or a fixed rent we shall be ruined by these delays at payment'. [29] But good advice is not always heeded and soon there were more financial problems.

Watt was fairly optimistic when he wrote to McGrigor in August 1779:

> Mr. B. said to Mr. Glassford that when all the engines we then had in hand were completed they ought to bring in a revenue of £4000 p year but it will be at least a year before they be in that state and it is also supposing all to happen for the best - Our present income ought to be considerable if we could obtain punctual payments but that is sometimes attended with difficulties; though on the whole our affairs are very prosperous, and we have made considerable progress, in getting all our several annuities properly settled and now find no difficulty in making decisive bargains for the engines we undertake which could not be done untill our fame was established. [30]

While this letter points to the possibilities, it also highlights the difficulties, particularly actually receiving the payments. In fact, as the price of copper fell, concessions had to be made over the premiums such as that in the autumn to secure an order for an engine at Dolcoath. Watt pointed out to Boulton that they must, 'on making this concession, insist on our having a good pay master to pay regularly once a month & not to be obliged to go like beggars to their accounts'. [31]

Another problem arose with Lowe and Vere. The engine business was financed through both Boulton and Boulton and Fothergill. Watt certainly thought of it as a separate business and there was some form of separate accounting. As it started to produce an income, Boulton saw it as a way of underwriting his debts. Lowe and Vere obviously welcomed this but made no distinction between the two concerns. They wanted security based on income from the engines. Watt reluctantly agreed to let Boulton take the entire income from specified engines provided he was allocated the same proportion from others not mortgaged. [32] But Lowe and Vere would not agree and wanted the inventor to be included in their settlement.

The whole affair reached another crisis in May 1780 after Watt had been asked again to enter into a fresh mortgage bond. That to Wiss was still outstanding. The size of the bond terrified Watt who was afraid that he would lose everything he had gained as well as being liable for all Boulton's debts. He told Boulton that he could not:

> ... bring myself to a fresh bond of mortgage – I will not add to your vexations by explaining my feelings on this subject – let it suffice that I will assign any quantity of engines you please provided that sufficient is left [to] secure L. V. & Co. and that I am bound in no bonds.
>
> I am sorry my dear Sir to prove in any such shape refractory at what you desire but my quiet, my peace of mind, perhaps my very existence, depends on what I have told you – I am unhappy in not having any person I can advise with on this subject and my own knowledge of it is insufficient... It grieves me much to write to you such a letter. [33]

Lowe and Vere were demanding all of Boulton's title deeds. [34] Two more similar letters followed shortly afterwards. Watt felt very isolated with his main psychological prop, Boulton, now leaving him in the lurch. But he agreed to stand security for another personal bond. This was the time when Watt's copying machine was being developed which needed capital investment.

The autumn of 1780 again shows the ups and downs of the partnership. A massive order for five engines was received from Wheal Virgin which Watt expected would take about a year to finish. It would bring in a good round sum but 'untill they come to pay we shall remain poor people after that better things is to be expected'. [35] In fact the engines were not set to work until probably September 1782 which delayed Boulton and Watt's financial recovery. [36]

The downside in the autumn of 1780 was that a surprised and shocked Watt found:

> ... that after all our remittances and payments last year & this summer that our ballance due to B. & F. is not diminished but is £300 greater than it was in the 1777 though almost £4000 which ought to have been profits has been paid in that time... It is to me amazing what has been the sinking fund but it will take some time to find it out, found out and stoppt - otherwise all we do is vain. It will be in vain we labour at the pump while there is such a leak. The ship will sink inevitably without a remedy is found. If there were any great sums of money due to us for goods at present I could account for it but that is not the case nor do we seem to have encreased our stock. Nevertheless if we check it now some good may be done and therefore I did not explain myself to you while under the dominions of the dark blue prince of terrors nor should have done it now only to show how necessary a reform is become and to beg your stay to be as short as can manage the business in hand... and to assist in getting our books examined and business put on a proper footing. [37]

It may be that this discovery caused Watt to take a greater involvement in the financial affairs of Boulton and Watt. He examined the books and discovered 'that Cornwall has hitherto eat up all the profits we have drawn from it and all we have got by other places and a good sum of our own money to the bargain'. [38] So up to this time, Cornwall, the jewel in the crown, was more of a liability than an asset and it seems to have been the need for cash by Boulton and Watt which was causing at least some of the problems.

Watt looked again at the books over Christmas and found to his horror that the amount owing to Lowe and Vere had increased. Watt thought that this debt would be reduced to £12,000 but found it amounted to £18,682. This was in spite of £3,000 being paid in from the engine business.

State of Accounts on the 18th December 1780

Owing Lowe Vere & Co for drafts accepted to pay on and Before the 31st December	920. 6. 8
Lowe Vere & Co for Notes payable by the end of Dec	946.13. 6
	1867. 0. 2
Owing Lowe Vere & Co for Drafts payable after the 31st December	11815. 3. 6
	£13682. 3. 8
Lowe Vere & Co have accepted to Wm Matthews for wch they have Counter drafts on Matthews	5000. 0. 0
	£18682. 3. 8
Amount of Notes payable at Lowe Vere & Co after the 31 Decem 1780	2105.12. 8
The ballance of Acct with Mr. Willm Matthews on the 18th December 1780	2844.17. 9
So that the Ballances would stand	
To Lowe Vere & Co for drafts & Notes	£13682. 3. 8
To do for Notes payable in 1781	2105.12. 8
To Willm Matthews (Ballance of acct)	2844.17. 9
	18632.14. 1 [39]

'I cannot tell you how much I am mortified and vexed at this state of the transaction – it appears that all my efforts are vain and that all the Engine Company can do will be swallowed up to their total ruin without producing any good effect', wailed Watt. [40] He was at a loss what to do but wanted to insist that he should receive all the engine money so that he could see it would be used to pay off the capital of the mortgage he had entered into with Lowe and Vere. He told Matthews, 'I hope your candour will excuse me in taking every step necessary to clear myself from an affair which renders my life unhappy to the last degree'. [41] Boulton may also have recognised that he needed to pay more attention to the financial state of his companies for he told Fothergill that he must either 'quit this business here (Cornwall) in favour of Soho (and I must say I do not think Mr. Watt has either health or spirits to encounter with the wrangles we do and must constantly meet with) or I must quit Soho in favour of the engine business'. [42] Boulton's patience with Watt may have been wearing a little thin for after all he had been the mainstay of the engine business.

Watt continued to try and escape from the burden of this bond and sought an ally in Matthews. Perhaps we should feel sympathy for Watt as the bond was for £14,000, which must have seemed astronomical to a person with his frugal, careful attitudes. He wanted Matthews's help to see that the money earned by Boulton and Watt should be used to repay the principal and that this should be acknowledged by Lowe and Vere. 'Therefore I trust you will do your best for me... You may safely say [to Lowe and Vere] that our affairs are in a most prosperous situation now, and more orders offer than we can execute but it requires time to gett money'. [43] At this juncture, Annie pleaded with Boulton to try and relieve her husband from this bond. Her letter reveals a great deal about Watt's character:

> I know the goodness of your heart will readily forgive me for this freedom, and your friendship for Mr. Watt will, I am sure, excuse me for pointing out a few things that press upon his mind. I am very sorry to tell you that both his health and spirits have been much worse since you left Soho. It is all that I can do to keep him from sinking under that fatal depression. Whether the badness of his health is owing to the lowness of his spirits, or the lowness of his spirits to his bad health, I cannot pretend to tell. But this I know, that there are several things that prey upon his mind as to render him perfectly miserable. You know the bond that he is engaged in to Vere's house has been the source of great uneasiness to him. It is still so, and the thought of it bows him down to the very ground... There is another affair that sits very heavy on his mind; that is, some old accounts that have remained unsettled since the commencement of the business. They never come across his mind but he is rendered unfit for doing anything for a long time. A thousand times I have begged him to mention them to you... Believe me, there is not a person on earth who is dearer to him than you are. It causes him pain to give you trouble. The badness of his constitution, and his natural dislike to business, make him leave many things undone that he knows ought to be done, and, when it is perhaps too late, to make himself unhappy at their being neglected... In his present state of weakness, every ill, however trifling, appears of a gigantic size, while, on the other hand, every good is diminished. Again, I repeat, that from the certain knowledge I have of his temper, nothing could contribute more to his happiness and make him go on cheerfully with business than having everything finished as he goes along, and have no unsettled scores to look back to and brood over in his mind. [44]

Boulton seems to have regarded his partner in the light of a permanent invalid and wrote to Fothergill regarding an order, 'The matter must be managed with some delicacy respecting

Mr. Watt, as you know that when he is low-spirited he is vexed at trifles'. [45]

Then in the spring of 1781, prospects seemed brighter. Watt wrote in a letter that he thought the following year would be profitable since he expected that income would double but expenses would not increase. Up to then any profits had been absorbed in discharging debts but now they were gaining. [46] Boulton tried to obtain Watt's release from Lowe and Vere but found himself obliged to 'postpone that business for a few weeks' since Vere was 'not in very good humour... I therefore beg you will not vex yourself about it for if you do you will vex me'. [47] Hamilton was surprised that Boulton and Watt were still having financial problems for he wrote to Watt:

> I observe you are erecting a considerable number of Engines in Cornwall. I would fain hope that by the time these are finished, your old scores will not only be cleared off but the annual revenue will be very great - indeed I expected before this time all your original sin would have been purged off. [48]

The potential for a prosperous future was bright but pitfalls lay ahead.

Boulton and Fothergill

At the same time as Boulton was trying to placate Watt, he was receiving complaints from his other partner, Fothergill. For many years Fothergill had been concerned about the indebtedness of the Soho Manufactory. He thought that Boulton was spending too much time on the engine business and neglecting theirs. Then he became ill in the spring of 1781. Watt wrote to Boulton, 'My principal reason for writing to you tonight is that Mr. Fl. grows weaker daily, is ordered to Bath whether he will not go until your return and the Doctor says it is his only chance - Indeed I fear there is no chance - However you will be blamed if he does not go'. [49] Watt's pessimism was proved wrong but Fothergill must have expressed his desire to terminate his partnership with Boulton.

Boulton must have tried to obtain a further loan from Lowe and Vere during the summer of 1781 which was met by a further outburst of complaints and pleadings from Watt who was in Cornwall. He saw all the profits from his Act going to Lowe and Vere and that both he and Boulton were so bound to the bankers that they would be acting criminally if they reduced the guarantees and security given. He moaned to Boulton:

> When I executed the mortgage to them my sensations were such as were not to be envied by any man who goes to death in a just cause nor has time lessened the acuteness of my feelings in other way than producing the means to satisfy some part of the demand - I considered upon the one hand the obligations I lay under to you your friendly treatment of me and your pressing necessities - on the other hand I thought I was resigning in one hour the fruits of the labour of my whole life that if any accident befell you or me I left a wife & children destitute of the means of subsistence by throwing away the only jewel Fortune had presented me with My feelings for you and the thoughts of the consequences which might ensue prevailed and made me swallow the bitter draught. [50]

In 1770, when he had to choose between his unfinished steam engine at Kinneil and a paying career constructing the Monkland Canal, he confided to Small, 'If I refused this offer I had little reason to expect such a concurrence of favourable circumstances soon. Besides, I have a wife and children, and saw myself growing gray without having any settled way of providing for them'. [51] (See Vol. I, p. 290) The same fears were roused again but Watt needed Boulton as much as Boulton needed Watt to get them out of their difficulties. However Boulton could not have been pleased to receive yet another of Watt's rather cringing letters.

Fothergill gave Boulton notice on 2 November 1781 that their partnership would cease at the end of the year. Boulton had to make a fair settlement with him and thought that he should receive some benefits for the risks he had taken and decided to allow him part of his own share of the potential engine profits. [52] This incensed Watt, partly because Fothergill had declined being a partner after they had obtained the Act of Parliament in 1775. All this occurred at a very difficult time when copper prices were falling due to competition with more ore being produced from the Parys Mountain on Anglesey and the need for more capital to support experiments on rotative engines at Soho.

Watt was away in Cornwall and sent another of his long letters to Boulton. He thought that Boulton had been very imprudent to offer sharing the profits from the engine business until they amounted to £20,000, 'which is a sum I fear they never will reach and they never can but by unwearied application and attention to both our parts'. He thought that either Boulton must devote his whole time to it or find an active partner for he, Watt, could not for long stand the load which he was carrying. If Fothergill were to have a share, then that ought to be in proportion to the outlay and risk Boulton and Fothergill had stood, although, as Watt pointed out, any risk would have fallen on Boulton alone. While he, Watt, was exceedingly sorry for Fothergill, he did not feel that this excused him demanding so much more than his fair entitlement. 'In truth I am so much vexed that I scarcely know what I write and may appear to be too harsh in my sentiments'. [53] Perhaps Watt thought he had gone too far for in his next letter to Boulton he told him he had written the earlier one which Boulton could show to Fothergill's friends but there is no doubt that he was very annoyed with the proposal that Fothergill should in any way benefit from the engine business. [54] Another of Watt's fears was that the 'Engine business cannot be a permanent one and I am sure it will not in any case prove so lucrative as you have flattered yourself. therefore hope that you will take every step you can to free me from L.V. & W.' [55] Watt could not foresee the success of his rotative engines which would soon generate additional profits to those from Cornwall.

Reorganisation of Boulton and Watt

The dissolution of the Boulton and Fothergill partnership gave Watt the opportunity to disentangle the affairs of Boulton and Watt and Boulton and Fothergill so he wrote to the clerks, James Pearson and John Buchanan, instructing them how to do this, something well overdue:

> By our contract of partnership all expenses whatsoever attending the Business prior to the passing the Act of Parliament (I believe in June 1775) were to be placed to his [Boulton's] Account solely.
>
> And also expenses which might or may ensue in making Experiments for the improvement of the Engines were to be charged to him alone.
>
> And in return such Engines models etc which were made for the purpose were to be and remain his property or part of his stock.
>
> Consequently he ought to be debited for the costs of the several Engines which have been erected for the use of Soho Manufactory, and also for the maintenance of the said Engines in coals and labour excepting an annual allowance for coals and attendance when the Engines were wrought on purpose to exhibit them to our customers or friends.

B. & F. ought to be debited to M.B. for the value of the said Engines which were applied to the use of the manufactory and also for their maintenance.

All expenses which were incident to the making of Engines for sale should be charged to the respective Engines.

General expenses should be debited for Interest of money, Rents, Clerks salaries, travelling expenses, etc. in prosecution of the Business. My personal acct. is clearly stated, except some claims agt. Mr. B. prior to 1775 of no great ammount which I cannot give in untill I return – whether Mr. B. ought to pay the late charges of patents I leave to himself.

All articles which you cannot with precision ascertain the place of to be placed in suspended acct. to be determined by Mr. B. and me when we meet.

Two thirds of the Profits and stock in trade to be placed to Mr. B.'s credit and one third to J.W.'s

As money has been advanced by Mr. B. through B. & F. at various periods and has been applied promiscuously to the purposes of experiments & Trade - It will be necessary to deduct the first from the latter according to their dates before the interest can be charged.

It may perhaps appear hard to charge Mr. B.s part with the whole expense of experiments etc. but when it is considered that upon that account he is allowed $^2/_3$ of the profits & that it is according to the tenor of our Bond of partnership, I expect that appearance of hardship will vanish. At any rate I am ready to forego any claim which does not in the present state of the business appear Just and reasonable.

It is only with Mr. B. that I am in partnership. That he is bound to admit no other partner in his share without my consent under my hand and seal. That I never did consent to his dividing any part of the profits of this business with Mr. F. any longer than Mr. F. and he should remain in partnership, and that I agreed to on consideration of Mr. F.'s enabling Mr. Boulton to give a greater share of his time and attention to B. [Watt] and his business than he has in reality done, and in consequence of which so much of the business has fallen to my share to do, that I have been rendered very unhappy - the want of Mr. B.'s assistance has been a loss of some thousands to B. & W., and I cannot pretend to see all the consequences. [56]

Extensive records must have been kept from which the accounts could be separated but they have not survived, thus losing much valuable early information about the engine business.

At the same time, Watt wrote to Lowe and Vere, pointing out that Boulton and Fothergill's debt had increased again and how he wanted every assurance that every sum he paid in would count towards its discharge. When it had been reduced to £10,000, he wanted part of his property securing the bond to be released. He also made his own position and lack of responsibility for Boulton and Fothergill quite clear which gave Lowe and Vere an opportunity of complaining about Boulton's handling of the transactions. Annie kept Boulton's letter to Watt upbraiding him for writing to Lowe and Vere from her husband for a few days. [57] Watt was discouraged by the situation in Cornwall where the mines were seeking abatements for their premiums through the competition from Anglesey. He felt they 'must seek another county of Cornwall for this will not be as good as it promised I think however that it may still be worth £5000 p anm but little more'. [58] Watt wrote to M. Walker, clerk at the Soho Manufactory:

I am driven to the brink of desperation... My mind is so distracted with anxiety that I cannot apply to my business and as Mr B. seems to be in the same situation and can give me little help our business must fall to ruin and produce no profit to

> anyone... With what spirit can I engage in any undertaking when I have reason to believe that all the money I can acquire for years must go to defray debts I had not the smallest share in contracting & which I may never live to see repaid. [59]

The surprising outcome of all this was that Boulton agreed that Watt should supervise the finances of Boulton and Watt, in a complete reversal of the roles normally associated with these two. But it must be remembered that Watt had run his own various businesses and engineering contracts in Scotland so had considerable experience of accounting practice. He informed Pearson on 7 March 1782, writing from Cornwall:

> As Mr Boulton has desired me to take the management of the Engine Compys cash, it is necessary that I should be acquainted with the present state of it vizt what cash is in Mr Mathews hands and what demands are likely to come against us; which state I will be obliged to you to furnish me with – It will be impossible for me while I remain here to take the absolute charge of it but I shall from time to time furnish you with such general directions as will I hope enable you to prevent misapplications untill can get home – at present the following are what occur to me. Let such cash as you may receive be lodged in our name in the Birmingham Bank – unless Mr B should have any good reason against it – Lett no money be paid by us on the acct. of Messrs B & F nor any be paid by them on our account without being immediately placed to their credit.
>
> As our finances in this County are anticipated for sometime by certain transactions of Messrs B & F it will be necessary to examine what moneys can be obtained from other quarters in order to supply the demands of our trade at Soho, and also to take care that no person belonging to us, either Masters or men receive any moneys in advance or more than is strictly due to them, and that every person receiving money for travelling charges give in an account of the same at return from their journey or at longest in 3 months otherwise such money to remain for ever at their own debit.
>
> It is highly necessary that this article be attended to with strictness, as we are continually subject to losses from the neglect of it, and after accounts which are reluctantly paid – also that Invoices be dispatched as soon as possible after the goods are sent off – The present state of our finances by no means admits of much generosity or unnecessary advances of money for others. In consequence of which you may perceive that I have confined myself within very narrow bounds since I came here, that I might show a good example if I could do no more; but my family economy will signify little if that of the business is not attended to.
>
> When I meet Mr. Boulton or perhaps before I shall settle what he and I are respectively to draw from the business and let you know that confusion may be avoided in your accounts.
>
> In the meantime I trust in your known integrity and accuracy, that our general interest will be attended to and that you will inform of such things as militate against it – I should have sent my monthly account but have not received Mr. Wilsons acct of cash paid out by him for me. [60]

The new broom was determined to make a clean sweep and put the house in order. Attention was to be paid to the minutest detail and economies made wherever possible. Watt told Boulton what he had done and offered to leave Cornwall and go to Birmingham to help sort out their affairs. It is possible that his taking over the financial management was one reason why he did not go back to Cornwall for some time after this trip. He was soon telling Boulton not to issue any circulating bills in favour of Boulton and Watt, 'as I am convinced no business whatsoever can support that way of borrowing money'. [61] He wanted to set up a proper scheme for organising their business. Engineering work carried out at Soho was also to come under review.

Probably it was fortunate for the firm of Boulton and Watt that Watt had separated their finances from those of Boulton and Fothergill because Fothergill died in June 1782. Boulton of course had to make a financial settlement for his wife as well as to wind up finally their partnership accounts. But this turn of events certainly simplified the relationship between Boulton and Watt. While money was needed to finance the construction of the prestigious Albion Corn Mill in London, this was more than offset by better prospects in Cornwall. That same month, Watt told William Craig in Glasgow that low copper prices had delayed setting the new engines to work at Wheal Virgin and Poldice mines, 'Our other business in Cornwall goes on very well and I hope will soon free us from debt which is a desirable situation'. [62] A little later he could report to Hamilton that large sums had been drawn from the engine business to liquidate part of Boulton and Fothergill's debts:

> The clear income of the engine business is above £3000 pr annum and has a chance of being £2000 greater but may also be less or nothing as we shall be able to defeat our opponents. All debts of the engine business to B. & F. are more than paid but I believe J.W. is a little in debt to B. & W. though not much. [63]

The principal opponent at this time was Jonathan Hornblower junior who was trying to have his compound engine accepted by the Cornish mine owners.

The Bond to Lowe and Vere

While therefore the internal debts between the two major parts of Boulton's empire had been settled, the millstone of the bond to Lowe and Vere still hung round Watt's neck. He pointed out to Hamilton how this was affecting him in spite of the brighter Cornish prospects. 'I have ever since I was obliged to enter into this security been a prey to the most cutting anxiety. I have been able to enjoy no pleasure... I have not been able to attend with perspicuity to any business'. [64] He was afraid that he might meet with an accident so asked Hamilton to sell his property in the Delftfield Pottery in Glasgow in order to refund his creditors in that city. Boulton was able to repay Lowe and Vere £10,000 in May 1783 and the bankers in turn agreed to release Watt when the account should be reduced to £8,000. [65]

Lowe and Vere recognised that Watt was more likely to repay them than Boulton and so were determined to keep him in their clutches for as long as possible. While Boulton was battling against claims over Fothergill's debts, Watt was hopeful that he might reduce the bond to the magic figure in the autumn of 1783. Cornwall was not only paying well but orders were coming in so that he expected £8,000 per annum from there. His own account with Boulton was nearly settled and considerable balances were being received. 'The debt to L V is reduced to £12,000 and they promise on receiving 4000 more to release C [Watt] of this since C has £2000 in his possession and can get the other by Oct.' [66] But Watt's hopes were dashed once more because in January 1784 he discovered that Boulton's debts had increased to £14,440. [67] Watt felt obliged to take steps to settle the matter which he had refrained from doing earlier out of respect for Boulton's feelings. Boulton and been shattered by the discovery of Anne, his second wife, drowned on 11 July 1783 in Soho Pool, the ornamental lake below the Manufactory. [68] How the accident happened was never discovered. Her death was a great blow to him since he was very fond of her. His health was so much affected that he felt he must get away from Soho and all its business affairs so he made a long tour through Ireland and Scotland. [69]

Watt suggested that Matthews should hold in trust Boulton's share of the engine income in order to pay off Lowe and Vere. He was unwilling to pay any more than he was bound because 'there is no reason why I should be obliged to lend him all I am worth and be deprived of the use of my own money, which is the case at present, and likely to continue so if I do not take some steps to prevent it'. [70] He wondered about taking advice from counsel to see whether Lowe and Vere had acted improperly by not accepting that what had been paid to them should be considered as reducing the principal of the bond. He tried to see if they could change bankers but this large sum could not be handled by Matthews. Watt feared that Boulton would continue to use these funds to pay off his other debts.

In April 1784 Lowe and Vere were still refusing to alter the terms of the bond but Watt was able to reach agreement that the account had been reduced to £9,000. [71] He was desperate to be free because he thought Boulton and Watt would have to go to law with Hornblower junior to protect their patent. He was afraid that Lowe and Vere 'may ruin us by stopping credit at a critical conjunction. I therefore entreat you [Matthews] to do what you can to gett the account removed to the other house as soon as possible & to gett the proper assignment from Mr. B.' [72] It was not until nearly Christmas before matters were finally arranged with Lowe and Vere and the account transferred to Matthews. [73] Watt finally wrote to Hamilton in May 1785:

> I have at last got clear of the B.A. [presumably the bond] but have been obliged to submit to considerable advances to accomplish it and have not yet got the deeds though I have a receipt in full which will clear me of any future claim. My share of B. & W. profits ending Sept 1784 was £2870 – besides £191 for interest my whole stock there 6600 but a great part of that stock is vested in mines, another part has been swallowed up by the sinking fund of the B.A. and this Albion Mill will drain us horribly – However the knowing I am in a manner independent gives me an ease of mind I never possessed in any former period of my life which has contributed much to clear me of headaches though I have still the same weakness of constitution. [74]

At last Watt was freed from his 'Hell on earth', [75] principally due to the payments from Cornwall. Hamilton replied saying that he hoped that matters had been arranged so that Watt would have no more trouble; 'this weight off your spirits will I hope re-establish your health, so as to prosecute your other business to your mind, & I would fain suppose a very few years will put it in such a train as to bring you a comfortable income without any trouble'. [76] Hamilton's prophecy was proved right.

Personal Debts

At the same time, Watt was able to clear off his remaining Scottish debts. While his partnership with Boulton settled his debts with Roebuck, Watt had also borrowed from his family and friends in Scotland. He was anxious that these people should not suffer as a result of any failure of Boulton and Watt. In 1781 he apologised to Black:

> I am very much ashamed of having been so long your Debtor, but I could not avoid it without putting myself to some inconvenience as our Business never defrayed its own current charges untill last year and the product of that was swallowed up in a very large pay[men]t: several P[aymen]ts we were obliged to make at Christmas. I have therefore been always in debt to the partnership but am now clear or nearly so

> As to being rich that is out of the question and to make you wait till then would be trespassing too much on your patience which I have encroached on too much already. let it suffice that it is not now inconvenient to make the payment and that I hope you will accept of my sincere thanks for the time you have indulged me - Whatever Ball[an]ce may be still due you I will either settle with you at meeting or order you pay[men]t of immediately as is most agreeable. [77]

Watt sent Black £40. Black sent Watt a statement of the account, leaving according to him, £3.3s.6 $^1/_2$d. [78]

In September 1782, Watt told Hamilton that his family commitments were bonds of £210 to Agnes and Betty Millar and to Betty a bill of £40 or £50, about £100 to William Craig and a further sum to Mr. Muirhead, some of which he was hoping to repay. [79] The amount owing to Muirhead turned out to be nearly £300. [80] Hamilton commented, 'You say affairs go on prosperously in Cornwall – I rejoice at it but wish you could make a point of having all the money that is drawn from the Engines applied to pay off the old engagements - Till that is done you will never be easy & happy'. [81] Watt must have been relieved when these debts were settled in the spring of 1783. This left the debt to Black. It was not until June 1784 when he paid off the outstanding amount here:

> I send you inclosed my Draft on Gilbert Hamilton for £18.8sh.6d being the ballance due by me to you (for I reckon nothing on the organ which you placed to my credit) It has grieved me much to have been obliged to have been so long owing you; but it is very lately that the sun began to shine on me, and I could not think of paying one friend by borrowing from another. [82]

At last Watt could look forward to a prosperous and secure future. But this was after the date when the 1769 patent should have normally expired in 1783 so that, had this patent not been extended, Watt would have derived no profit from it.

Administration

At the same time as the finances of Boulton and Watt were being sorted out, the administration also began to take the shape that would continue until 1800. We have seen that Henderson had been brought in to help Watt with the erection of engines. His position was not filled after he left in 1783. (See p. 108) Possibly he was replaced in Cornwall by Murdock while the other engine erectors were becoming more experienced, needing less supervision. But somebody was still needed to help both Boulton with the Soho Manufactory and Watt with the engine business.

The choice fell on Keir. [83] Watt had met him while staying with Small in 1768 and continued to correspond with him subsequently over producing alkali from salt. (See Vol. I, p. 158) When Boulton went to join Watt in Cornwall in October 1778, Keir was brought in to take charge of the works at Soho with a view to his becoming a partner. [84] The letter that Keir wrote to Boulton at the end of that month shows that he was involved both with administering the erection of engines as well as giving advice about arranging loans from Wiss. [85] Watt noted on Sunday 18 April 1779, 'all day with Messrs. B. & Keir going over agreements and our contract of partnership'. The partnership was executed two days later before Boulton left for Holland. [86] It is not clear what this partnership was. It is very late for that between Boulton and Watt and too early for the one between these three for the copying machine. It may be that Keir actually joined a partnership with Boulton and Watt in the engine business only to withdraw shortly afterwards when he discovered their true financial state. Keir wrote later:

> I then visited Mr B. & Soho daily. We examined together the details of some of the business & saw that considerable amendments in the economical part were requisite. Mr. B. consulted me in many subjects relative to his business & to his correspondences. I perceived that there were great financial embarrassments. Mr Fothergill told me the amount of their discount or bill account which appeared quite alarming. Mr Walker [clerk at the Soho Manufactory] confirmed it for I never looked into their books of accounts. It appeared evident that the business had been carried on with great loss, as it was afterwards proved to have been by the insolvency of Mr Fothergill at his death – Under these circumstances, the intended partnership could not with any propriety be compleated, & it was never mentioned by either of us. [87]

Whatever happened over the partnership, Keir continued to assist at Soho until the middle of August 1780 when Watt told Hamilton:

> We have been all along in great want of a proper person to superintend the mercantile part of our Engine business and that want is made more sensible just now by Mr. Keir's going to carry on the Alkali business on his own bottom at some distance from hence and consequently Mr B. loses the benefit of his attention. [88]

Watt was left to soldier on with the engine business but in the meantime Keir had accepted a partnership in the copying machine enterprise while Watt helped him over his alkali patent. The firm of Boulton and Watt did not seek another administrative supervisor, partly perhaps through the growing competence of their office staff.

A Scot, James Pearson, had been engaged as book-keeper and cashier as early as 1775. [89] Surprisingly Watt complained in 1780 that 'Pearson does not understand our business' [90] but in spite of this Pearson remained at Soho until 1817 when he retired on a pension. Another Scot, William Playfair, had arrived by early 1778 when he was helping with drawings for the Ketley engine. [91] Copying drawings was a task he continued to do although he made a mistake over an eduction pipe. [92] Watt told Boulton that 'Playfair does not understand bookkeeping nor method' which is scarcely surprising considering his other abilities. He left in late 1781, much to the regret of Boulton who held a high opinion of him. [93]

In the spring of 1781, Watt complained to Boulton, 'Some draughtsman must be got or we must retrench our business. I cannot get forward I have so much bad health and distracted with other things'. [94] But it was not until the end of August that Boulton suggested John Southern:

> He seems good-humoured and very obliging. He is now with his brother as a surgeon, but says that though he is keen in the study of surgery, yet he had rather been employed under you as a draughtsman and assistant, and if you wish to have him I know he will gladly come. He draws tolerably neat. [95]

Watt gave a guarded reply from Cornwall:

> If you have a notion that young Southern would be sufficiently sedate, would come to us for a reasonable sum annually, and would engage for a sufficient time. I should be very glad to engage him for a drawer; provided he gives bond to give up music; otherwise I am sure he will do no good, it being the source of idleness. [96]

Southern started with Boulton and Watt in June 1782 at the age of twenty-four. In spite of continuing with his violin playing, he proved to be a very valuable assistant, relieving Watt of a great deal of work. He also helped Watt with his scientific research and was probably the inventor of the steam engine indicator. He filled a managerial role and was made a partner in 1810, receiving one-sixth of the profits as well as a salary. After some illness, he died suddenly in 1815 while still working for Boulton and Watt. [97] Watt was very distressed and thought that Watt junior would be overwhelmed with business, so indispensable had Southern become.

Notes

1 Muirhead, (1858), p. 301.
2 Ibid, p. 302.
3 Ibid.
4 Dickinson, (1935), p. 90 and Lord, (1923), p. 110.
5 Smiles, (1878), p. 218.
6 Hebert, (1838), Vol. II, p. 719.
7 Muirhead, (1854), Vol. II, Letter 136, p. 79, J. Roebuck to J. Watt, 12 Nov. 1774.
8 Goodison, (2002), pp. 28 – 9.
9 Lord, (1923), pp. 112 – 13 and Roll, (1930), p. 100.
10 Goodison, (2002), p. 29 from Kenneth Morgan, Ed., *An American Quaker in the British Isles*, Oxford, (1992), p. 255.
11 Dickinson, (1936), p. 209, and Lord, (1923), p. 111.
12 Goodison, (2002), p. 60.
13 Boulton 348.57, M. Boulton to J. Watt, 24 Feb. 1776.
14 Schofield, (1963), p. 149.
15 JWP, C3/3, 'Journal', 23 May 1778.
16 Boulton 348.92, J. Watt to M. Boulton, 6 July 1778.
17 Boulton 348.93, J. Watt to M. Boulton, 8 July 1778.
18 Ibid.
19 Boulton 348.95, J. Watt to M. Boulton, 16 July 1778.
20 Boulton 348.99, J. Watt to M. Boulton, 3 Aug. 1778.
21 Smiles, (1878), p. 192.
22 Roll, (1930), p. 102, M. Boulton to J. Watt, 25 Sept. 1778; B. & W. Col.
23 Tann, (1981), p. 206, Doc. 84, M. Boulton to C. Vere, April 1779.
24 Lord, (1923), p. 116.
25 Boulton 348.124, J. Watt to M. Boulton, 20 Jan. 1779.
26 Boulton 348.125, J. Watt to M. Boulton, 22 Jan. 1779.
27 Boulton 348.121, J. Watt to M. Boulton, 22 Jan. 1779.
28 Tann, (1981), p. 207, Doc. 84, M. Boulton to C. Vere, April 1779.
29 Boulton 348.136, J. Watt to M. Boulton, 10 May 1779.
30 JWP, W5, J. Watt to J. McGrigor, 2 Aug. 1779.
31 Boulton 348.158, J. Watt to M. Boulton, 16 Oct. 1779.
32 JWP, 3/8, J. Watt to M. Boulton, 13 April 1780.
33 JWP, 3/8, J. Watt to M. Boulton, 18 May 1780.
34 JWP, C3/6, 'Diary', June 1780.
35 JWP, W5, J. Watt to G. Hamilton, 28 Oct. 1780.
36 JWP, 4/43, G. Hamilton to J. Watt, 9 Sept. 1782.
37 JWP, 3/8, J. Watt to M. Boulton, 20 Sept. 1780.
38 JWP, 3/8, J. Watt to M. Boulton, 31 Oct. 1780.
39 JWP, 6/21.25, W. Matthews to J. Watt, 'State of Accounts on the 18th December 1780'.
40 JWP, 3/8, J. Watt to W. Matthews, 28 Dec. 1780.
41 Ibid.
42 Lord, (1923), p. 123, M. Boulton to J. Fothergill, 11 Dec. 1780; Tew Mss.

43 JWP, W5, J. Watt to W. Matthews, 5 March 1781.
44 Smiles, (1878), p. 217, A. Watt to M. Boulton, 15 April 1781; Boulton Mss.
45 Ibid, p. 213.
46 JWP, W5, J. Watt to 'Dear Sir', 27 May 1781.
47 Tann, (1981), p. 298, Doc. 139, M. Boulton to J. Watt, 27 Aug. 1781.
48 JWP, 4/43.7, G. Hamilton to J. Watt, 15 Aug. 1781.
49 JWP, 3/8, J. Watt to M. Boulton, 24 April 1781.
50 JWP, 3/8, J. Watt to M. Boulton, 21 June 1781.
51 Muirhead, (1854), Vol. II, pp. 2 – 3, Letter 83, J. Watt to W. Small, 9 Sept. 1770; Boulton 348.20.
52 Dickinson, (1936), p. 110.
53 JWP, 3/18, J. Watt to M. Boulton, 11 Feb. 1782.
54 JWP, 3/18, J. Watt to M. Boulton, 13 Feb. 1782.
55 JWP, 3/18, J. Watt to M. Boulton, 20 Feb. 1782.
56 JWP, 3/18, J. Watt to J. Pearson and J. Buchanan, 20 Feb.1782. See also Lord, (1923), pp. 127 – 28.
57 Lord, (1923), p. 129.
58 JWP, 3/18, J. Watt to M. Boulton, 23 Feb. 1782.
59 JWP, 3/18, J. Watt to Mr. Walker, 6 March 1782.
60 JWP, 3/18, J. Watt to J. Pearson, 7 March 1782.
61 JWP, 3/18, J. Watt to M. Boulton, 21 March 1782.
62 JWP, Letter Book 1, J. Watt to W. Craig, 2 June 1782.
63 Ibid, J. Watt to G. Hamilton, 11 July 1782.
64 Ibid.
65 Lord, (1923), p. 131, M. Boulton to J. Watt, 31 May 1783; Tangy Mss.
66 JWP, Letter Book 1, J. Watt to G. Hamilton, 12 Aug. 1783.
67 Ibid, J. Watt to W. Matthews, 24 Jan. 1784.
68 Ibid, J. Watt to J. Fry, 22 July 1783.
69 Dickinson, (1936), pp. 118 – 19.
70 JWP, Letter Book 1, J. Watt to W. Matthews, 9 Jan. 1784.
71 Ibid, J. Watt to G. Hamilton, 30 April 1784.
72 Ibid, J. Watt to W. Matthews, 20 July 1784.
73 Lord, (1923), p. 132, M. Boulton to W. Matthews, 16 Dec.1784; Tew Mss.
74 JWP, Letter Book 1, J. Watt to G. Hamilton, 8 May 1785.
75 Ibid, J. Watt to W. Matthews, 29 Jan. 1784.
76 JWP, 4/43.38, G. Hamilton to J. Watt, 15 May 1785.
77 Robinson & McKie, (1970), p. 108, Letter 79, J. Watt to J. Black, 9 April 1781.
78 Ibid, p. 110, Letter 80, J. Black to J. Watt, 23 April 1781.
79 JWP, Letter Book 1, J. Watt to G. Hamilton, 29 Sept. 1782.
80 Ibid, J. Watt to R. Muirhead, 18 Feb. 1783.
81 JWP, 4/43.20, G. Hamilton to J. Watt, 12 April 1783.
82 Robinson & McKie, (1970), p. 143, Letter 99, J. Watt to J. Black, 6 June 1784.
83 For a biography of Keir, see Smith, Barbara M.D. and Moilliet, J.L., 'James Keir of the Lunar Society', *Notes and Records of the Royal Society'*, Vol. 22, Nos. 1 & 2, 1967.
84 Smiles, (1878), p. 191.
85 Tann, (1981), pp. 174 – 75, Doc. 67, J. Keir to M. Boulton, 29 Oct. 1778.
86 JWP, C3/4, 'Journal', 18 and 20 April 1779.
87 Dickinson, (1936), p. 101.
88 JWP, W5, J. Watt to G. Hamilton, 14 Aug. 1780.
89 Dickinson & Jenkins, (1927), p. 284.
90 JWP, 3/8, J. Watt to M. Boulton, 20 Sept. 1780.
91 JWP, C3/3, 'Journal', 3 Feb. 1778.
92 JWP, 3/8, J. Watt to M. Boulton, 1 May 1780.
93 JWP, 4/81, J. Keir to J. Watt, 12 Nov. 1781, 'Playfair to settle in London'.
94 JWP, 3/8, J. Watt to M. Boulton, 24 April 1781.
95 Dickinson & Jenkins, (1927), p. 285, M. Boulton to J. Watt, 31 Aug. 1781: B. & W. Col.
96 Muirhead, (1854), Vol. II, p. 131, Letter 196, J. Watt to M. Boulton, 1 Oct. 1781.
97 JWP, C6/3, J. Watt to J. Weston, 1 Aug. 1815.

8
Problems in Cornwall

After reading about Watt's purchase in 1781 of their new house at Cosgarne, Hamilton replied, 'The description of your House & situation made me think you had found some Fairyland in the dreary County of Cornwall'. [1] In fact Cornwall was already proving to be a bed of nails. Mining prospects were suffering through the increasing production of ore at Anglesey which forced down prices and made Cornish mines less profitable. Mine owners sought to reduce costs and so sought reductions in the premiums being paid to Boulton and Watt. Hostility began to manifest itself as people saw how much wealth was being taken out of their county by a firm from far away Birmingham. Some decided to challenge the 1775 Act of Parliament. Then Hornblower junior patented his compound engine in 1781. It was thought that this might rival and even excel Watt's engine in economy so that Cornishmen used it as a bargaining counter to try and beat down Boulton and Watt's dues or delayed reaching agreements over existing engines. But the mines could not manage without the more efficient Watt engines and were obliged to install them if production were to continue.

At first, prospects for Boulton and Watt in Cornwall seemed reasonable, particularly as the number of their engines in that county increased. It was not an unduly pessimistic Watt who wrote to Black in the autumn of 1780:

> I am sorry I cannot give you so good an account of our gains as you have heard. Our income if regularly paid would perhaps be about £3000 pr annm but it is not so. there are many deductions to be made and the expense of carrying on the business is great, as are our debts however our prospects are brilliant, our income increases yearly, will have a considerable increase soon in Cornwall, and a much greater one when Wheal Virgin setts to work about a year hence they are to pay £2500 per annm and Poldice which will go on at the same time 1500 My situation hitherto has been of the most uneasy Sort, and I am so habituated to disappointment that even these splendid prospects cannot raise my spirits to par. [2]

These splendid prospects were shattered by the falling price of copper, causing closure of mines and much distress for the miners and their families.

Fall in Copper Price

In the summer of 1778, Watt had pointed out to Boulton how ore prices were falling daily and that Chacewater mine might not be workable. He feared others might close. [3] He told Hamilton three years later how the miners were disheartened as the price of ore was still sinking. [4] However there was a brief recovery over the New Year of 1782 but then in the following June prices had 'sunk to so low a price that many of the mines could scarcely live'. [5] This was due to increased production from Anglesey. The Cornish mine owners were not blameless either because they had increased copper production there to try and balance costs with greater sales at lower prices which exacerbated the situation. The following figures reflect these problems.

Average Output of Copper in Cornwall and Devon in Five-Yearly Periods.

Year	Ore (tons)	Metal (tons)	Standard £ per ton	Value at mine £ per ton (metal)
1770-75	28,750	3,500	76	50
1775-80	27,600	3,300	77	52
1780-85	33,160	3,990	73	48
1785-90	35,500	4,220	65	45
1790-95	40,240	4,670	-	-
1795-1800	49,900	5,200	108	84 [6]

The cost of coal had risen at the same time. Watt was relieved in June 1783 when Parliament refused to grant the Parys Mine Company relief on coals to be imported there especially as they were reported to have made a clear profit of £60,000 in one year. [7] He wrote to Wilson in January 1784:

> The greatest objection is the extravagant rate at which the Anglesea Compy are getting ores, which on any demurr in sales must overstock the markets and infallibly lower the price of copper so much that nothing but loss can be got by it, which indeed makes one tremble for the business in general, but to do these gentlemen justice they are indefatigable in seeking out foreign markets, which will throw the backwater of sales in some of the foreign mines and probably stop some of them entirely. [8]

It was only around 1790 that the production of the Parys Mine Company began to fall and the price of copper started to rise again. [9]

Anglesey

Competition from copper extraction in Anglesey caused the deterioration of mining prospects in Cornwall, which affected the position of Boulton and Watt there. While there was some renewed mining for copper on Parys Mountain in the early 1760s, it was not until March 1768 that a great mass of copper was discovered, pointing to the potential richness of the area. However the west part of the mountain was divided between two owners, Sir Nicholas Bayly of Plas Newydd and originally William Lewis of Llysdulas. There followed complex and lengthy negotiations and lawsuits between these families. The Lewis estate was settled on Lewis's niece who married a clergyman, Edward Hughes. To represent them, the Lewis faction sought legal advice from the attorney, Thomas Williams of Llanidan. The deaths of Lewis, his widow and sisters left Hughes and Bayly to effect a settlement. Bayly decided to abandon mining here on his own account and signed a lease for his part of the mountain for twenty-one years to the London banker, John Dawes, on 12 November 1778. Having obtained the lease, Dawes joined Hughes and Williams to form the Parys Mine Company, thus uniting the whole of that area under one control. The active partner in the concern was Williams. [10] The first two large Boulton and Watt engines in Cornwall had been working for only three months so had had little time to prove their greater efficiency and help reduce the mining costs there. Williams was quick to exploit the advantages of Anglesey. Not only was Parys Mountain only about a mile and a half away from the port of Amlwch so that his costs of transporting the ore ready for shipment were much lower than the central Cornish mines, but Parys Mountain could be worked as an open pit which was much cheaper than the deep Cornish mines. A writer in 1827 said:

> The annals of mining exhibit no instance of a mine so productive as the Parys copper-mine has been, accompanied with so little expence in working. The labour consisted in quarrying an immense mass of ore which rose to the surface of the ground on the summit of a hill of moderate elevation. Here no hydraulic machinery was necessary for the drainage of the mine, a solid body of copper ore of prodigious extent was exposed to the glare of day, tempting the labour of the miner. There was no useless rubbish to remove, no toil in descending, no risk of life; nor was it possible after many years employed in excavating to calculate its extent or possible duration. [11]

Legal obstacles now no longer prevented the full exploitation of the mineral wealth. Yet Williams quickly realised that he would be unable to gain maximum profit from the ores raised through the existence of a smelter's cartel. Already as early as 1775, Hughes had declared his intention of erecting smelting works on Anglesey. He petitioned Parliament in 1776 for the removal of duty on the coastwise shipment of coal. The Parys Company tried again in 1779 as well as 1782 and 1783 but the Bills were thrown out, partly through opposition from Cornwall. Eventually an Act was obtained in 1786 for the remission of duty on coal used in smelting or for steam engines up to £1,500 per annum. But Williams realised that freight charges for the 30 to 40 tons of coal needed to produce a ton of fine copper would be too high so did not set up a smelter in Anglesey. The same situation existed in Cornwall. But Williams did think it worth importing a limited amount to calcine his low-grade ore, so reducing its bulk by getting rid of some impurities and so lowering transport costs.

In the meantime, in the autumn of 1779, the Parys Mine Company leased land and signed a coal contract for large works at Ravenhead in Lancashire where a smelter was set up. A 40-ton sailing barge could leave Amlwch and work its way finally up the canal to moor almost at the door of the smelting house. Coal, building materials and fire clay could be obtained with virtually no costs for transport. By 1782, Williams had purchased the existing smelting works at Upper Bank, Swansea, in addition. He completed his copper empire by establishing a manufacturing plant at Holywell, Flintshire, powered by the then abundant water flowing from St. Winifred's Well. Thomas Pennant related how:

> Thomas Williams, esquire, that useful active character, with unparalleled speed, covered the lower part of the stream, or that next to the sea, with buildings stupendous in extent, expense and ingenuity of contrivance. These great works are under the firm of the Parys Mine Company. The buildings were completed in the year 1780. [12]

There Williams sent his smelted copper where the plant consisted of a rolling mill, a forge and a wire mill so that he could manufacture sheet, strip or wire. He had set up a vertically integrated industrial empire, with units situated at the best sources of raw materials, copper ore on Anglesey, coal in Lancashire and waterpower at Holywell. He could be, and was independent of all other competitors.

In 1785, Lord Uxbridge, Bayly's heir, handed over the entire direction of his mining affairs to Williams which included the Mona mine at the eastern end of Parys Mountain. The establishment of this great industrial organisation represented a vast increase in the power at Williams's disposal so that within five years he had the copper trade of Britain at his mercy. Anglesey could produce copper at £50 per ton or even lower while that from Cornwall was more in the region of £80. His aim was to produce as cheaply as possible and sell at whatever price would bring himself and his partners the largest profit. [13] Williams, with his ruthless ambition, caused many problems for Boulton and Watt in the ensuing years. One way Boulton tried to mitigate the effects on Cornwall and on Boulton and Watt was to help organise what was in effect a cartel, the Cornish Metal Company, the intention of which was to buy up all the Cornish ore to control the price.

Hostility in Cornwall

One result of the fall in price of copper was that the Cornish mine owners wanted an abatement of the premiums they paid to Boulton and Watt. Much earlier in 1776, when Sir Archibald Hope in Glasgow had been enquiring about an engine for use where coals were cheap, it was suggested that he pay a single lump sum based on the amount of half the fuel savings calculated over ten years. [14] Perhaps Boulton remembered this when he visited Cornwall with the object of settling agreements with various mining companies. He discovered that there were many complaints against the tables for calculating the premiums which Watt had so laboriously drawn up and which he thought would be absolutely fair. Boulton told Watt, 'The fewer columns, and the fewer figures, the less they will be puzzled and the less room for cavilling'. [15] There was temptation to evade payment by the mine owners who found these tables so complex that they suspected Boulton and Watt of trying to cheat them. They were also suspicious of the counters on the engines.

Watt reported that Wheal Virgin would order engines if they could beat down Boulton and Watt's and their landlord's terms. They wanted to pay an annual premium which Watt strenuously opposed, saying it was contrary to their normal manner of proceeding and he could not agree to it without his partner's consent. He told Boulton:

> This is no light matter for besides being a firm precedent for all the county it will be at least £2,400 p annum to us & may be more... now I am hardened I will be for taking all I can get fairly - and I am sure that the allowances made on the two engines at poldice ought to have satisfied reasonable men but the more is given the more will be required and I find a disposition to fool us out of all if they could. [16]

Watt recognised that he was unable to drive a hard bargain with these Adventurers and in his next letter pleaded with Boulton, 'I repeat you must come and come quickly – If you had been here and had gone to that meeting with your chearfull countenance and a good heart perhaps they would not have been so resolute'. [17] Boulton came and reached a draft agreement with Wheal Virgin that they should pay £2,500 per annum for the use of five engines. Poldice agreed to pay £1,500 per annum for their two engines. [18] Watt reluctantly consented but commented:

> These disputes are so very disagreeable to me, that I am very sorry I even bestowed so great a part of my time and money on the steam-engine. I can bear with the artifices of the designing part of mankind, but having myself no intention to deceive others, I cannot brook the suspicions of the honest part, which I am conscious I never merited even in intention far less by any actual attempt to deceive. [19]

Watt found these disputes and haggling over premiums very difficult to handle, partly through his ill-health. He had to point out to Boulton how:

> In the situation of health I then found myself I could not go to be worried by 30 people united against me I was afraid that either through weakness I might have betrayed our cause or through passion have said things which might have excited personal animosity, for I was well informed of their sentiments and that their design was merely to beat us down as a precedent for Wheal Virgin. [20]

Watt deserves sympathy for having to face a room full of hostile men but we are reminded of the instance in his surveying career when he told Small how he 'would rather face a loaded cannon than settle an account or make a bargain. In short I find myself out of my sphere when I have anything to do with mankind'. [21] (See Vol. I, p. 218) While Watt's original method of seeking a premium based on the amount the engine was used may have been more reasonable, the set annual premium was simpler and easier for reckoning. Even so there were more attempts to obtain reductions.

Some Cornishmen considered challenging the validity of Watt's patent. This led to a further outburst from Watt which is quoted at length to show his reactions:

They charge us with establishing a monopoly, but if a monopoly, it is one by means of which their mines are made more productive than ever before. Have we not given over to them two-thirds of the advantages derivable from its use in the saving of fuel and reserved only one-third to ourselves, though even that has been further reduced to meet the pressure of the times?... The invention [of the engine] has been the product of my own active labour and of God knows how much anguish of mind and body... No person has a greater aversion of going to law than I have but when I see these gentlemen pursuing methods which if they are allowed to persevere in must end in our total ruin, at least in mine, I am forced to exert myself to prevent it and to show that though I may be cajoled out of what I esteem my right, I will not be kept out of it by obstinacy to give it no worse name. In regard to what you have been told of the county intending to apply to Parliament to abolish our Act, which is so great a loss to them as this monopoly of ours, I agree with you that the proposers of such a thing have not given themselves new time to think - Nevertheless for my own part whenever they chuse to present such a petition I will join issue with them and shall refer all my right willingly to the justice of Parliament You have had occasion to know my sentiments in that head from the beginning and know that in place of this Act which is a grievance to them I would willingly have taken £7000 and made the invention free to all men but neither Parliament <u>nor any body else</u> would then give me even that sum, though by the by I should not have put much of it in my pocket, yet I should have been much richer than I am now after that we have spent 6 years of a most laborious and anxious life and have spent many thousands of pounds in thoroughly establishing the powers of the engine to the conviction of all men, so that they are come to believe that we shall get some reward for our labour or at least get some of our money back again. then the people who receive the greatest benefit from the invention to whom we freely resigned 2 parts of the profitts out of 3 <u>These</u> people propose to petition parliament to take away the remainder, why don't they petition to take away Sir Francis Bassett's mines? He acknowledges having received great profits from them, which we cannot with truth say we have got from ours, for it appears by our books that Cornwall has hitherto eat up all the profits we have drawn from it and all we have got by other places and a good sum of our own money to the bargain - We have no power to compell any body to erect our engines? What then will parliament say to any man who comes there to complain of a grievance he can avoid? Will Parliament give away our property without an equivalent? Will they adjudge them to pay us any less sum than it has cost ourselves? Will they not further add some reward for the quantity of life which has been lost in the pursuit of what is evidently for the advantage of others but hitherto has not been for our own. Lastly will Parliament compell us to work for anybody without a reward adequate to our experience? or will they oblige us to labour for anyone without our consent? We are in the state of the old Roman who was found guilty of having better crops than his neighbours, and was therefore ordered to bring before the assembly of the people his instruments of husbandry and to tell them his art which he complied with and when he had done said These O Romans are the instruments of our Art, but I cannot bring into the Forum the labour, the sweat, the watchings, the anxieties, the cares which produced these crops. Everyone sees the reward which we may properly receive from our labours; but few consider the price we have paid for that reward & which nevertheless is by no means a certain annuity but of the most precarious sort To put an end as far as lies in my power to all disputes with the people of Cornwall let them pay my debts and give me a reasonable sum for time I have lost and I will resign my part in their favour and think myself well off by the bargain. Or if you can find any man who is agreeable to yourself, I will sell him my share on reasonable terms and like the sailor I shall promise to contrive no more fire Engines.

In short my dear Sir with a good cause in hand I do not fear going before Parliament or anywhere. I am sure that if they did anything they would put us in a better situation than we are in. I beg pardon for enlarging so much on this subject but I thought it might be proper to point out to you some of the arguments in our favour and to enable you to show our friends what are my real sentiments on the subject, that neither my hopes nor desires are exorbitant and that though I am not disposed to return the ill offices which are done us, in kind, yett I have a proper sense of them. [22]

Perhaps Watt should have followed a literary career. We see his strong feelings that the steam engine with the separate condenser was his own invention and that he had struggled long and hard to bring it to a successful working state. But this was written at a time when debts were mounting and he was held in bond to his bankers. Also he was being overworked and so under considerable stress. This particular threat to try and annul the 1775 Act of Boulton and Watt abated at this time.

But Boulton took a very different attitude. Cornwall was their most important market, both actual and potential. He was prepared to support the mines by investing in them because he realised that, if the mines stopped, the engines would stop and they would 'gain nothing'. [23] Watt agreed so that in April 1780 Boulton and Watt purchased a share in the Marazion mine, 'it being our interest to support the mine'. [24] By the end of the year, they held shares in five mines. [25] For example, they held a 1/16th share in Wheal Union and a 1/32nd in Pool mine. [26] Later they took a 1/32nd in Wheal Virgin to help set that mine to work which was expected to cost £30,000 initially and £3,000 monthly. [27] It was a policy which was followed in ensuing years. In 1784 they took a 2/32nd share in North Downs mine and one right in the New Chacewater Company. [28] Watt had £6,600 tied up in mine shares in 1785. (See Vol. III) One advantage was that it gave the right to attend meetings of the Adventurers and therefore influence policy.

With the continuing fall in the price of ore, the demand for reduction in premiums continued. In March 1782, the Adventurers of the United Mines asked for an abatement of one third of their dues for their engines at Ale & Cakes and Poldory. [29] Instead, Boulton and Watt agreed to rescind payments for six months to keep the mines running. [30] Other mines also requested reductions. It was a time of unrest in the county:

> The Tinners have threatened to pull down some of the Engines at Wheal Virgin if they do not go to work this summer & Capt Paul thinks they will keep their words. he is very uneasy lest they should mob him as indeed he runs a great risque As to myself I think I am not safe as every effort is used to prejudice them against us, by the Horn[blow]ers and other infernal agents – I am therefore determined silently to withdraw as soon as possible and leave the ferment to subside of itself. [31]

This hostility may be another cause of Watt's not returning to Cornwall. Later Boulton and Watt's representatives had to be withdrawn for their safety.

Boulton and Watt's Royalties

Rough figures based on a list compiled by Boulton in September 1782 will show why Cornwall was so important to Boulton and Watt. It covered most engines built, their full premiums and what was actually being paid.

Average Premiums Received, 1782

Place	Number of Engines	Present Average Income	Actual Average Income per engine	Total possible Income	Average per engine
Cornwall	c. 20	£2,690	£169	£8,560	£428
Elsewhere	15	£1,184	£ 78	£1,468	£ 98
Not listed	11				
		£3,874		£10,028 [32]	

Those for Wilkinson and the Birmingham Canal Navigation were not listed because special arrangements had been agreed. Although the amount received from the Cornish mines was much lower than it might have been, it still eclipsed that received from the rest. Each engine unit there was more profitable than in the rest of the country. Even with reduced premiums, each engine paid more on average. While the total income should have been £10,028, owing to reduction in premiums or engines stopped, it was actually only £3,874 of which Cornwall contributed the greater share. The amount received from Cornwall up to the end of 1785 was £36,351.10s.2d. but, if the full one third value of the savings had been paid, it was estimated it should have been £47,719.14s.7d.

Royalties Received by Boulton and Watt from Cornish Engines, 1778 - 1785

Year	1/3 of Fuel Savings	Received
1778	468. 15. 0	291. 13. 4
1779	1,741. 2. 2	1,250. 4. 0
1780	3,198. 10. 10	2,441. 12. 5
1781	3,902. 16. 2	3,211. 18. 5
1782	6,694. 1. 0	4,192. 17. 0
1783	11,397. 7. 0	8,224. 15. 3
1784	10,312. 6. 6	8,227. 12. 3
1785	10,004. 15. 11	8,510. 17. 8
	47,719. 14. 7	36,351. 10. 2 [33]

While these figures show how much was received from Cornwall during the early years, there must have been a comparable amount from the engines erected elsewhere, but no figures have been discovered. In January 1783, Watt wrote to his friend Charles McDowal in Scotland:

> We have altered all the engines in Cornwall but one, and in many other parts of England, but do not acquire Riches as fast as might be imagined, the expenses of carrying on our business are necessarily very great and have hitherto consumed almost all our profits but we hope to do better by containing our attention and exertions and by multiplying the number of our workers. [34]

When in the same year he was preparing the draft of their case against Hornblower junior's patent, he wrote:

> ... nothing less than the most unremitted endeavours to please on the part of B & W and the undeniable merit of their invention could have brought their Engines into the general use they now are in Cornwall. General may be safely said when out of 22 engines which are at work in the county 21 have been constructed by Boulton and Watt. [35]

Boulton and Watt engines had acquired a commanding position in Cornwall through their excellent performance. Faced with such a monopoly controlling the vital steam pumping engines, it was only natural that Cornishmen should seek an alternative source of power.

Engine Performance

Figures from Cornwall are the main source of the performance of the pumping engines. Almost as soon as he had made his crucial invention of the separate condenser, Watt realised that his 'perfect' engine would be more efficient than earlier types. He wrote to James Lind on 29 April 1765, 'If there is not some devil in the hedge, mine ought to raise water to 44 feet with the same quantity of steam as theirs does to 32'. [36] He claimed in his first patent that he had 'after much labour and expense, invented a Method of lessening the Consumption of Steam and Fuel in Fire Engines, which he apprehends will be of great Public Utility'. [37] This claim was renewed in the application to Parliament in 1775 for the extension of this patent. It was confirmed by Roebuck who stated in his evidence, 'That this Engine will not consume Half the Steam that a common Fire Engine does, probably not the Third Part, and produce an equal Effect; that it will do double the Work of a common Fire Engine at the same Expence'. [38]

The Watt engine, size for size, was more powerful because Watt aimed to 'make a compleat vacuum' as well as using steam a little above atmospheric pressure. These gave an advantage of his engine over 'the Common Engine, in the degree of 12 to 7'. [39] This greater power was later dramatically confirmed in practice when Watt rebuilt the Chacewater engine and placed his cylinder of 63 ins. diameter inside that of Smeaton's 72 ins. The smaller engine drained the mine effectively which the large one had failed to do. (See p. 96) Then in 1779, Watt replaced two common engines at Poldice, one of 66 ins. and the other of 60 ins., with a single one of his own at 63 ins. diameter. By this time, he had had practical experience with running these engines but his claims for their performance still needed to be supported by actual tests.

There are few surviving figures of the performance of Savery's engine which can be compared with those of either Newcomen's or Watt's. None have been discovered where such engines operated with both the suction and forcing cycles. In 1774, Smeaton tested two engines erected by Joshua Wrigley at Joseph Thackery's cotton spinning mill in Manchester where they raised water with the vacuum part of the cycle only to power a waterwheel. The duty was given as 5¼ and 5½ millions. [40] In 1793 – 4, Peter Kier erected another with a similar arrangement in London. 17,325,000 lbs. duty was claimed for the steam part alone without the waterwheel. This was reduced to 11,550,000 lbs. for the power delivered at the wheel shaft. Kier claimed that this result, which was equal to the best Newcomen engines, was achieved by a valve which let air into the receiver, so separating the steam from the water. [41]

Smeaton tested many ordinary Newcomen engines. In 1769, he produced figures for the performance of fifteen in the Newcastle-upon-Tyne area, which he averaged at 5.59 millions, with the lowest at 3.22 millions and the highest at 7.44 millions. [42] Ten years later, he tested a further fifteen large engines in Cornwall and recorded better results, perhaps reflecting the higher price of coal there and so better maintenance and firing. He found the average was 7.1 millions, with the lowest 5.3 millions and the highest 7.79 millions. [43] These figures probably reflect the normal performance of the ordinary or common type of Newcomen engine.

However Smeaton showed that this performance could be improved through systematically testing a small engine he built at his home in Austhorpe, near Leeds. [44] Although this engine recorded only 2,919,017 lbs., [45] later engines constructed through the experience gained on it showed dramatic improvement. A worried Watt recognised the potential competition and wrote to Small that Smeaton 'told me of an experiment he had made on the New River engine which seems to say that common engines can be made to do more than I imagined'. [46]

Performance of a steam Engine of 40 HP.	On Newcomen's principle; Atmospheric	On Mr Watt's principle Without expansion	1.518 times expansion
Coals consumed per hour	(7.56bush. =) 635lbs	(4bush. =) 336lbs	(2.98bush. =) 250.5lbs
Coals consumed per hour by each horse-power	15.87lbs	8.4lbs	6.26lbs
Pounds of water raised 1foot high by each bushel (= 84lbs.) of coals	10.5 million	19.8 million	26.6 million
Water evaporated from the boiler, per min.	1.33cub.ft	.647cub.ft	.501cub. ft.
Cold water injected per min. temp. 52°	13.66cub.ft.	19.3cub.ft.	14.35cub.ft.
Height to which a cubic ft of water is raised, by each cubic foot of steam	9.31ft.	18.18ft.	24.43ft.

This was proved by Smeaton replacing a common engine at Long Benton colliery with one of his new designs. Both had cylinders 52 ins. diameter. He tested both engines and found that he had raised the duty from 4.6 millions to 9.1 millions on his new one. [47] For its day, this was a staggering improvement and he anticipated that his Chacewater engine would reach 10 millions. Curr's engine at Attercliffe Common colliery, Sheffield, achieved 9.38 millions in 1790. [48]

As anticipated, Watt's first engines showed even higher duty figures. Smeaton tested the Spon Lane 20 ins. diameter engine on the Birmingham Canal in 1778 where he recorded 18 millions. He followed this in 1779 with the one at Hull, 22 ins. diameter, which gave 18½ millions. [49] Farey commented that these were small engines which might do less than Boulton and Watt expected but the 58 ins. diameter Hawksbury or Bedworth engine tested in 1779 also gave a result of only a little below 19 millions. However this was working at a disadvantage through slow speed. In 1778, Boulton had laid down a rule for computing the consumption of fuel which resulted in a duty of 21,772,800 lbs. [50] Yet other figures show that the consumption of a bushel of good Newcastle coals (84 lbs.) with a non-expansive engine would yield a duty of 19.8 millions. One bushel per hour would supply an engine of 10 horsepower, or one horsepower required 8.4 lbs. of such coal per hour. [51] With such formulae, Watt could judge the performance of his engines and see when they were falling below standard. He must have felt well justified when he wrote to McGrigor in August 1779, 'As to the engines themselves, they answer fully as well as we expected and give satisfaction to our employers. By several comparisons that have been made lately bet^{n} the common engines and ours, ours exceed them from the proportion of 3½ to 1 to the pr[oportio]n. of 4 ¼ to 1'. [52]

With increasing numbers of Boulton and Watt engines, it became necessary to establish a standard performance of the common Newcomen engine against which the savings of fuel could be judged and their premiums reckoned. Accordingly the fuel consumption, the strokes made and the water raised by the two eastern engines on Poldice mine were recorded during August and September 1778. The 66 ins. engine gave a duty of 7,027,495 lbs. and the 60 ins. a duty of 7,070,392 lbs. [53] Therefore it was agreed that a duty of 7 millions should be taken as standard. The Boulton and Watt engine that replaced them gave 26½ millions in 1783 and 1784 but it may have been renewed by then. [54] During August 1791, its duty was 32.8 millions. [55] Watt was careful to safeguard his position and retain the Poldice common engines as standard by pointing out:

> If it should be alleged that there are some common engines that are better than those of Poldice, we answer, That if there are such, they are not proper engines to fix a common standard by, and therefore if we were asked to compare with such extraordinary engines, we should require a larger share of the savings. [56]

Watt was not going to accept Smeaton's improved engines as the standard.

The introduction of the expansive engine gave further increases in the duty figures. Farey wrote:

> The practice of Messrs. Boulton and Watt has always been to employ steam a little more elastic than the pressure of the atmosphere, and to cut off the supply at one-half, or two-thirds of the stroke, according to the circumstances under which the engine works. In this case, the decreasing pressure in a large engine, is not much greater than to counteract the acceleration; and the energy of the heavy working lever, pump-rods, and the rising column of water, produced a nearly uniform motion. [57]

Nowhere has a figure for the cut-off been recorded and it is possible that a higher figure may have been the result of a shorter cut-off but less smooth running. Farey gave figures of 26.6 millions for a 48 ins. cylinder worked expansively compared with 19.8 millions non-expansively and 10.5 millions for one of Smeaton's improved engines with a 52 ins. cylinder. [58] The figure of 26.6 millions seems to have been a little higher than the average for Boulton and Watt's Cornish engines. In 1803, Watt wrote to Baron de Prony in Paris, pointing out how the Wheal Maid double-acting engine in August 1791 returned 24,783,395 lbs., a saving of 70% compared with a common engine. He gave annual figures for the duty of this engine.

Duty of Wheal Maid Engine, 1788 - 1791

Year	Duty in lbs.
1788	25,302,671
1789	25,237,261
1790	25,069,939
1791	24,894,256

These show a steady decline but no reason was given. Watt continued to Prony:

> It is needless to give details of other Cornish engines, let it suffice that their average effects vary from 22 million to 27 or 28 million according to the encumbrances with which they were loaded & perhaps according to the good order in which they were kept. It does not appear from these accounts that the single Engines consume more coals in doing the same quantity of work than the double ones do, cateris paribus. [59]

The comment about the double- and single-acting engines performing equally may point to the reason for the lack of popularity of the double ones with their complicated pump work and why the later Cornish engines, in which the duty was raised so greatly, were the single type.

Boulton thought that a greater duty might be obtained if the rate of strokes was increased to 12 or 14 per minute. For example, Poldice New Engine, which in 1786 was returning from 24 to

28 millions, might give $31^{1}/_{4}$ millions. 'This effect will be produced by B. & W. Engine whether single or double with good coal & the engine in good order'. [60] Boulton showed figures for Ale & Cakes in March 1786, 30 millions and that July, 31,338,850 lbs. He later claimed that Ale & Cakes had been above the 30 millions figure 'for months together'. [61] Even better figures could be obtained under trial conditions. In 1792, Watt thought that Poldice double engine, which had been showing 26,730,960 lbs. duty, could be raised to 'more than 30 million on a days trial', taking account of stoppages, cleaning, etc., in normal running. [62] The engine at Wheal Butson was given a trial on 17 September 1792 'in the presence of Capt. Buzza' and 'produced an effect of 33,001,632 lbs.' [63] This is the highest reliable duty figure for a Boulton and Watt engine discovered so far. However a further trial was held on 10 November when only 28,466,782 lbs. duty was recorded. [64]

Other figures show that engines installed in waterworks, where the weight of the pumping mechanisms was less as well as the height to which the water had to be raised, gave similar performances. The figure for a test on the Shadwell engine in March 1779 surprised Watt and must be doubted. 'Our engine at Shadwell I have not seen but it is said to raise 600,000 cubic feet of water one foot high with a bushel of coals... I scarcely believe this more than the other but have better evidence as Mr. Boulton says he saw it'. [65] While seeing is believing, this would have given a duty of over 37 millions, a result not matched until possibly 1815 when a Woolf compound engine recorded 52 millions. [66] Yet this first Boulton and Watt engine at Shadwell did perform well for it was tested by Rennie in probably 1788, returning $26^{1}/_{2}$ millions when a new one there gave $30^{1}/_{2}$ millions. That at Chelsea in 1788 showed 33 millions, 'but all the circumstances could not be so well ascertained as they were in the Shadwell experiments'. [67]

Conditions varied greatly from engine to engine but Boulton and Watt kept records to calculate premiums due to them which could be checked against the official returns for the drawback of the duty on coal so their figures ought to be reliable. However other figures published by Davies Gilbert in a paper to the Royal Society in 1830 show lower results than Boulton and Watt claimed. In 1793, the average performed by seventeen Watt engines in Cornwall was 19,569,000 lbs., which exceeded 'the standard experiment on the two [Poldice] atmospheric engines in the proportion of 2.78 to 1'. [68] The duty of twenty-three engines was recorded in 1798 but it was not stated whether they were all made by Boulton and Watt. Against the two highest, one at over 29 millions and the other 28 millions, it was noted, 'It was believed at the time that some inaccuracy must have occurred in the communications respecting these two'. The next highest was 27,503,000 lbs. There was only one other over 20 millions at $24^{1}/_{2}$ millions. Two double engines both returned over 18 millions. But a third small one with a $14^{3}/_{4}$ ins. cylinder showed only 6 millions. This might have been a rotative winding engine in which case this was quite a respectable figure. The next lowest was a single 20 ins. cylinder at 10 million. The general average was only 17,671,000 lbs., a result which shows a decline in performance towards the end of the century. To compare the later development of the Cornish engine, Gilbert gave the average duty in 1820 as $41^{1}/_{2}$ millions and the best engine in December 1829 as 75,628,000 lbs. [69] Watt had declared in 1798 on the tests at Wheal Harland which reached $27^{1}/_{2}$ millions that further improvement could not be expected. [70] Even if he did not understand the thermodynamic importance of higher boiler pressures and shorter cut-offs, he had dramatically improved the steam engine and laid the foundations for future enhancement.

The Hornblower 'Compound' Engine

In the early 1780s, Watt was faced with two competitive designs of steam engines emanating from the West Country. One was the rotative engine of Matthew Wasborough which will be examined later and the other was Jonathan Hornblower junior's double cylinder 'compound' engine. In 1788, Hornblower junior wrote in Howard's *New Royal Cyclopaedia:*

> In the year 1776 I determined to make a small steam-engine for my own amusement, but wished to perform something new. Accordingly, the first idea of consequence, that occurred to me, was, a possibility of causing the steam to perform more than one operation, instead of destroying it, in the manner peculiar to all other steam-engines.
>
> I made a model, which succeeded so well, that I recommended it to my father as a matter of great consequence; whose judgement, from long experience as an engineer, was sufficient at that time of life, to bias me, either to continue my pursuit, or relinquish it. He was of opinion, that as I proposed a lid to the steam vessels, and a rod of iron to move through a collar of tow in the lid, there would be more friction than could be dispensed with; and also, that as the pistons would require frequent repair, the cylinder lids must be unscrewed as often as that was done. I therefore had no other thought at that time about it, than to amuse some of my friends by a working model. [71]

He reiterated the same claims to his uncle, Josiah Hornblower, in 1789. [72] There the development of his engine remained until after his brother Jethro had seen the Wheal Bussy engine going, about which he exclaimed, 'Pshaw! its a bauble I wouldn't give 2 ½d. for it'. [73] The news of Watt's engine revived interest in his own:

> In 1778 Mr. Watt came into this country, with his much improved steam-engine; and seeing him make use of one cylinder lid, and piston rod, in the same manner as I had proposed to apply them, the matter was again revived, and a friend directing me where to find a representation, by a copper plate of thirty years standing, of a steam-engine with a lid to the cylinder, and a rod moving thro' it connected to the piston, I therefore completed another model, on a larger scale, in 1781, obtained a patent, and have since proved it to be capable of producing nearly double the effect of any other steam-engine, with the same quantity of fuel. [74]

Hornblower junior submitted his patent specification on 13 July 1781 and it was passed on the twenty-third. [75]

News of the engine spread quickly. Watt wrote an indignant letter to Hamilton on 23 July, already threatening legal action:

> To crown all we are pestered by the envy of every body numbers of schemes of new engines arise which are boasted of for a while as superior to ours most of them have hitherto hurt nobody but their owners but now we have a formidable opposition by a set of people in this county (some of whom are under the highest obligations to us) who pretend to have invented a new Engine much superior to ours & have taken a patent for it which makes much noise at present and indeed gives me some pain – I know not what it is but conceive it to be a base attempt to evade our patent; which I expect the law will not countenance them in, at least it shall at any rate cost them dear. I have some good hopes that nature all puissant will prove the Lord Chancellor & condemn them to costs & infamy for attempting to infringe her laws – This I am sure of that the inventors are insolent ignorant puppies; but fools may have fortune. [76]

Watt could never believe that anyone else could invent a steam engine which would out-perform his perfect one. He could not recognise that others would seek to follow his example and also try to make their fortunes.

In this case he felt very hurt because Boulton and Watt had given assistance to the younger Hornblowers by employing them as engine erectors. Their father had died by now. Boulton was equally incensed in his reply to Watt:

> When I consider your generous Conduct towards that family in being Engineer for them at Poldice & giving them the reward of that office as well as the battles we have fought for them on every occasion, I say when I reflect upon these things I am fired with resentment & therefore I would expell them root & branch from ye Engine trade in Cornwall. [77]

Another letter followed a couple of days later, with more rather stronger language about the Hornblowers:

> If we do suffer our selves to be pissed upon by that family we deserve to be sh-t upon. For Gods sake dont loose a moments peace upon that head but proceed, get the Engines erected & the Money in our pockets & not doubt whether we have a right solely to press down the piston by Steam or condense in a Separate Vessel.
> doubt the Sun is fire
> doubt all ye powers of sight
> doubt truth to be a Lyer
> But never doubt our right
> I am sure you want assistance shall I send Harrison or will you send to Scotland for Wright to come directly. [78]

Boulton recognised that the best way to defeat the Hornblowers was by out-performing them not only with their engines but also in the service offered.

Watt wrote to Boulton that he understood that Hornblower's engine was on some new principle, and the only novelty which he could think of was a form of hot-air engine. He therefore asked Boulton to make all the inquiries he could as to the bulks and prices per 1,000 feet of all possible kinds of air in their most expanded state. He continued, 'I am much vexed by this affair. Jabez does not want abilities: the rest are fools. If they have really found a prize, it will ruin us... Bankruptcy might ensue to both. But I don't fear getting my bread independent of engines, though much easier with them'. [79] Watt was always afraid that someone else might follow his example and find a new principle for driving engines.

It was not until November 1781 that Watt learnt from Robert Cameron that 'the Horner's is certainly the double cylinder'. [80] It could be that Cameron had seen an engine being erected to drain a colliery at Radstock. Watt kept his ear close to the ground for he discovered in February 1782 that Hornblower junior had formed a partnership with John Winwood, the Bristol ironfounder, whom Watt described as 'a sneaking fellow that supports the Hornblowers'. [81] Watt had more respect for the design of the Hornblower's main beam for the engine which he recognised 'will make a very strong beam and I am sorry to see they have so much genius among them'. [82] An early drawing shows a conventional beam made from a single baulk of timber but later ones show a space-frame design in which the arch-heads were strengthened top and bottom with angled braces and the tops were linked with an additional horizontal beam. This construction utilised thinner sections of timber which were more readily available.

Watt was afraid that, if the Hornblower engine worked tolerably, the Cornish mine owners would seek to drive hard bargains with Boulton and Watt. Delicate negotiations with Wheal Virgin were in progress at the time. [83] Watt instructed Richard Cartwright to return to

Birmingham via Radstock to try and find out about the rival engine. [84] Watt expected that the Radstock engine would be set to work by the middle of March. [85] It seems that the engine was running by 21 March, when his worst fears about the reactions from Cornishmen were being realised for he told Boulton:

> Almost the whole country is against us and look upon us as Oppressors & Tyrants from whose power they believe the Horned Imps of Satan are to relieve them – I have therefore resolved to make no further abatement to the United Mines, and none at all to the Virginians but will agree to lett them go by the Tables if they will. [86]

Watt was not to know that this would turn out to be a wise policy through the failure of the Horn-blower engine. He also pointed out to Henderson the 'unreasonable-ness of the mines. The hopes all true patriots have of the success of the Hornblowers & of their hearty prayers for the destruction of B & W – Confound their Politicks'. [87]

Watt was determined to make a strong stand against any reduction in premiums even if they could not defeat the Hornblowers. He realised that any reduction for one would lead to 'needless discussions' and 'perpetual demands' for abatements because 'The minds of most of them seem to be more intent upon ruining us than upon the good of the county'. [88] It was pointed out to Watt that the future prospects for Boulton and Watt in Cornwall 'depended on our Engines proving superior to the Hornblowers as otherwise our attempting to stop theirs by law would be resisted by the whole county'. Watt replied that he would defend his rights 'in a legal manner' against those 'base enough to espouse the cause of such scoundrels'. [89] He was convinced that 'The hand of all men is against us thanks the Hornblowers who I hope will be duely rewarded for it by their patron Belzabub in his good time'. [90]

However by April 1782, Watt had received a description of the engine and, although not fully understanding how it worked, thought:

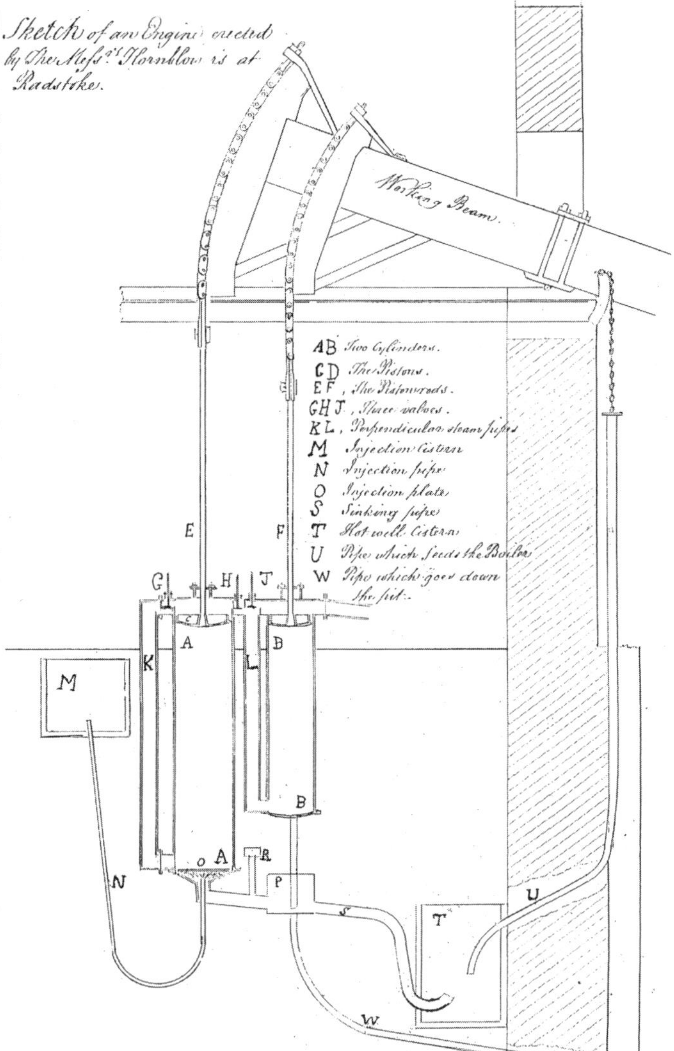

The Radstock engine, showing the smaller high pressure cylinder on the right. The large second cylinder has a jet of water to condense the steam at the bottom. Condensate flows down a pipe into the mine shaft.

> ... the contrivance ingenious but not faultless and I think all its ingenuity has been borrowed from us and also that in strict justice it is within our specification in all the principal points. However if they are protected in it, we may give up Engine making as the difference between what may be done in this way and the best is not worth minding... I think this scheme will not work. [91]

He was proved right by the performance of the Radstock engine during the summer so it was with relief that he wrote to Hamilton in September:

> The Hornblowers who have caused us so much uneasiness by the lies they have told in Cornwall have never been able till lately to get their engine to work at all and now it performs as badly as one could wish it being very much worse than a common engine, notwithstanding they have uniformly asserted in Cornwall that it performed beyond their most sanguine hopes, which by the way must have been very humble and that it was the very best engine ever made - We had a person there last week who informs us that when they have got a very strong steam it will make 21 strokes in three minutes but then it comes to rest and must stand 5 minutes before it gets strength enough to take another stroke and all the while they must fire away as hard as ever they can otherwise it will not work at all. To add to their triumph a Cornish Gentleman whose knowledge candour and abilities give him influence there has seen it in this good order and has borne witness in Cornwall of its performance much to the Horners comfort and humour as you may suppose. Mr. Boulton is now in Cornwall & will not fail to take proper advantage of our enemies defeat. [92]

But Watt's feelings of triumph were to change quickly once he heard of the hostility against Boulton and Watt which was being generated in Cornwall:

> Mr. B. writes me from Cornwall that the Horners notwithstanding the real bad performance of their engine (which I have no reason to doubt of) actually made proposals to several of the mines we are interested in to alter our engines to their construction at a small expense and to serve them afterwards and allow the mines the use of their engines for a small Salary - And the Cornish people believe them and some of the principal are determined to support them against us and in short have set the whole County against us in such a manner as to render the situation of Mr B. most disagreeable - you may conceive that we have now no choice but to enter into a lawsuit against them - and the event of this God only knows we have right law & justice on our side, we are highly injured, yet who knows what the chicanery of the law may suggest as the rage of an English jury against monopolies? Yet if our right cannot be supported by law or custom it is time we knew it before we have wasted more of our lives improving machines for the benefit of scoundrels. [93]

The Hornblowers were trying to undercut the premiums of Boulton and Watt by charging less, a move which delighted the Cornish mine owners. In view of the later patent lawsuits, it should be noted how Watt recognised at this time that going to law had its risks. Yet he seriously thought of applying to the Chancellor for an injunction to stop the Hornblowers building engines.

The colliery owners at Radstock had been warned that they were infringing Watt's patent. They said that they were unaware of this. Their engine never went regularly, sometimes five or six strokes a minute and once or twice might have gone ten for a short time. It burnt one hundred bushels of coal in twenty-four hours to perform the same task which Watt reckoned one of his engines would do with ten, 'if the coals were tolerably good'. Watt put an advertisement in the Bristol paper warning that people would be prosecuted if the patent were infringed. Although the Hornblowers continued to brag about their engine, Watt thought 'their day of retribution is near. by calculation their engine is 8 times worse than a common fire engine'. [94] Setting the first of the engines to work at Wheal Virgin began to convince people in Cornwall of the superiority of the Watt engine. 'The clamour in Cornwall is somewhat abated, and the people there begin to believe that the Horners engine cannot serve their turn in defeating us. But these devil trumpeters still go on blowing their own tune and blowing lies'. [95]

Watt prepared a case for counsel. The advice was not to go to law because judges and juries were biased against monopolies. [96] In early 1783, the Hornblowers were keeping quiet and business in Cornwall was going on reasonably well. Watt told Joseph Fry who had supplied information about the Radstock engine how, 'Many of those who favoured them there have thought proper to change sides and there is scarcely anybody who will own that they have patronized them. We ought not however to go to sleep as if the ?hoard? of rogues has failed; but should keep a watchful eye upon them'. [97] The Hornblowers continued to be a potential competitor to Boulton and Watt. In 1783:

> The Hornblowers have made another attempt to get a small mine in Cornwall to erect one of their engines but their proposal was not agreed to Their engine is still going at Radstock but very indifferently. I have no doubts but we shall have some further plague with them, if they can get any fools to listen to them. [98]

Another was set up at Penryn in 1784 and a further one at Timsbury close to Radstock in the following year. They also succeed in making the Radstock engine work tolerably. [99] After 1790 they erected several quite successful ones in Cornwall. [100]

The Hornblower 'compound' engine was a double threat to Boulton and Watt. We have seen how Hornblower junior attempted to capture the lucrative Cornish market, which he did to a considerable extent only in the 1790s. This would have become a reality earlier had his engine fulfilled the claims he made for the greater performance of his double cylinder engine over Watt's and had the copper mining industry in Cornwall been more prosperous. Improved performance was the second threat. Here we must contrast Watt's development of his engine with that of Hornblower junior. Watt had carried out many experiments from which he learnt about the nature of steam, its relationship between its temperature, pressure and volume. From this he could calculate the proportions of his engines, based on scientific and mathematical principles. He had started with small models to prove his theories and developed these gradually into actual working engines. He knew full well the difficulties in translating a model into an actual machine and had not wanted to make the jump from the small engine at Soho to the large one for Bloomfield. He had had experience of designing and constructing both Newcomen and his own engines. In contrast, Hornblower junior moved straight from a couple of models to the Radstock machine. He had, of course, been trained by his father on the Cornish Newcomen engines and assisted with the construction of Boulton and Watt engines there but it is doubtful whether he had ever designed any engine before that at Radstock. He certainly did not have the mass of records of engine performance which Watt had accumulated on which to base any design so it is hardly surprising that his first attempts were unsuccessful. Yet the concept behind his engine showed great ingenuity in an attempt to circumvent Watt's patent.

The Radstock engine was a reciprocating beam engine with two vertical cylinders in line, 19 ins. and 24 ins. diameter with strokes of 6 ft. and 8 ft. respectively. The cylinders were surrounded by a wooden casing 'stuffed with hair or some other body which transmitts heat slowly'. [101] The pistons in both cylinders moved in unison. The tops of the cylinders had covers through which the piston rods passed. Steam from the boiler entered the top of the smaller cylinder and was allowed to fill it as the piston descended. There was no inlet valve so that the steam remained at full boiler pressure. At the end of the stroke, the steam passed through a second valve down a pipe to the lower part of the cylinder to fill that at full pressure as the piston rose. At the start of the next stroke, that steam was transferred up the pipe, through a third valve to the top of the large cylinder 'permitting it to expand itself' as that piston descended. Finally the steam was taken through a third valve and down a second pipe to the bottom of the large cylinder. Here it was condensed originally 'by causing it to pass in contact with metal-line surfaces while water is applied to the opposite side'. This metal surface was perforated and was soon abandoned after it had been broken by the piston descending too far through over-stroking. A horizontal jet of water across the bottom of the cylinder was introduced instead. A perforated plate might separate the area of the jet from the higher part of the cylinder, so avoiding cooling the main part of the large cylinder. At first the condensate drained away in a long tube, 'constructed... on the principles of the barometer, the upper end having open communication with the steam vessels, and the lower end immersed into a vessel of water'. This was similar to the eduction pipe on a Newcomen engine which had a flap valve on its lower end immersed in water. Air was discharged 'into a separate vessel, whence it is protruded by the admission of steam'. [102] The patent contains no drawings but a drawing is preserved in the Boulton and Watt Collection which does not show any mechanical air pump to remove the condensate.

In April 1782, Watt was able to send Boulton a sketch and assessment of the Radstock engine which he concluded was 'much more formidable than we have thought it'. The engine started with a vacuum below the large cylinder, steam from below the small one pressing on the top of that piston while the small one was in equilibrium. As the pistons descended and the steam expanded from the bottom of the small into the top of the large with the top of the small cylinder being filled with fresh steam, more power was developed in the small and less in the large. The engine was being considered as acting through the pressure of the steam. Watt made two attempts to calculate the power and decided that 'Its maximum seems to be when Cyl A is 1/2 of B, without they applied an equalizing Beam'. He thought they would be 'perplexed with valves which they will find troublesome to construct and keep in order'. On the later engine for Tin Croft in 1791, the valves were flat discs with two holes. They were turned round to align the holes with others on the casing, allowing the steam to pass through. The flat surfaces would have been difficult to keep steam-tight. The steam passages were small and convoluted so would have partially obstructed the free flow of the steam.

From his scientific knowledge, Watt knew that, through using a cold water jet in the bottom of the cylinder, it would be impossible to create a perfect vacuum because 'part of the water formed by the last [condensation] will indeed come alive again by the partial vacuum but then will always remain a defect'. He thought that water would always remain on the top of the large piston which 'will be in a fluctuating state between evaporation and condensation'. He had not discovered how they purged the air out of the system but thought it might be blown out of a valve by using steam from the small cylinder controlled by a cock, not a very satisfactory solution. [103] He assessed that the Hornblower engine failed to meet the high standards he set for his own and doubted whether it would become a serious rival.

But Hornblower junior continued to improve his engine for the account, in Howard's *Royal Cyclopaedia* 1788, shows that he had developed it along lines much closer to the Watt engine. An inlet valve was added to regulate entry of steam to the small cylinder. The gland round the piston rod of the larger piston was fed with steam from a small pipe to exclude air. The jet of condensing water was forced in by atmospheric pressure as the supply tank was on the same level, 'owing to a constant partial vacuum within'. No claim was ever made for a perfect vacuum. The condensate and air passed along a pipe into a receiver through a valve. The receiver sat in a cistern of water. At the top of it was another valve to allow in water, possibly to expel air. The water was drawn off from the receiver by what Hornblower junior carefully described as a 'water pump' worked by the beam. Rennie saw a form of bellows as the pump. The description in Howard of how the engine worked was used subsequently at least by Farey and Galloway. [104]

The advantages of a compound engine should be that the steam is expanded in two distinct stages, lessening the pressure and temperature drop in each cylinder and reducing the extent to which each needs reheating to accept the next charge of steam. The reduced pressure in each cylinder also lessened steam passing around the piston as well as giving a more equal average pressure than expanding the steam in one cylinder only. Ideally steam should enter the high pressure cylinder, be cut off at the appropriate point and expand with temperature falling. That expanded steam should be taken out of the cylinder into a receiver with the exhaust valve open as long as possible. From the receiver, the steam would pass through the inlet valve in the low pressure cylinder to be cut off again and allowed to expand once more to reduce the temperature and pressure as close as possible to that of the condenser. In this way, there is no temperature/pressure overlap between the cylinders, so using the heat most effectively.

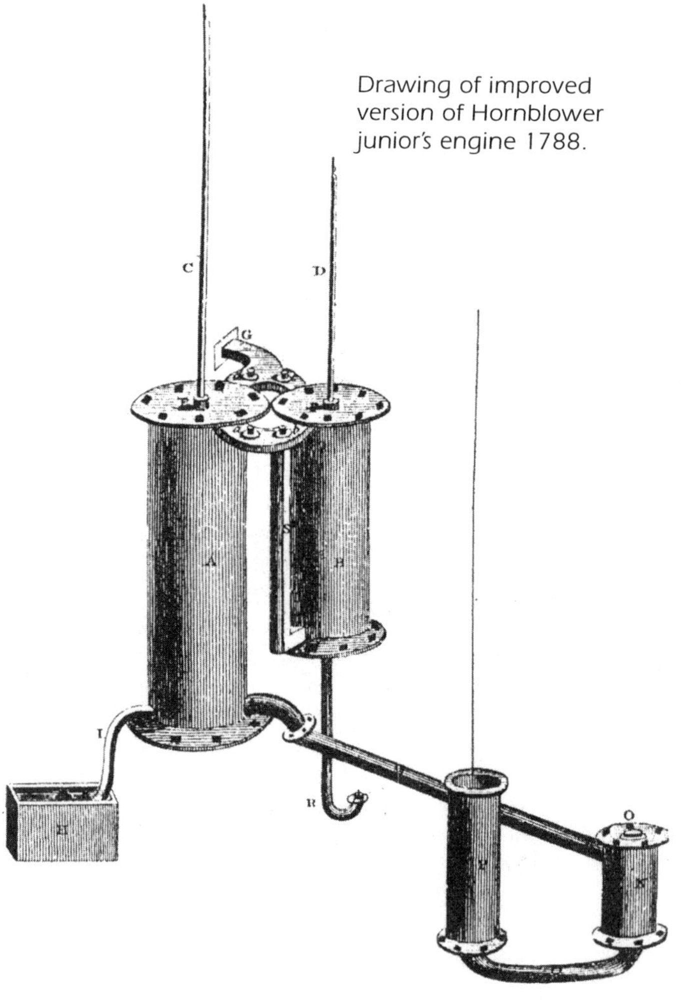

Drawing of improved version of Hornblower junior's engine 1788.

But in the Hornblower engine there was no expansion in the small cylinder so that steam remained at or near the admission temperature. That steam passed first below the small piston and then into the large cylinder so the incoming steam to the large cylinder was virtually at the same temperature as that in the boiler. As the steam expanded from the small into the large cylinder, the temperature would fall in both, necessitating both having to be reheated. Taken as a pressure engine, Hornblower's had an advantage in practice 'because the combined effect on the two pistons, approaches more nearly to a uniform action, than could be done by the same extent of expansive action of the steam, when operating in only one cylinder on Mr. Watt's system'. [105] So it had one but only one of the advantages of a compound engine. Steam pressures at the time were not high enough to derive benefit from compounding and the extra friction and complication of two cylinders and their valves made such an engine even less attractive.

It is possible that the Hornblower engine could have had another advantage over the Watt engine through the lower final pressure and temperature of the steam. While Watt created a near perfect vacuum in his condenser, the pressure in the cylinder at the end of the stroke seems to have remained high because he stated that this steam rushed into the condenser. The Hornblower engine might have achieved a lower terminal pressure and temperature in the cylinder which could have given a thermodynamically better engine than Watt's in spite of creating only a partial vacuum. In actual practice, it seems that certainly at first the Radstock engine failed to achieve as high a pressure in its cylinders as even in the Newcomen types for Fry told Watt that 'The owners are convinced she does not work four pounds on the inch with very very much coal'. The barometer pressure gauge stood at 17 or 18 ins. [106]

Watt approached the colliery owners and pointed out that, in his opinion, they were infringing his patent. He called to see the Mayor of Bath but had to wait until three o'clock because the Mayor had gone out hunting. [107] An advertisement was prepared for the Bristol newspapers pointing out six features of his 1769 patent: 1, cylinder with closed top; 2, piston pressed down by steam; 3, steam case, or non-conducting casing to the cylinder; 4, separate condenser; 5, air pump; 6, piston kept tight by oil or grease. [108] Watt submitted a long report for consideration by counsel, outlining the history of his invention. He finished by pointing out that the Hornblowers were using steam as the acting power on the piston instead of the atmosphere, they were using grease to keep the piston steam tight and were not condensing 'the steam in ye steam vessel itself but in some vessel under it the nature of which B & W have not yet received proper information of'. They also encased their cylinders with insulating material. Watt had clear evidence concerning these infringements but asked for clarification about the layout of the condensing apparatus:

> ... whether condensing the steam in a distinct vessel but always communicating with one part of the steam vessel; or, condensing it in a part of the steam vessel itself which is always kept exhausted and cold and which part of the steam vessel the steam never enters except to be condensed is not an infringement of the second article of the specification...
>
> Whether condensing the steam in a part of the lower end of the steam vessel itself but separated from the other parts by a diaphragm ought to be considered as condensing in a distinct vessel and how far the law in such cases will protect evasions. [109]

Boulton and Watt never took Hornblower junior to law probably because they realised that they could not prove he was condensing the steam in a separate condenser. While he was infringing other parts of their patent, it would be difficult to prove that he was infringing the crucial section concerning the separate condenser. But Watt had not heard the last of Hornblower junior who became a serious rival.

NOTES

1 JWP, 4/43.7, G. Hamilton to J. Watt, 15 Aug. 1781.
2 Robinson & McKie, (1970), p. 99, Letter 73, J. Watt to J. Black, 25 Oct. 1780.
3 Boulton 348.89, J. Watt to M. Boulton, 27 June 1778 and Boulton 348.95, J. Watt to M. Boulton, 16 July 1778.
4 JWP, 3/8, J. Watt to G. Hamilton, 23 July 1781.
5 JWP, Letter Book 1, J. Watt to W. Craig, 2 June 1782.
6 Barton, (1961), p. 94.
7 JWP, Letter Book 1, J. Watt to G. Hamilton, May or June 1783.
8 Wilson Papers DDX 318/1, J. Watt to T. Wilson, 27 Jan. 1784.
9 Hamilton Jenkin, (1927), p. 157.
10 Harris, (1964), pp. 18 – 35 for an account of these disputes.
11 Hamilton Jenkin, (1927), p. 157, quoted from *Transactions of the Royal Geological Society of Cornwall*, iii, p. 284.
12 Harris, (1964), p. 39, quoted from T. Pennant, *A History of the Parishes of Whiteford and Holywell*, (1798), pp. 203 – 4.
13 See Harris, (1964), pp. 36 – 40.
14 Boulton 348.60, J. Watt to M. Boulton, 8 July 1776 and Boulton 348.62, M. Boulton to J. Watt, July 1776.
15 Roll, (1930), pp. 79 – 80, M. Boulton to J. Watt, 2 Oct. 1780.
16 Boulton 348.161, J. Watt to M. Boulton, 28 Oct. 1779.
17 Boulton 348.162, J. Watt to M. Boulton, 30 Oct. 1779.
18 Barton, (1961), p. 31 and Smiles, (1878), p. 212.
19 Smiles, (1878), p. 213.
20 JWP, 3/18, J. Watt to M. Boulton, 5 April 1782.
21 See Muirhead, (1854), Vol. II, p. 34, Letter 102, J. Watt to W. Small, 24 Nov. 1772; Boulton 348.25.
22 JWP, 3/8, J. Watt to M. Boulton, 31 Oct. 1780.
23 Roll, (1930), p. 88, M. Boulton to J. Watt, 19 April 1780; B. & W. Col.
24 JWP, 3/8, J. Watt to M. Boulton, 21 April 1780.
25 Roll, (1930), p. 89.
26 P. Watts-Russell, 'A "copper revolution" cut short: the Cornish Metal Co. (1785 – 1792)', typescript, 2004.
27 JWP, 3/8, J. Watt to W. Craig, 10 Sept. 1781.
28 JWP, Letter Book 1, J. Watt to G. Hamilton, 30 May 1784 and J. Watt to W. Matthews, 5 June 1784.
29 Ibid, to Adventurers of United Mines, March 1782.
30 Ibid, J. Watt to Capt. T. Mitchell jun., 27 March 1782.
31 Ibid, J. Watt to M. Boulton, 21 March 1782.
32 Lord, (1923), pp. 155 – 57.
33 From Wilson Papers DDX 318/11, Mine Accounts. For slightly different figures, see B. & W. Col., Box 9 IV (1), quoted in Tann, (1995 – 96), p. 41.
34 JWP, Letter Book 1, J. Watt to C. McDowal, 3 Jan. 1783.
35 JWP, 4/1, 'Draft of a Case for Counsel', 1783.
36 Muirhead, (1854), Vol. I, p. 3, Letter 1, J. Watt to J. Lind, 29 April 1765.
37 Ibid, Vol. III, p. 1, Patent 913, 5 Jan. 1769.
38 *Journal of the House of Commons*, Vol. XXXV, (1775), p. 168.
39 JWP, 4/53, 'Copy Minutes on Recommitment of Mr. Watt's Engine Bill', see p. 45.
40 Farey, (1827), Vol. I, p. 125 and Rees, (1819), article WATER, c. 1818.
41 Hills, (1989), p. 48.
42 Farey, (1827), Vol. I, p. 234.
43 Ibid, Vol. II, p. 92.
44 Skempton, (1981), p. 182.
45 Farey, (1827), Vol. I, p. 116.
46 Muirhead, (1854), Vol. II, p. 9, Letter 30, J. Watt to W. Small, 30 Sept. 1770.
47 Rees, (1819), reprint 1972, Vol. 5, p. 91, gives 5,044,158 lbs. for the old and 9,636,660 lbs for the new. This may have been the London bushel of 88 lbs. See p. 94
48 Farey, (1827), Vol. I, p. 206.

49 Ibid, p. 338.
50 Ibid, pp. 338 – 39.
51 Ibid, p. 367.
52 JWP, W5, J. Watt to J. McGrigor, 2 Aug. 1779.
53 Wilson Papers DDX 318/11, 'Proposals'.
54 Ibid, DDX 318/4, M. Boulton to T. Wilson, 23 Aug. 1790.
55 Dickinson & Jenkins, (1927), p. 329.
56 JWP, 4/69, 'Proposals to the Adventurers'.
57 Farey, (1827), Vol. I, p. 352.
58 Ibid, p. 368.
59 JWP, 3/18, J. Watt to M. Prony, 30 April 1803.
60 Wilson Papers DDX 318/4, M. Boulton to T. Wilson, 23 Aug. 1790.
61 Ibid, M. Boulton to T. Wilson, 3 Dec. 1791.
62 Ibid, J. Watt to T. Wilson, 24 July 1792.
63 Ibid, DDX 318/5, T. Wilson to Mr. Printer, 25 Sept. 1792.
64 See Dickinson & Jenkins, (1927), pp. 329 – 30 for more figures.
65 JWP, W5, J. Watt to J. Black, 3 March 1779.
66 Hills, (1989), p. 106.
67 JWP, 3/18, J. Watt to M. Prony, 30 April 1803. See also JWP, 6/30.14 from T. Barnes, 10 March 1797, for figures of a Watt engine achieving 28 million pounds compared with that of a common engine at only 11 millions.
68 Gilbert, (1830), p. 125.
69 Ibid, p. 126 - 7.
70 Farey, (1827), Vol. II, pp. 240 – 1.
71 Howard, (1788), p. 1192.
72 Torrens, (1982 – 83), p. 191, J. Hornblower to Josiah Hornblower, 12 March 1789.
73 Tann, (1981), p. 81, Doc. 26, M. Boulton to J. Watt, 28 July 1781.
74 Howard, (1788), p. 1192.
75 Patent 1298, 13 July 1781 and Tann, (1981), p. 57, Doc. 15, M. Boulton to J. Watt, 30 July 1781.
76 JWP, 3/8, J. Watt to G. Hamilton, 23 July 1781.
77 Tann, (1981), p. 81, Doc. 26, M. Boulton to J. Watt, 28 July 1781.
78 Ibid, p. 57, Doc. 15, M. Boulton to J. Watt, 30 July 1781.
79 Smiles, (1878), p. 236.
80 Muirhead, (1854), Vol. II, p. 134, Letter 200, J. Watt to M. Boulton, 1 Nov. 1781.
81 JWP, 3/18, J. Watt to T. Handley, 2 Feb. 1782.
82 Ibid, J. Watt to M. Boulton, 13 Feb. 1782.
83 Ibid, J. Watt to M. Boulton, 20 Feb. 1782.
84 Ibid, J. Watt to M. Boulton, 2 March 1782.
85 Ibid, J. Watt to M. Boulton, 9 March 1782.
86 Ibid, J. Watt to M. Boulton, 21 March 1782.
87 Ibid, J. Watt to L. Henderson, 1 April 1782.
88 Ibid, J. Watt to M. Boulton, 1 April 1782.
89 Ibid, J. Watt to M. Boulton, 28 March 1782.
90 Ibid, J. Watt to M. Boulton, 5 April 1782.
91 Ibid, J. Watt to M. Boulton, 7 April 1782.
92 JWP, Letter Book 1, J. Watt to G. Hamilton, 22 Sept. 1782.
93 Ibid, J. Watt to G. Hamilton, 29 Sept. 1782.
94 Ibid, J. Watt to G. Hamilton, 24 Sept. 1782.
95 Ibid, J. Watt to G. Hamilton, 4 Nov. 1782.
96 Ibid, J. Watt to G. Hamilton, 18 Feb. 1783.
97 Ibid, J. Watt to J. Fry, 1 March 1783.
98 Ibid, J. Watt to G. Hamilton, 28 Sept. 1783.
99 Ibid, J. Watt to G. Hamilton, 25 Aug. 1784.
100 See Rhys Jenkins, 'Jonathan Hornblower and the Compound Engine', T.N.S., Vol. 11, 1930 – 31, pp. 128 – 54 and J. Tann, 'Mr. Hornblower and his Crew: Watt Engine Pirates at the End of the 18th. Century', T.N.S., Vol. 51, 1979 – 80, pp. 95 – 109.
101 JWP, 'Draft of a Case for Counsel', 1783.
102 Patent 1298, 13 July 1781.
103 JWP, 3/18, J. Watt to M. Boulton, 5 & 7 April 1782.
104 Farey, (1827), Vol. I, pp. 385 – 86 and Galloway, (1830), pp. 97 – 9.
105 Farey, (1827), Vol. I, p. 390.
106 JWP, JW/23, J. Fry to J. Watt, 2 Oct. 1782 and JWP, C3/6, 'Diary', 16 Nov. 1782.
107 JWP, C3/6, 'Diary', 17 Nov. 1782.
108 Dickinson & Jenkins, (1927), p. 305.
109 JWP, 4/1, 'Draft of a Case for Counsel', 1783.

9
Natural Philosopher

Various Interests

> I know that disease too well having been the victim of it almost all my life. The irresolution, the procrastination, the diffidence, the pusillanimity, and dread of mankind which it inspires would have totally ruined me, if I had not been saved by an order & enthusiasm in the pursuits of science, that at times forced their way through the languor to which I was subjected & commanded exertion. Lucky circumstances seconded this favourable disposition otherwise I should have sunk under my afflictions and indeed without an energy of mind of some kind or other which nothing could continually depress & a strong sense of duty, nothing could have enabled me to persevere in a course of life for which I am naturally not qualified and which is and always has been to the last degree disagreeable to me... I have not an atom of philosophical news – our business encreases in making Mill engines but my cares encrease in the same ratio. Our profits from these machines are not great in proportion to the trouble but there is some pleasure in success. [1]

Thus Watt wrote to his friend De Luc in October 1786 after years of intense activity not only with his steam engine business but with scientific investigations as well. It is suspected that he would rather have been involved with natural philosophy which he found more interesting and stimulating than engineering. He had pioneered the development of steam engines, although assisted by others, but in his scientific explorations he was more often than not taking up discoveries of others. Entries in his Common Place Book show the broad range of his interests, only a few of which can be mentioned. There are few connections with steam engines but some of the experiments were carried out in anticipation of evolving further profitable industries. Frequently the links came through stimulation from other members of the Lunar Society. Darwin provided the catalyst for Watt's copying machine which he was developing for over a year from the middle of 1779. Keir helped with that as well as further investigations into making alkali during the summer of 1780. The profit motive may be seen in Watt's designing a steam-heated machine for McGrigor in early 1781 to dry linen cloth more quickly. There was the cement for the steam joints in engines in April 1782; his metric system for weights in November 1783, which originated with John Whitehurst, another Lunar Society member. Watt's interest in ballooning about the same time was another venture typical of people interested in natural philosophy. This was true also of his work on the composition of water which was strongly influenced by Priestley. Of these, the copying machine, the drying machine and the jointing cement proved to be important but their inventor was forgotten.

The Delftfield Pottery

Throughout his time at Birmingham, Watt remained a partner with the Delftfield pottery in Glasgow although he did not visit Scotland from 1776 until probably Christmas 1787.[2] (See Vol. I, pp. 162 – 76) Hamilton became the senior trading partner and took control of the concern from 1782 until his death in 1808. He frequently sought Watt's assistance. Earlier, Watt's chemical and technical advice had helped to put the products on a firm basis, so contributing to its prosperity. While in Birmingham, he harboured thoughts of becoming involved again in this industry but, beyond carrying out a few further experiments on clays and glazes, he never actively participated in running any business of this type there.

However he was able to give some assistance to the Delftfield pottery. His motto was 'Observare' and he closely observed the countryside through which he travelled to see what minerals or other substances might be exploited. While erecting engines in Cornwall, he investigated sources of ball clay, kaolin and petuntse, all crucial for cream coloured earthenwares and porcelain. Samples were sent not only to Glasgow but also to Josiah Wedgwood at Etruria, another member of the Lunar Society. Watt described to Hamilton details about the method of grinding flints under water which had been developed in Staffordshire. While in Glasgow, he had improved the glazes used at Delftfield but, in spite of much prompting from Hamilton and Muirhead, he was unable to find his recipes for a white glaze. He did not carry out further experiments to rediscover the secret. However he did test cobalt found at a Cornish mine but the mine did not pay. He also sent Hamilton a description of transfer-printing on pottery. To what extent this type of advice proved to be of any value cannot be determined but it shows the breadth of Watt's knowledge.

Through his links with Wedgwood, Watt was able to send Hamilton some lathes for turning pottery. It was through these wider contacts that Watt gave more assistance. Boulton and Fothergill had trading links with Russia through which Watt secured outlets for the Delftfield products. He was able to secure another agency in Paris during his visit there with Boulton over New Year 1786-7. He saw some of the latest fashions on visits to London, samples of which he purchased and sent to Glasgow. His guests dined off plates made by Delftfield. He commented on the design of some of the Delftfield ware which he received.

The Delftfield concern generally paid a dividend until after 1800 which may have been a great help to Watt's hard pressed finances during his early years at Birmingham. But, after Hamilton's death, the business declined. The other partners had little knowledge of this trade and Hamilton's son, Archibald, who succeeded him, seems to have been incompetent. Watt must have been saddened to see the Company which he had done so much to promote in his early years running into difficulties. In his final letter to Muirhead on 24 March 1819, Watt wrote:

> The Delftfield concern seems hopeless of doing good under the present circumstances and as we seem unable to mend them I agree with you that it will be prudent for some of us to dispose of the concern, or of our shares as soon as we are able to find a purchaser for like you I have no desire to engage with new partners and my son has enough on his hands. [3]

Watt died in the following August. The Delftfield pottery struggled on until closure in 1826.

Alkali Manufacture

During the early part of 1766, Watt had formed a partnership with Roebuck and Black to produce the alkali, soda (sodium carbonate), from sea salt (sodium chloride) and lime (calcium hydroxide) according to Black's process. (See Vol. I, pp. 155 – 9) Perhaps they hoped to achieve a second successful industry similar to that which Roebuck had built up with his oil of vitriol or sulphuric acid process. However they were not to know that this reaction does not work. It needed a sulphate instead of the chloride. The sparse results obtained, which encouraged Watt to continue experimenting, were the result of some sulphate in his sea salt. [4] When Watt stayed for a couple of weeks with Small at Birmingham in 1768, it is likely that he revealed too much of his secrets to Keir who was also staying with Small and was experimenting to make alkali. Watt's experiments were discontinued in 1769. Keir had a lime-sulphate process which was successful so they corresponded but no partnership resulted. In 1771, Keir entered a caveat at the Patent Office against the granting of any patent for producing alkali from salt and there this group left the matter in abeyance.

In 1780, there was a great flurry of activity because Alexander and George Fordyce petitioned Parliament to grant a drawback on salt duties. George Fordyce alleged that he had discovered a method of manufacturing alkali and marine acid (hydrochloric acid) from common salt. Keir decided to petition Parliament that he should be allowed a similar privilege. Watt heard about this at the end of May and decided to petition Parliament as well on behalf of himself and Black because:

> ... such a law if passed would confine the exemption to him [Fordyce] no other person could afterwards attempt that process on an equal footing and as I have not totally laid aside all thoughts of the process we carried on for that purpose – and as Mr. Keir had invented a similar process about the same time we were at work upon it and is determined to go on with it upon his own account, we have judged it proper, separately, to present petitions to the house of Commons praying to have the like exemption as Alex[ande]r Fordyce. [5]

Keir intended to take out a patent. Since Watt was in no position to offer financial help, Keir would grant him and Black licences, or Keir thought Black might wish to take out a patent.

Watt went to London in early June for the first of several visits to attend the Parliamentary enquiries which were disrupted by the Gordon riots so that the House of Commons was adjourned. [6] There were half a dozen other petitions in addition to those from Fordyce, Keir and Watt. Submission for a patent continued at the same time with Keir suggesting that Watt should have 'half the property', half of which Watt in turn offered to Black. [7] Watt complained to Boulton, 'My being so long at London has been an immense loss to our business'. [8] In the middle of August, he reported to Hamilton that he had been 'in London during the summer at Parliament about a bill for fosil alkali. We lost the bill but are promised success next session'. [9] Hamilton replied that he thought that the high price of ashes would make it 'a losing business'. [10] Perhaps it was this, or the competition from others as well as Roebuck who 'was constantly ferreting about in every public office' which made Keir and Watt reconsider taking out a patent application. 'We thought it prudent to delay taking out the patent untill he [Roebuck] should take himself away which has only happened very lately and now we are cool both Mr. K. and myself doubt whether it will be worth while to take out a patent at all'. [11]

Watt recognised that it would take many more experiments, which he had no time to undertake, to determine which would be the best method. Keir decided to leave Boulton and Watt in order to set up his own chemical works at Tipton, eight miles from Birmingham. He had discovered a method using potassium and sodium sulphates which were waste products from vitriol and aqua fortis manufacture, so avoiding salt with its heavy taxation. [12] Watt was asked by Keir in 1781 if he would bear half the costs of a patent but he replied that he 'could spare no time from my present business to do anything in it'. [13] In May 1782, he wrote to Fry who set up his own works to produce alkali for his soap manufactory, 'I have almost entirely dropt thoughts of the Acaline business now I hear you are to be saddled with Excise men... I will stick by the Engine business while it sticks to me, but we have got so many pretenders now that I fear they will make us little people if we let them'. [14]

Watt had played a leading role in the Parliamentary negotiations in his own right and no longer as second fiddle to Boulton. He was beginning to have influence in government circles, writing directly to some of the leading politicians. Although his methods of alkali manufacture were unsuccessful, this episode shows that he, and other members of the Lunar Society, were alert to the possibilities of scientific investigations opening up new fields of exploration for industrial applications and commercial operations. In addition, it shows the mutual trust which had been formed between this group of people centred on Birmingham.

Steam Heated Drying Cylinders

In February 1781, Watt sent to McGrigor drawings of a machine using the heat of steam for drying long lengths of linen during the bleaching process. [15] The wet cloth was wound round a roller from which it was taken off and passed over and under three steam heated copper cylinders in succession before being rewound onto a second wooden roller. Watt recommended trying one cylinder at first to see if that would be sufficient. He also showed how the condensate would be lifted out of a cylinder by a false end-plate into which the water would pass and be raised to a pipe connected to the hollow centre axle from where it would flow out as the cylinder revolved. Because only low pressure steam was used, the joints could be made with soft solder. [16]

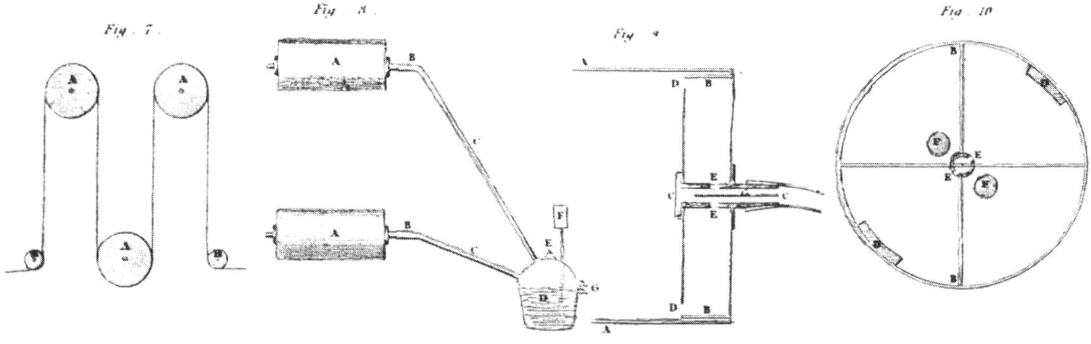

Watt's steam-heated drying cylinders.

Watt recommended that 'one of my old servants J. Gardiner or Murray Osborne may be of use to you in fitting the several parts'. McGrigor had the machine made by Gardiner under the supervision of Dr. Irvine and Hamilton. [17] In 1814, Sir David Brewster asked Watt to describe it for the *Edinburgh Encyclopaedia*. Watt thought that his was the original 'from which such machines were made, and which I believe, is claimed by somebody else'. [18] So by that time, his machine had been copied by others. It would have been very useful to Thomas Bell with his engraved roller calico printing machine with which Watt had helped in 1770. (See Vol. I, p. 135) It may be that Bell introduced drying cylinders to Lancashire because his printing machine was used at Masney near Preston, England, by Messrs. Livesey, Hargreaves, Hall & Co. in 1786. [19] In 1820 when T.B. Crompton, Farnworth, Lancashire, was considering how he could dry the continuous lengths of paper produced on his paper making machine, it is said that he took a roll of wet paper to a textile mill where he had it run over the drying cylinders. This showed him that it was practical to dry paper in this way so he took out a patent. [20] It was only after the addition of these drying cylinders that the papermaking machine developed its full potential, perhaps through Watt's invention. If so, the origin has never been acknowledged before.

Cement for Joints on Steam Engine Cylinders

The continued expansion and contraction of the various parts of steam engines which were subjected to the heat of the steam caused their joints to fail. Movement through insecure assembly might have contributed as well. Failure resulted in loss of steam or, in the case of a vacuum, entry of air and so loss of efficiency. Watt had put his chemical knowledge to practical use when he told Boulton in April 1782:

> I am trying some experiments on a new cement for joints, as I have lost faith in putty which always fails in the long run in all the vacuum joints owing to the repeated exhaustions & repletions. I have several sorts in view, a fine powder of iron in a metallic state mixed with substances which may dispose it to rust and some mucilaginous matter which may give it elasticity and keep it in place till it rusts and a new substance which is neither soluble in water, [spirits of wine] nor oils but can be softened by water to the consistence of Caoutchou & which when mixed with earthy substances becomes as hard as stone, and can be had cheap, being the product of an English vegetable. What continued heat may do it experience must determine in other respects it would make a hole of an inch wide quite tight in one minute after applied and adhere to dry substance most viciously caustic alkalis or acids destroy it or weaken it. [21]

Watt does not say what he used to make the iron rust. He was aware of this property in sal ammoniac (ammonium chloride) which had been manufactured in Edinburgh from the soot of bituminous coal from about 1756 by James Davie and James Hutton. [22] Hutton was an intimate friend of Black from either of whom Watt may have learnt about it. Watt was well aware of the rusting effect of sal ammoniac because he had recommended its use in 1779 when he wrote his instructions for engine erectors. He directed that the end of a screwed stud should be wetted with it to fix the stud firmly in tapped holes in cast iron. [23] Perhaps he did not think it suitable for the cement.

Later Boulton wrote to Watt from Cornwall in November 1782 saying:

> What a pity it is that all the joynts will in a little time give way. W[m] Murd[k] says that Trevascus is the only one (lower joynt) that will stand it being put together with a ring of soft Spanish Iron anointed wth Salamoniack that it was rusted together before steam was let in, so that it was solid as if no joynt at all. [24]

Apparently Murdock had recognised the rusting properties of sal ammoniac through some being spilt in his tool box where it spoilt the tools. [25] From Boulton's description, it sounds as if Murdock was not making a paste with iron filings but had a sort of iron gasket which he caused to rust.

The idea was not taken up immediately since it was not until 8 July 1784 that Boulton wrote to Watt, 'The method of using salamoniack in the making of joints seems to answer so well that an engine, being once well put together may be carryd about as one solid piece without spoiling the joints'. [26] Watt replied on the 20th:

> I have got a new composition for Cylr. Joints viz sulphur and iron. It is vastly preferable to the sal amon[c] which by the by was my own proposition together with the whole process by which the Trevaskus bottom joint was made. Should this new compos[n] be specified or not, I fear it cannot be concealed, its smell is so atrocious. [27]

It is now disputed who was the first to use sal ammoniac and iron filings for the cement, Murdock or Watt. But it is clear that the standard cement used by Boulton and Watt was this mixture with the addition of sulphur to make it set quicker, the sulphur certainly being added by Watt. This cement remained steam-tight for years and contributed to the excellence of Watt's engines. [28]

The Philosophical Pound

John Whitehurst, a clockmaker and early member of the Lunar Circle, had been investigating a method for determining an invariable standard for a unit of length. A proposal for a uniform system of weights and measures was suggested at a meeting held at the Royal Society on 1 February 1781. [29] This encouraged Whitehurst to continue with his experiments and soon other Lunar Society members were considering the same problems. Watt had experienced difficulty when comparing the experiments of Pierre De La Place and Antoine Lavoisier with his own. So in a letter to Richard Kirwan in November 1783, he proposed the standardisation of weights in a metric system:

> My proposal is briefly this, let the philosophical pound consist of
>
> | | | | |
> |---|---|---|---|
> | | 10 ounces or | 10,000 | grains |
> | the ounce of | 10 drams or | 1,000 | grains |
> | the dram of | | 100 | grains |
>
> Let all elastic fluids be measured by the same measure of water by which the valuations of the different cubic inches will be avoided and the common decimal tables of specific gravities will immediately give the weights of these elastic fluids. [30]

He suggested that the Amsterdam or Paris pound be taken as the standard, since that was used throughout Europe and dividing that. Priestley agreed with this proposal and Watt asked Kirwan to fix upon the weight of the pound if he approved of the system. Watt also wrote to other people such as De Luc and Magellan as well as Whitehurst. Nothing further came of these proposals while the metric system had to wait for its introduction under the French Revolution. [31]

A Calculating Device

In December 1785, Watt wrote to De Luc about a calculating machine he was planning:

> I have been turning some of my idle thoughts lately upon an Arithmetical machine, how I shall succeed I know not not having made it yet. Its properties are to be that when you want to multiply you first turn up one figure of the multiplex, you then turn up in their order all the figures of the multiplicand and the machine will show the product by that multiplier, you then turn up the second figure of the multiplier and beginning one place towards the left hand you turn up again all the figures of the multiplicand & the machine shows the product by these two figures already added & so on for any number of figures: and it will perform division nearly as easily without the least calculation or burthen to the memory, other than to take the figures in their order beginning always at either end you like. I intend to make an attempt at making it. I say an attempt, for though the machine is exceedingly simple yet I have learnt by experience that in mechanics many things fall out between the cup and the mouth. [32]

Watt did attempt to make his calculator for there is a device with star wheels in his Garret Workshop. Presumably he did indeed find the actual manufacture too difficult.

Calculating device in Watt's garret workshop.

THE COPYING MACHINE

The copying machine was one of the few of Watt's inventions besides the steam engine which he developed into a commercial business and hence is worthy of treatment in greater depth. It is also an instance where members of the Lunar Society helped each other. In addition, it shows Watt's search for perfection against considerable difficulties. He had to create a separate, new industry for producing all the equipment needed. Not only did he have to find a papermaker capable of manufacturing the correct quality of paper but special ink had to be prepared at the Soho Manufactory for sale to customers. Then special machines and tools had to be developed for production of the presses, a significant early example of specialised production engineering.

An instrument for writing many copies of the same thing at once had been patented by Sir William Petty in 1648 but it was found to take more than twice the time to produce two copies than it did to write one. – 'Our story starts towards the end of 1778 with Darwin assuring Charles Greville that he had 'incessantly labour'd at the completion of the Polygrapher in its triple capacity'. [2] This was a form of writing device with multiple pens. Darwin was satisfied with the way the paper was held and could write as easily with the instrument as with an ordinary pen. 'I flatter myself I shall be able to send you one that will perform all its parts to considerable perfection'. But Darwin had to apologise in the following May that he had been unable to send Greville the machine because 'it remained many months at Birmingham before my mechanical friends return'd from fire-engine building in Cornwall'. [3] It was complete as far as Darwin was concerned and he did not intend to patent it. It never came into commercial production, perhaps through the difficulty of keeping the pens filled with ink. Much later, a similar, but much more complicated, device was used to prepare the multiple engravings for calico printing rollers on pantograph machines but these had cutting styluses and not pens.

When Watt returned from Cornwall, he set about improving it, as might be expected. Soon he had constructed one to his own design. [4] In the middle of January 1779, he 'tryed some experiments with Dr Darwins and my own writing machines the Doctors has least friction but some how the hand is confined by it so as not to have the freedom it has by mine which latter can be improved by being made lighter'. [5] He had given Darwin's device a fair trial but found it impracticable. A couple of days later, he 'made trial of a new writing machine which consists of two pens fastened to a frame which slides on a wire but found it to have a good deal of friction and also to confine the hand'. [6] He continued with more experiments, such as fitting a roller to support the 'tail part' which helped, but he 'found it still too heavy, unless one can be made a deal lighter the machine may be given up as it has too much friction'. [7] But he was unwilling to let the matter rest there. Once the idea of the possibility of making multiple copies had been suggested to him, he saw the advantages of having a system which would ease the burden of laboriously copying a second draft of all his voluminous correspondence which was now necessary for the steam engine business.

Watt's Early Experiments

While we can understand why Watt abandoned making copies with multiple pens, we do not know what made him investigate taking an impression off the original manuscript onto another sheet of paper. Unfortunately he did not tell Black the source of this inspiration through pressure of work:

> I should without Hesitation communicate the whole to you but cannot do it without writing at least a sheet of paper and recalling my experiments and Ideas on the subject which at present would be a hardship on me when my head is filled with other business, the weather and my spirits both very bad.
>
> In brief the first Idea was the forcing the ink through thin paper so as to appear on the other side, the second the improving the colour by wetting the paper with an astringent the third the depriving astringents of their colour, without depriving them of their effects as astringents – fourthly managing the operation so as to prevent the original being defaced or forced through to other side – But the greatest part resides in the Mechanical Manoeuvre – all this to yourself only, at present. [8]

Watt was unwilling to give too much away because he had not secured a patent for his invention at the time of writing in December 1779. He also pointed out that he was 'unwilling to have my inventions come into the world imperfect'. His idea was to write the original with gelatinous ink from which a copy could be taken by pressing onto it a sheet of damp unsized paper. The ink soaked through the copy paper so that it could be read from the other side. We will see that in his search for perfection, he carried out many trials both with the composition of the ink, the astringent in which the copy paper was soaked as well as the copy paper itself. He told Black, 'It gives me pain to be obliged to sell philos[ophi]cal discoveries but circumstances render it necessary as to the world in general, as to my friends they shall welcome to it so soon as I am secured'. He was assisted by the Lunar Society members, Boulton and Keir, but only Watt's archives survive for these experiments which show mostly his role.

Watt had been carrying out numerous experiments before the spring of 1779 when he went to Lichfield on 11 April with 'Wedgewood' to see Darwin's horizontal windmill. In his letter apologising to Darwin that he had been unable to send his comments on the windmill, he added triumphantly, 'I have fallen in a way of copying writing Chemically which beats your bigrapher hollow I can copy a whole sheet letter in 5 minutes – I send a copy of the other page enclosed for your conviction and I tell you farther that I can do still better than that copy'. [9] Although the date is missing on the letter, Watt's Journal shows that it was written on 12 May. This was the beginning of a long series of experiments to prove the idea and turn it into a commercial proposition.

Watt's Journal also shows that he may have started his experiments at the end of March when he noted 'tryed a mixture of gypsum and glue for white ink, but it wrought flaky and would not spread'. [10] By the end of April, he was copying papers concerning the Bedworth engine and 'Found the copy I had made of the answers much sunk by taking off the impression too wet – copied 2 sheets of it over again'. [11] During the second week in May, a series of experiments began which lasted almost a fortnight. On 19 May, he told Henderson that he had 'not brought it to its ultimate perfection though have copied all my letters that way for some time'. [12] Watt told Black, 'I have lately discovered a method of copying instantaneously, provided it has been written [?] or written within 24 hours. I send you a specimen [and will] impart the secret if it will be of any use to you. [It enables] me to copy all my business letters'. [13]

While at first Watt may have thought of it basically for copying correspondence, he soon saw the potential for copying engineering drawings as well. On 6 May he wrote to Hamilton, 'I intend to apply my method of copying to copying drawings in which will be exceeding useful. I can now do it better than the specimen I sent'. [14] He jokingly told Henderson that 'every gentleman engineer must now wear a pocket Rolling press and some sheets of Condum paper'. [15] During July, he instructed Playfair how to copy drawings for the Ale and Cakes engine. [16] Such copies had to be reproduced on thick drawing paper through which the ink could not penetrate and so were a mirror image of the original. Therefore they had 'REVERSE' written on them and dimensions had to be added on both original and copy after copying. Even so this system proved a great boon to the drawing office. In February 1781, Watt told Patrick Wilson, 'I find this machine of much more use in copying drawings, than in copying letters'. [17]

News of the copying system spread among Watt's friends who were interested in the process. Garbett called in June 1779 but Watt's attempts at the process were 'remarkably unsuccessful by being hurried and over zealous... I shall comply with your desire tomorrow and hope to be more successful'. [18] Watt should have realised that people without his extraordinary skills in chemistry and mechanics would find his process very difficult to operate. However he sent copies to Boulton in London as advertisements:

> I send you inclosed some of Mr Nobody's draughts with authentic copies of them; you will observe that that which appears full Black will not preserve even the Colour it has but will grow browner by keeping. The paler kind are nearly if not fully as white as the paper was before the operation and will stay so.
>
> I will be much obliged to you to procure me a quire or two of the most evenly & whitest unsized Cambric paper, and also specimens of the cheaper kinds with the wholesale prices – If you apply at Curtis's may perhaps be better served as we have dealt with him before – it is absolutely necessary that there be no size in the paper which may be known by touching it with a wet finger. [19]

There was a greater range of paper in London where Watt frequently sought supplies. His request should be noted for unsized paper which would act rather like blotting-paper.

His 'draughts' may have been in the form of banker's orders which Boulton showed to some of the Directors of the Bank of England. They were afraid that the invention would considerably facilitate forgery and were inclined to suppress it. [20] Watt's reaction was, 'I despise the Bankers more fully than they despise me; and like the whores, I will rather give it to those I love for nothing; than to those I despise for nothing always remembering the sour grapes'. [21] Black was quick to canvass the idea in Edinburgh during August 1779 and received a similar worried reaction:

> I much admire your Secret for copying writing – I shewed it last night to a Writer to the Signet who at first was much alarmed on account of the frauds which he thought it might give occasion to, he supposed that it needed only a little more address and ingenuity to enable you to copy writing of any age and in any manner. – I thank you for the offer of the Secret but have no particular occasion for it as I very seldom need to copy what I write – If it were peaceable times and money more plenty in the Kings pocket I think you might obtain some present or reward for communicating it to the Secretary of State etc. [22]

During the second half of 1779, trials continued to improve the copying processes and prepare a submission for a patent. It was agreed to set up a partnership consisting of Watt, Boulton and Keir who would form a separate concern, James Watt & Co. The shares were Watt one half, Boulton and Keir one quarter each. The expenses of the patent would be met by Boulton while Keir would manage the business. An agreement was executed in August but was not finally initiated until 31 March 1780 after the initial patent specification had been submitted. [23] At the beginning of November 1779, an affidavit was prepared for submission to the Privy Council. [24] Watt confidently told Black soon afterwards:

> In relation to the copying scheme the state of it at present is as follows – I have given in a petition for a patent and it is now in train, but as my occupations and health cannot permit me to follow that business my self I have taken in 2 partners – Mr. B[oulto]n who is to be at the expense of the patent and Mr. Keir who is to manage the business – Our proposition is to open a subscription for 1000 persons who are to be put in possession of the secret; a quantity of proper paper and materials with a press for taking off the impressions at the rate of $5^1/_2$ guineas each; no one to be put in possession untill the whole is subscribed, as the thing is so simple and easy that after divulging it to a number we might lose the rest – We have reason to believe that such a subscription will be immediately filled up, particularly as the price we have fixed is not much if any more than the value of the press itself which will be made of Iron with *lignum vitae* Rollers – The Iron work japanned Black – We cannot publish our proposals for the subscription untill the patent passes the seals the news of which I expect every day. [25]

This letter was written at the beginning of December but the date for the acceptance of the specification is 14 February 1780 in the text of the patent. [26] Watt was in Cornwall which may have caused this and later delays. Although he was writing the final specification with Keir in March, it was submitted to the Patent Office only on 22 May and enrolled on 31 May. [27] It is possible that Watt wanted to perfect his methods so that the application would be as complete as possible. Watt wrote the final specification with his usual thoroughness:

> Let the letter, or other writing that is intended to be copied, be written with the ink hereinafter described, or with any other writing ink fit for the purpose. Take a piece or pieces of thin paper which contain no size, or glue, or gummy mucilaginous matter, or which at least does not contain so much size or other matter as would make it fit for being written upon. Cut this paper or papers to the size and shape of the writing of which a copy is wanted. Moisten or wet the said thin paper with water, or other liquid, by means of a sponge or brush, or by dipping or otherwise. Having moistened or wet the thin paper, lay it between two cloths, or other substances capable of absorbing the superfluous moisture from the thin paper. When it has been slightly pressed between such thick spongy paper, or other substances, by the hand or otherwise, lay the said thin paper, so moistened and pressed, upon or under the side of writing which is to be copied, and in such manner that the one side of the said moistened paper shall be in contact all over the side of the said writing so intended to be copied; and that to the other side of the said moistened thin paper there shall be applied a piece of clean writing paper or cloth, or other smooth uniform surface. Lay the said writing intended to be copied with the thin moistened paper intended to receive the copy, placed respectively as above directed, upon the board of a common rolling-press, or that of which a description and drawing are hereunder written and drawn, and press them once, or oftener, through the rolls of the said press, in the same manner as is used in printing by copper plates;... subject them to any other pressure sufficient for the purpose, by means of which pressure, in whatever manner applied, part of the ink of the writing intended to be copied shall pass from the said writing into, upon, and through the said thin moistened paper, so that a copy of the said writing, more or less faint according to the quality of the ink and paper employed, shall appear impressed on both sides of the said moistened paper, viz. upon one of the sides in the natural or proper order and direction of the lines as they are in the original writing, and on the other side in the reverse order and direction. [28]

The patent marks a sort of mid-point in Watt's development of his copying process. Many experiments had resulted in a method which operated reasonably satisfactorily under his own capable hands. However sales to the public after the granting of the patent showed problems which he had to overcome. Although many of his notes have survived, he used code names and symbols for which the keys have disappeared, so it is now very difficult to determine precisely what he was doing and trying to achieve.

The Ink

When Black first received his machine, he had difficulty in producing good copies. Watt blamed the ink. [29]

> That subject is attended with more difficulties than you can be aware of – The following qualities are required – 1st that the Ink be capable of being made up in the form of a powder readily soluble in water 2dly that it be fitt for use in an hour or two after it is made – 3dly that it be thin and flow readily from the pen 4thly that it do not grow thick or mould by keeping; nor yet deposite much sediment 5thly that it is capable of giving an impression without diffusion soon after it is wrote with, 6thly that it shall be capable of yielding an impression at the end of 24 hours after it is wrote with and if would do it a longer date would be still more valuable, lastly that the Copy as well as the original keep its Colour.
>
> We have occasionally made ink with all or most of these properties but I am sorry that we cannot always succeed which is the more mortifying as we do not know the causes. [30]

The ink had to be capable of being penned onto the original in reasonable quantity without smearing. Some of this dried ink had to become fluid again when in contact with the damp copy paper under pressure but not to the extent that it would flow laterally. Watt wrote this in October 1780 after he had been experimenting for over a year. More experiments followed to make satisfactory ink which could be used by ordinary people.

Watt selected as the basis for his ink the type that was used for writing with a pen and not that used by letter-press printers. It was:

> ... commonly made of copperas and galls, and gum arabic; but astringent plants may serve the same purpose; such as oak-bark, red roses, logwood or sumach...
>
> For many years the ink most used by European writers has been the infusion of galls and other astringent vegetables, containing gallic acid, rendered black by sulphate of iron and thickened by the addition of a little gum or sugar. [31]

The paper on which the original was written had to be 'very well sized, that is such as the ink does not sink into'. [32]

Paper in Watt's day was made principally from linen rags, which he would have known from his visit to the paper mill at Ruthven in 1770 when he was surveying the line of a canal from Perth to Crieff. [33] The cellulose fibres in paper would react in the same way to receiving colours as would those in linen or cotton cloth when it was being printed. He had studied dyeing so may have been expecting that he could adapt similar techniques for his copying process. (See Vol. I, p. 148) A pointer in this direction may be found in his letter to Black when he outlined his ideas on the subject and wrote, 'the first Idea was the forcing the ink through the paper'. [34] In calico printing with hand blocks, the colour was forced into the cloth by hitting the back of the block with a small maul. Keir, with his interests in chemistry, had also studied dyeing and in particular mordants for making the dye adhere better to the fibre. [35]

Watt's journal for May 1779 records the earliest surviving details about his experiments:

> May 6 In afternoon tryed experiments for proper ink for copying by counterproof common ink strikes through best but becomes too purple if the paper which take it off be wetted with a weak solution of verdigris it makes the original Blacker and takes off fully as good a proof – if the thin paper be first done over with Galls and then dryed and wetted with verdigris it takes a very good thorough proof.
> May 7 Copied some of these letters by counterproofing with water only but found it does not come through distinctly and is apt to spread about.
> May 8 Took off these proofs with a weakened solution of Galls in which suspect there is a little Iron find that the dryer the paper the neater the impression and that if more pressure could be aplyed [sic] and Blotting paper on each side it would be neater still – The verdigris wetting is only for drawings.
> May 9 Tried some more experiments abut the copying I find that by mixing fine powder of whiting with the solution of galls and laying it on with a brush that the ink does not spread – If some Gum arabic is added it rather improves the neatness and gives more strength to the paper though it seems to hurt the whiteness of the ground. [36]

Verdigris was a green pigment obtained by treating copper with vinegar and may have helped to intensify the colour but its addition was soon given up. [37] Gum arabic was obtained from some species of the acacia tree. It was very adhesive and hardened without losing transparency. More important, it easily softened again when moistened with water. [38] It was mixed in ordinary ink to thicken it and help to prevent it spreading.

The crucial chemical used by Watt was 'Galls' or gallic acid. This was prepared from pounded gall nuts which grew on a type of oak and were imported from Aleppo. Blueish ones were preferred by him as they gave the strongest solution when treated in water but the infusion had a brownish colour. [39] Gallic acid was used by the dyer to make various shades of black and to improve or fix other colours. [40] It may be that Watt and Keir thought it would help the ink to adhere to the copy paper through acting as a mordant but its real importance lay in its reaction on copperas. Copperas, sometimes called green vitriol, was a natural combination of iron with sulphur, which turned black when treated with galls. This black might precipitate out of the solution which was another reason for adding gum. The gum, when dry, would also act as a sort of protective varnish. But this composition is liable to fade, chiefly due to the deficiency of galls which 'are the most perishable ingredient, the quantity of these, which gives the greatest blackness at first (which is about equal parts with the vitriol) being insufficient to maintain the colour, that for a durable ink the quantity of galls cannot be much less than three times that of the vitriol'. [41] It was also known that, as long as the sulphate of iron remained in the paper, the colour of the writing could be restored by washing it with a fresh infusion of galls. [42] Keir found that the ink had to contain excess tannic acid in order to penetrate the unsized paper while containing the optimum amount of gum arabic in order to give the correct balance between penetration and sharpness of the letters. [43]

Just how much Watt knew of the chemical reactions at this early stage is not clear but he was probably aware that ink could be moistened again by the water with which thin copy paper had been wetted. He may have hoped to obtain a further chemical reaction on the copperas with the addition of more galls either by an initial soaking of the copy paper, drying it and then adding more water, or by wetting the paper with a solution of galls and placing that directly on the writing to be copied. But he was dealing with chemicals and paper which varied in their composition. The earliest copies have paper which has been turned distinctly brown. On one or two, the ink has burnt holes through the paper. Penetration varies and the ink has tended to diffuse. [44] Whatever caused the holes was quickly corrected. To produce a viable commercial product which could be used by the general public, he needed to improve all the techniques to achieve simplicity and consistency.

Only a few of the experiments and different substances can be mentioned. In June 1779, Watt pointed out to Boulton how 'The copy will continue to grow blacker as other writing does and I fancy you will find the originals rather blacker than they were before copying and as far as I can judge not in the least defaced'. [45] Watt could not have known that this was caused by oxidation of the iron compounds. The solution of gallic acid and water with copperas (ferrous sulphate) at first gave a colourless liquid which oxidised in air to give a blue-black complex which is insoluble in water. He investigated many ways to stabilize and improve the colour. He suggested that Keir should 'try solutions of iron in other acids than the vitriolic'. [46] He found that a white soap, 'a sort of Prussian alkali gave a pretty black impression'. Oak bark he found to be 'too astringent and does not strike through'. [47] At the same time, he was experimenting with different solutions of galls, alum and alkalis to find a satisfactory preparation to treat the copy paper.

During the second half of January 1780 while he was in Cornwall, Watt undertook a further series of experiments, probably in preparation for the patent submission. One quite common addition was lime water probably to neutralise the acid but later this was found to hinder the process. Sulphuric acid and another acid from tartar have been identified. Alum was tried as well. This is composed of sulphuric acid, alumina and potash. It would act as a preservative against mould and was used as a mordant but the sulphuric acid will damage paper. He found that alum 'resisted penetration more than it did diffusion. Ink with no gum spread much but was very black'. [48] These trials continued into February until he returned to Birmingham.

Watt gave instructions about the composition of the ink in his patent specification. He took:

> Four quarts, ale measure, of spring water; one pound and a half, avoirdupois weight of Aleppo galls; half a pound of green copperas, or green vitriol; half a pound of gum-arabic; four ounces of roach allum; pound the solid ingredients, and infuse them in the water six weeks, or two months, during which time the liquor should be frequently shaken. Strain the liquor through a linen cloth, and keep it in bottles closely corked for use. [49]

Preparation of the ink took considerable time and equipment. J.F. Stansfield experimented with this composition in 1966 to prepare samples for the Lunar Society Bicentenary Exhibition that year. His copies were very faint.

Watt continued to experiment with compositions for ink for many years after publication of his patent. He kept an 'Ink Book' [50] in which he revealed the component for one of his recipes called 'Newhall'. This was 6 quarts water, 24 oz. galls, 8 oz. vitriol, 8 oz. gum arabic and 2 oz. alum. He seems to have used more galls and gum arabic than conventional inks of the period. In August 1780, he told James Woodmason, his London stationer, he was altering the composition of the ink so that no preparation of the copy paper would be necessary but this was not successful. [51] He found that 'Ink which was very blk. when first wrote or copyed did not stand well in the copy and therefore have been at pains to make it pale and strong at the same time'. [52] Watt may have been moving towards obtaining a second reaction of the galls and vitriol in the copy paper to achieve darker copies. He also knew that exposure to air might darken ink. There does not appear to be any methodology in his experiments. He must have set up quite a large plant for the manufacture of ink because in May he ordered two hundredweight of galls and said in September that they could make 120 or 150 gallons of ink per week. [53]

During September 1780, comments first appear about two important developments. One was the production of ink in powder form which contained 'Extract [presumably galls] one pound, copperas, 5 1/3 oz, Alum 2 oz, Glauber Salt 2 oz, sugar 1 oz, to one gallon of water'. Watt instructed that:

> The utmost care should be taken to mix the ingredients equally and well as upon that the goodness of each particular parcel of powder [depends] – I therefore recommend weighing the ingredients in smaller parcels than the above by which means the mixture may be made more equal It must be put up in parcels for 1/2 a pint. [54]

A little later, Watt instructed Woodmason not to give out the new ink powders owing to sedimentation and other problems. Watt recognised that sediment forming in the ink was 'a great imperfection which we have been labouring to remove, and are in hopes that we have in a great measure succeeded within these two days. The number of faults to which ink is liable is so great that by remedying one fault we are apt to fall into another'. [55] Another problem was the ink turning green. Transport to London from Birmingham would have been much safer, easier and cheaper of ink as a powder than a liquid.

In the Garret Workshop there is a parcel containing two packets of powder for making up ink. The instructions say that the larger packet had to be boiled for an hour or longer and then, when it had cooled to blood heat, the contents of the second could be added. While this concoction could be used immediately, it was better left to stand for a couple of days during which time the colour would develop. Watt explained this to Woodmason:

> We think that the reason of the powders not doing so well is that all the ingredients are put together and we have accordingly changed our method... Each paper for 1/2 a pint is divided into two little parcels, the bigger is to be thoroughly ?dispersed? in hot water, and by keeping the solution near a fire till there is no sediment, and then the little parcel of ingredients is to be added. Stirred, upon which in a few minutes the ink will be made... The ink will scarcely have the colour of ink at first but the writing becomes black very soon. If the ink is left exposed to the air a little while it will become blacker. [56]

It sounds as if this powder was prepared from copperas and gallic acid in an anhydrous form. Sediment still remained a problem. This was in October 1780 and Watt continued to improve the powders for he asked Hamilton a year later, 'Have you tried the Ink powders I fancy you will find them better than the ink sent you first as we have much improved our Ink about 2 months ago'. [57]

The other important development was the addition of sugar which prevented the ink from drying too quickly. Sugar became a common ingredient for copying inks during the nineteenth century when it was stated that 'The powerful affinity of water to sugar causes the damp paper to immediately absorb the ink by the powerful aid of the copying machine'. [58] Logwood or sumach used in black dyes might be added to improve the colour and Watt believed 'that in all the old specimens none have stood so well as those which have had logwood in them'. [59] With the state of chemistry at that time, his achievement in developing a copying system that proved to be so enduring was truly remarkable. Yet in November 1780, he wrote to Keir in some despair:

> I have wrought at ink making until I am quite bewildered and have discovered very little – ?vitriol? seems to prevent the curdly precipitation but if overdosed makes the ink thin & impedes copying – the Refined extract gives the blackest copies and at the longest date but precipitates more than any other. Gum remedies the precipitation, but makes the ink too thick – sugar does not seem to prevent precipitation, though it thickens the ink and disposes extract inks to spread. [60]

Experiments on ink continued intermittently into 1782 with some even in 1784.

The 'Preparation' for the Copying Paper

In the earliest surviving copies of Watt's letters, the paper has turned brown through having been wetted by some solution probably brushed on. He had realised very quickly that moistening the copy paper with plain water was insufficient either to draw the ink through so it could be read the correct way round on the reverse or to prevent lateral diffusion. He had to develop some preparation which could be applied to the copy paper. On 13 May 1779, he recorded that he 'copied them with the chalk vinegar and Galls mixed a larger proportion of Galls than usual'. [61] Lime water and alum were other ingredients tried but lime was found to be deleterious. However, at one attempt, he perceived 'the outside sheet growing Black occasioned by the precipitation of the iron in the Chalk by the Galls – so was obliged to separate the sheets and dry them speedily'. [62] He found alum helped to remove the colour given by the galls.

It was difficult to apply the preparation in the correct strength for the paper on which it was being used. In addition, the paper had to be damped correctly. In October 1779, Boulton was reprimanded, 'I must mention to you that in copying your letter you have made the copy paper too damp so that is miserably struck through so as to be scarcely legible – I believe that it may in part be caused by the copy paper which is too strong prepared'. [63] While he was in Cornwall at the end of 1779 and into the New Year, Watt continued experimenting with different mixtures to try and remove the colour given by the galls; vinegar, lime, oak bark, alkali, alum, tartar and vitriol were tested. He continued after his return to Birmingham. At one time he thought that the addition of sugar improved the colour but later changed his mind. [64]

The main component of the preparation remained the solution of galls. He gave recipes in his patent:

> Take of distilled vinegar two pounds weight, dissolve in it one ounce of sedative salt of borax; then take four ounces of oyster shells, calcined to whiteness and carefully freed from their brown crust, put them into the vinegar, shake the mixture frequently for four and twenty hours; then let it stand until it deposits its sediment. Filter the clear part through unsized paper into a glass vessel; then add to the said mixture or solution two ounces of the best blue Aleppo galls, bruised, and place the liquor in a warm place, shaking it frequently for twenty-four hours. Then filter the liquor again through unsized paper, and add to it after filtration one quart, ale measure, of distilled or other pure water. It must stand then twenty-four hours, and be filtered again if it shows a disposition to deposit any sediment, which it generally does. [65]

Watt suggested alternatives in place of the vinegar and oyster shells, adding the caveat:

> It may be found necessary to add more or less water in the preparation of the above liquor to be used for moistening the thin paper, or to vary the proportions of the other ingredients according as they are more or less perfect or strong, or as the impression is required to be more or less deep coloured. [66]

How the person purchasing a copying machine was expected to know this is not explained.

When he was in London, Keir complained that some of the preparation turned green which discoloured the paper badly. Watt thought that he had discovered the cause and wanted to use something called 'Satan' but 'Mr Keir has great objections to any dealings with the devil'. [67] Unfortunately we do not know what Satan was but there appears to have been nothing of any great danger in a recipe of July 1782, unless it was the brandy:

> Preparation for copying letters – Best Galls one pound, Water one gallon, or rather 2 Gallons and boil the Galls therein for 12 hours, then filter or Clarify the liquor, and evaporate until only one Gallon remains, add pounded cloves _ oz, Brandy 4 oz – each half pint of this liquor to be diluted with 30 gallons of water and used for preparation. [68]

There had been many experiments in the meantime to reach this composition.

At first Watt brushed the preparation onto the thin copy paper but it must have taken a long time to treat a number of sheets. Then in May 1780:

> At Soho prepared some paper by soaking some flannels in the preparation and laying one on each side of the quire of paper, so repeatedly squeezing gently and slackening alternately the paper was penetrated quite through but seemed crumpled Left it in the press all night to see if that would level it. [69]

The following day, he announced to Boulton, 'I have spent these two afternoons on preparing paper without opening the quires [144 sheets] and find I can do it but it wants proper tools yet'. [70]

At this time, Woodmason in London was supplying paper to Watt in Birmingham where it was prepared and returned to London with waste of time and unnecessary costs. One reason was that different batches of paper needed different treatment. Watt informed Woodmason that he had 'Almost been thrown on my back to find a preparation to make this new paper answer tolerably I think I have now accomplished it though not without a good deal of trouble... I can easily prepare it sufficiently quick to answer any demand'. [71] Watt realised that the papermakers had facilities for drying paper and so recognised that the task of preparation should be given over to them. Therefore he sent Woodmason parcels of the preparation to be forwarded to John Bigg. [72] To overcome variations in paper quality, Watt tried to avoid giving it any treatment by altering the composition of the ink but came to the conclusion that 'the preparation of the paper is the surest means of having permanent copies'. [73]

The instructions sent to Bigg were to mix 'half a pint of the strong preparation to 30 gallons of water, for if used stronger the writing will spread most prodigiously. I am anxious until I hear how he succeeds as it is a very ticklish matter at first beginning'. [74] Watt did not reveal how he determined the strength of his preparation nor how he assessed the quality of the paper. However he was 'glad to hear that Mr Biggs succeeds so well with preparation and with the paper'. [75] So it seems that, from the autumn of 1780 onwards, Woodmason was supplied with treated copy paper directly from the mill, the preparation having been produced in Birmingham.

Anyone wanting to copy a letter had to moisten this special paper before placing it on top of the writing. This must have been quite a delicate task because wetting any paper weakens it, especially thin unsized sheets. The method outlined in the patent was to use a sponge or brush, or dipping to apply the water and then absorb the excess by placing the sheet between two cloths and gently pressing by hand. Soon wetting books were introduced. These consisted of sheets of 'bibulous' paper like blotting paper. The copy paper was soaked and placed between the leaves of the wetting book to absorb excess moisture. To make a copy, the original was placed on a sheet of 'painted' paper on top of a cloth-covered board. The damp copy paper was placed on the original and covered with a sheet of oiled paper then paste board and finally a piece of cloth. This pile was passed through the press rollers. [76] On removal, the copy was placed between the leaves of a drying book.

It appears that at first each sheet of paper was moistened separately for copying. Watt's friend, Patrick Wilson, explained how he had moistened 20 or 30 sheets, put them in groups of four in the wetting book for a couple of hours and then stored them for up to a month wrapped in tinfoil between wooden boards. [77] During February and March 1781, Watt improved Wilson's method by making a mahogany box half an inch deep of a suitable size for the paper. A close-fitting lid dropped into the box. The wooden surfaces were lined with tinfoil to prevent damp penetrating the wood. Paper prepared in the wetting book and placed in the box for a week had proved to be satisfactory and Watt thought that it would keep for months but he was worried that it might turn mouldy in the summer. [78] He then realised that moistening the paper could be done in these boxes by putting a layer of well-soaked sponge paper above and below the stacks of sheets, placing the top on and leaving them to soak for five or six hours 'by which time you will find that the moisture has penetrated the whole paper'. [79]

Because these damping boxes made the final preparation of the paper so much simpler, Watt started to manufacture them commercially but ran into difficulties. He told Keir at the end of March 1781:

> We have very few of the damping boxes but the charge of them at Soho runs so high that without a very high price vizt 10/- pr pair we shall be no gainers by them I have sent out to try what can gett the boxes made for in Bir'm as our workmen seem to be of the very laziest kind and I do not think that Mr Playfair stickles enough with them – They take 3 hours to cover one of them with tin which I did with my own hands in about one hour but intend to make another trial & note the time exactly – When things are apparently dear as is the case with these boxes if to be sold at 10/- pr pair it prejudices people against the sellers of them & am therefore sorry we ever meddled with them at all unless we can get them done cheaper. [80]

This is a very early instance of a time and motion study as well as determining the production costs of an article. It shows that Watt was also aware of sales psychology regarding costings. Yet the price which was charged for a pair sold to Boulton and Fothergill in June was even higher. They purchased a press for £8.8s. and the pair of damping boxes for 15s. [81] In the Garret Workshop there is a pair of these boxes. One has what looks like a plaster of Paris bottom which would have been soaked in water to damp sheets of paper placed on top. The second box, lined with sheets of tin, was the store for the damped sheets. Yet in 1806, Watt told De Luc:

> I have given up the use of the damping boxes in any private practice... they do very well in a country house where the consumption of paper is great, but when it is slow they are subject to the faults you mention. I wet the paper as I use it by means of a large brush about 1¼ inches diameter close to the socket or tying. [82]

Watt had reverted to the method he had tried at first.

The Copying Paper

It took a long time before a suitable paper was found for the copying process, reflecting Watt's quest for perfection. A thin paper was necessary for the ink to penetrate right through so that the writing could be read the correct way round. But the paper had to be tough enough for handling when wetted. Watt's notes and letters mention the names of five papermakers, Whatman, Ruse, Lea, Taylor and Bigg. This Whatman was the younger James Whatman whose Turkey Mill was on the outskirts of Maidstone, Kent. The mill which Joseph Ruse may have operated at this time has not been identified. Later in 1799 he was a papermaker on his own account at Upper Tovil Mill, Kent. The Lea family were running Hurcott Mill at Kidderminster from at least 1779 to 1790. Taylor was Clement Taylor the younger who took over Upper Tovil Mill from his more famous father, running it from 1772 to 1797. The Bigg family had a long association with Iping Mill in Sussex. John Bigg took over the business from his mother Susannah and developed it into a large concern for that time with six papermaking vats when he retired in 1800. [83]

Whatman and Ruse were soon out of the running. Ruse is mentioned only once; 'The specimens you have sent me of Ruse papers are not sufficient to form a judgement by as nothing less than a whole letter is proof'. [84] Watt was well acquainted with Whatman paper since he used it both for his letters and drawing paper. [85] However he found it unsuitable for copying and told Keir:

> Taylors wove paper, at least the spec[ime]n sent me was too thick and seems to participate the faults of Whatmans that is though a good ink and a good preparation will strike through it yet it will in those cases be difected [sic] – I have today tried it again over again with all the care I could but find the same faults. [86]

In a copy of a letter to Hamilton in April 1779 on Whatman laid paper, the ink has penetrated in blotches and in a couple more a year later to Annie and Capt. Marr the ink is very faint. [87] On one of the small test samples found in the Garret Workshop, Watt wrote, 'Despond not, Whatman, perhaps it may do at last', but it never did.

The surviving Whatman press copies are on laid paper. This was formed on a hand mould which had the surface through which the water drained to leave behind the cellulose fibres made from a grid of parallel wires running from left to right tied at their correct spaces apart by chain lines running from front to back. These laid and chain lines left their impressions in the paper with thinner or thicker lines. In some thin laid paper in the Boulton and Watt archives, the paper at the chain lines is very weak indeed. This unevenness caused problems in copying, sometimes making words difficult to read. Spacing of the chain lines in Whatman paper was about $^3/_4$ ins. apart.

The elder Whatman had introduced for the first time paper called 'wove' with no laid or chain lines for the edition of Virgil's Aeneid published by the Birmingham printer, John Baskerville. The younger Whatman continued to promote this type of paper after his father's death in 1759. It was made on a mould with a cover of woven wire so that the paper was of equal thickness throughout. [88] Its smoothness made it attractive for Watt's copying process. It has been a surprise to find that other papermakers were also manufacturing thin wove paper at such an early date as 1780.

The enrolment of Watt's patent hastened the search for a copy paper that performed well. Some samples have very fine laid lines with chain lines spaced at only half an inch apart, giving a very smooth paper. On one is written 'Kidderminster' and on another dated 18 July 1780, 'Hurcott Paper'. This paper seems to be the same as some with the watermark

'AMSTERDAM' so it appears that all this paper was produced by the Leas. In fact the earliest copy paper that can be identified came from Hurcott Mill. On 9 April 1780, Watt wrote to E. Lea, 'I find upon trial that your paper answers very well'. He wanted more, half as thick again and pressed harder. [89] He commented to Keir about this paper, 'We should never have sent you a good impression without it. It has saved our credit'. [90] This copy is on paper watermarked 'AMSTERDAM' with narrow chain lines. While harder pressing would have given a firmer paper, its effects would have been partially negated when it was damped in the copying process. Watt thought he was:

> ... confirmed in the belief that very hard pressing makes better paper for copying for I took some of the worst of this cargo of Hurcott paper [a second batch], wett it with the preparation and past it 3 or 4 half sheets together very often through the Rolls hard sett wch answered in some degree the purpose of pressing though not equally with the papermakers press. [91]

This would have had the effect of partially restoring the initial pressing. Lea had made a second batch before the end of April about which Watt commented that the colour was not so good and thought it still needed more pressing. [92] He found that paper which was pressed more copied better than that which had not been. Lea had produced this second trial very quickly because more time was usually necessary to complete the various craft processes in hand made papermaking. Watt's comment on the colour shows the difficulty of quality control. At the same time, Watt enquired about woollen pasteboard 'as per sample'. True paper cannot be made with the protein fibres of wool. A small amount of wool can be added as a filler but this was unusual. Perhaps Watt was mistaken about the content.

The next papermaker to be identified is Bigg. Watt told Keir on 16 April 1780, 'Biggs thin reddish coloured paper does very well but is vastly inferior to Lea's and is made on a very coarse mould'. [93] Watt liked its texture and thought that if the price were right, it might do for the cheaper sort of copying paper. [94] He 'Re-examined Biggs's paper and find it to be the best fabric of any and quite of the proper thickness – If you have any quantities of these papers made they should be sent down here that we may sett about preparing some as that process must take time'. [95] This paper was laid with the watermark 'J BIGG' appearing in the copy of a letter sent to John Lean on 9 May. [96] Watt told Keir, 'In relation to the paper I approve of Mr Biggs being employed as his reddish paper was by much the best quality of any we ever had, and in my opinion, of the proper thickness'. [97] Soon Bigg was also supplying wove because on an undated sample in the Garret Workshop Watt wrote, 'This is written to try Mr Biggs new thin paper of which I have just received a ream from Mr Woodmason it seems very good as far as can be judged from appearances'.

By this time the third competitor had entered the scene. On 26 April, Watt reported to Keir that he had:

> ... mentioned to Mr Boulton that Taylors thin paper was exceeding good and so was Biggs's reddish kind. Taylors wove paper at least the spec[ime]n sent me was too thick and seems to participate the faults of Whatmans... N.B. I don't object to the wove mould, which I admire but to the texture of the paper which is too dense and close. [98]

This is the first definite mention of a wove sheet. Watt was impressed by the equal thickness of the sheet because he pointed out to Boulton how Taylor's laid was 'excellent saving its being almost cut through by the mould'. [99] These thin places were caused by the chain lines protruding into the layer of pulp on the mould.

However Watt remained suspicious of paper made on wove moulds and told Boulton, 'I am glad that so many objections have occurred to the wove moulds for the paper made on them, at least the specimen sent me, did not copy well which seemed to be owing to some bad arrangement of the fibres caused by the mould'. [100] Watt may have been justified in his criticism because early wove moulds had different drainage characteristics to laid ones and the vatmen had to learn how to handle them. Watt found fault with Taylor's wove paper because, while 'being very beautiful', it weighed seven pounds a ream instead of only five. He therefore rejected it for copying paper and suggested that it be sized and sold as writing paper. [101]

Watt had ordered some special wove paper with the watermark 'J WATT & CO PATENT COPYING' which he was using by 2 June 1780. This watermark appears in many more copies later but the manufacturer is unknown. He still remained sceptical about this type of paper so a trial was held by Taylor in July:

> Taylor's wove mould paper did not take good copies & that they did not stand so well as they did upon the Kidderminster paper... I suspected from something in the appearance of the paper that the texture of the mould might have some hand in it. On finding that these very thin papers of Mr Biggs's have the same fault I am more confirmed in this latter conjecture. My reasons for thinking that the being made on a wove mould may make the paper unfitt for our purpose was that the order Mr Taylor sent the first specimen of wove mould he also sent some exceedingly thin post paper which received very good copies and I was told that this thin post paper was made out of the same vat. [102]

Therefore Watt ordered Woodmason to see that no more paper was made on wove moulds.

This edict was also applied to wove paper from Bigg whose paper was not to have his name in it. In the middle of August, Watt said that Hurcott's was the best, followed by Bigg's and then Taylor's. [103] It may have been another batch of wove paper from Bigg which caused Watt to change his mind before the end of August when he told Woodmason, 'Wove paper is the best... I perfectly agree with you it will be best to adhere to the wove moulds... it is of a superlatively beautiful texture and evenness'. At this point, Watt was preferring Taylor's paper and wanted some manufactured immediately:

> It takes a good impression and will bear stronger preparations than Mr Biggs wove paper... I have a notion the Mr. Biggs paper does not differ from it so much from the coarseness of the mould as from the manner of cutting the rags. In good paper I apprehend that though the rags are long cutt they are also made very fine and silky which makes the paper closer and finer in the grain. It appears plainly by some trials which I have lately made that the smoothness and neatness of the impressions or copy depends much on the fineness of the fibres & on the thinness and solidity of the paper. [104]

Watt had discovered the truth of the saying that 'Paper is made in the beater'. Pulp preparation can be regulated to avoid cutting the rag fibres into short lengths since long fibres give strength. Then beating can follow to fibrillate or break open the fibre sheaths so that the fibrils on different fibres can bond together, again adding to the strength. At last Watt had found the type of copying paper he wanted. From September 1780 onwards, he does not record in his correspondence any comments about searching for better paper.

However there is a suggestion that, while he preferred paper from Bigg, he wanted to keep his options open to avoid reliance on a single supplier – 'Mr Biggs's water is much purer than Mr Taylors but this information you may keep to yourself as it may make Mr Biggs worse to deal with'. [105]

But there still remained problems with the other papers necessary in the copying process, particularly the wetting and drying books. Watt wrote to Woodmason in October 1780:

> The wetting book & drying book papers do not answer well. They should both be very moderately pressed in the ? and made of soft stuff and the drying book paper should be prest smooth in the second pressing. These articles are of more consequence to the well doing of the scheme than you can readily imagine for if the wetting book does not suck the paper sufficiently dry the copy diffuses and if the drying book paper does not touch every part of the copy so as to suck it immediately dry the same thing happens. [106]

Watt was still searching for perfection. Yet through all his struggles to find the right types of paper, Watt increased his knowledge of different types and their manufacture both in England and abroad. He was tempted to launch into papermaking himself which he thought might make a good living and told Mons. Reveillon, 'If I did not hate to have any concern with the Excise Laws, I should have made some attempts in that way myself'. [107]

The Copying Press

The patent contains drawings of two presses, screw and roller types. Screw presses were used in houses to press linen and in paper mills to press the newly couched sheets of paper between felts. Watt would have been aware of these as well as many other applications. The advantage of this type was that the pressure could be maintained so as long as it was left screwed down. The illustration shows a small cast iron example which later was commonly used in offices for copying. In May 1780, Watt prepared some paper and 'left it in the press all night to see if that would level it', which is the only direct reference discovered in his notes to this type. [108]

Watt preferred the press with rollers, either the 'common rolling-press, or of that of which a description and drawings are hereunder written and drawn... in the same manner as is used in printing by copper plates'. [109] Watt had seen such a press in Edinburgh printing engravings for his Uncle's Map of the Clyde as well as the rolls for calendering linen at Luncarty. (See Vol. I, pp. 92 – 7 & 134) These gave a much higher pressure at the nip of the rollers but for a short period only. It was probably this feature which he recognised was better for his copying process since it gave less time for the ink to transfuse and spread across the copying paper. In 1770, he had advised Bell about his invention of printing calico with engraved rollers which would have been similar. (See Vol. I, p. 135) Production at Birmingham concentrated on the roller press for it seems that the partners hoped to make a profit from sales of it to recoup expenses of the patent. Payment by the purchaser covered a press, paper and ink in a single transaction with no annual premium as in the case of the steam engine.

While Watt noted that he was 'Taking Proofs with roller press' in the middle of September 1779, [110] it was only early in December that he 'finished the drawing of the rolling press & Mr. B[oulton] wrote dimensions on it'. [111] One roll was placed on top of a second with their journals supported in bearings in cast iron frames. The bearings of the upper roll were fixed and this was the roll turned by a handle. Those of the lower roll were supported by springs so that the rolls could separate when the board with material to be copied was passed between

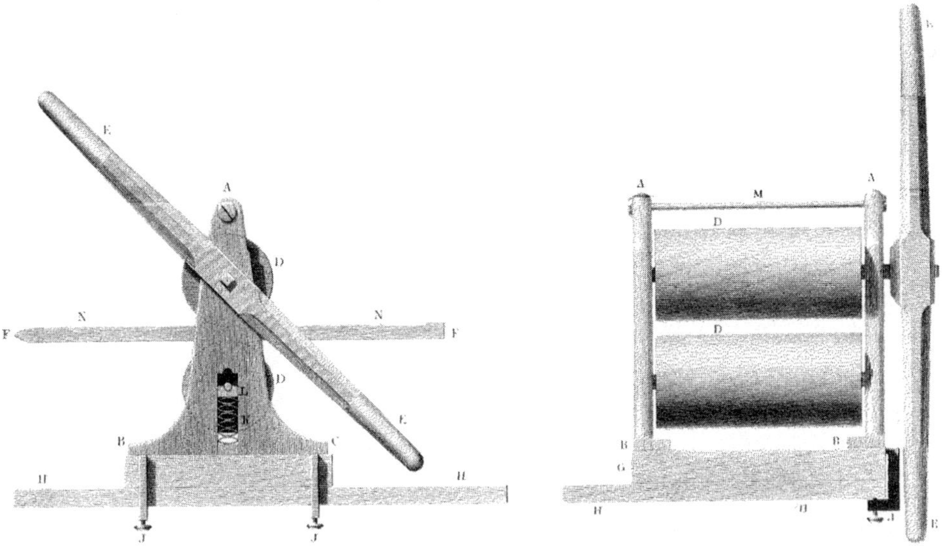

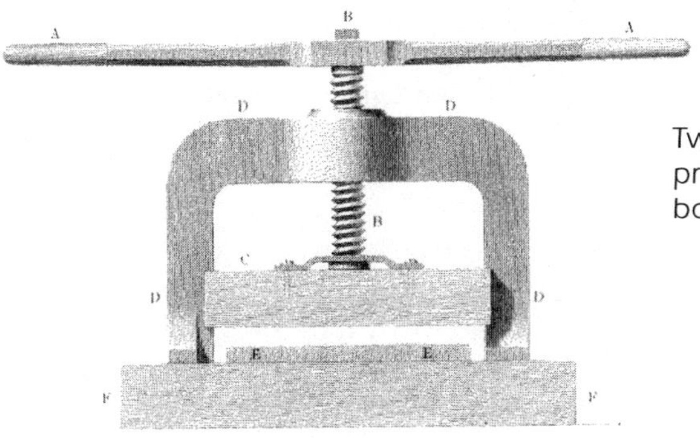

Two types of copying presses; top roller press, bottom screw press.

them. There was no method of adjusting the pressure. Iron journals were pressed into the ends of the wooden rollers. While those for the lower roll might be plain round ones, to drive the upper roll the part of the journal pressed into the roll had to be formed into either a cross or a square. [112]

The presses were manufactured at Soho in various sizes, with concentration at first on ones suitable for people's houses. Watt wrote to Boulton in May 1780 about 'his new plan of Roll frames' [113] but it is not clear whether he was referring to himself or Boulton. The frames gave no trouble, which was far from the case with the rollers. The first choice for their material was the very hard wood, lignum vitae, which might have been expected to withstand damp. At first, sets of rolls were ordered from outside suppliers. '100 setts of Rollers of 4 & 5 inches diam.' were ordered from a Mr. Taylor in December 1779. [114] Watt was in Bristol during the following February seeking supplies of lignum vitae and found some at 'Mr. Haywards Blockmaker which he offered to be cut in proper lengths'. [115] Generally this material was purchased from John Lean at Bristol who was the main forwarding agent for steam engine parts and much more. But around early 1780, the partners had decided to establish full manufacturing facilities at Soho.

Watt returned from Cornwall at the end of February and set about organising production but, presumably to tide them over until special tools were ready, ordered '50 pair of 4 inch rolls – also a few pairs of 2 feet long $5^{1}/_{2}$ to 6 inch dia.' from Lean. [116] Watt gave instructions about boring the rolls and a little later for making a lathe on which the rolls could be turned with special tools. [117] Boring the end holes continued to give problems so he had to alter the method because 'The machine for Boring Rolls is apt to bore them of one side the cause of which not yett found out, proposed to make the Roll turn round and the bitt stand still'. This is modern practice. The roll was to be turned by a driving dog at one end while the drill bit was fed into the other through a special guide. While this worked satisfactorily to bore round holes, problems were still encountered when squaring the holes with a taper to fit the drive shaft. He solved this partly by grinding the forged shafts more accurately as well as using squared punches followed by tapering with 'floats'. [118] At the end of April he reported to Boulton, 'The Roll turning after much labour is at last nearly in perfect order and we shall go on quickly'. [119] Watt had achieved this through designing special tools.

Watt standardised sizes of presses. In May 1780, he asked Black what size his Grace the Duke of Buccleugh wanted, 'Whether only for Common post paper which requires 12 inch Rolls, or larger. Our present apparatus will make them as far as 16 inches long which will take a half Sheet of Imperial paper – The price is $^{1}/_{2}$ a guinea an inch of the length of the Rolls'. [120] Different lengths needed different diameters. Four inch diameter rolls were the minimum for ones twelve inches long on a machine 'sufficiently portable'. [121] The more usual diameter was $4^{1}/_{2}$ ins. on this size. In June Watt ordered enough lignum vitae for 400 rollers of this diameter by 12 ins. long, for 40 rollers 5 ins. diameter 14 ins. long, for 20 rollers 5 or $5^{1}/_{2}$ ins. diameter 16 ins. long and for 20 rollers 6 to 7 ins. diameter 24 ins. long. This is likely to have been from Lean since Watt complained about the poor quality of some lignum vitae supplied earlier. [122]

While the machines for producing the different sizes of rollers were being constructed, Watt had to solve another problem, that of finding suitable springs to keep the rollers under pressure. In April 1780, he tried new steel springs which were 'immensely more elastic than the others and much easier made'. [123] Boulton had suggested bristles which at first Watt thought better but tried steel again. He 'adjusted steel springs in press found that the bad copies I had made were owing to the press not being hard enough sett on the side of the steel springs which I remedied'. [124] However a few days later he told Boulton he was sick of steel springs which broke, something that happened to spiral springs as well which in any event were too weak. He reverted to bristles tied up in small bundles with waxed thread which he found best. Eventually a dozen slightly curved steel springs 'were put together and then as many reversed... proved stiff enough and sprung about $^{1}/_{2}$ an inch. It took about 60 of them to pack one end'. [125]

Through these and other problems such as lack of lignum vitae, Watt sent a letter to Woodmason on 23 July apologising that he was 'ashamed we have been so dilatory in getting the presses finished yet we push them forward as fast as we can – Mr Boulton directs his principal part of his attention to them'. The first presses were despatched two weeks later on 7 August. Hamilton commented, 'As to the copying machines, I am glad they are at last coming to light. There have been so many delays to their birth that I was beginning to be afraid of ever seeing them'. [126] Hamilton knew his Watt.

Soon a design for a stand was prepared, engravings of it printed and sent to Woodmason in September. [127] In October, about 150 small presses were nearly ready which could be sent off at the rate of '6 pr day or thereabouts' while some larger ones were in preparation. [128] Some form of batch production had been organised for Watt told Woodmason, 'We have never yet made any large machines, otherwise one should have been sent to Mr Cooper. The making of them requires an alteration to be made in our tools, and therefore we wished to finish all the small ones that were subscribed for first'. [129]

In spite of all Watt's care, the presses were not entirely trouble-free in use. Hamilton complained that the board on which the material to be copied was placed sometimes ran to one side. This had happened to Watt who was 'certain it is not by the want of cylindricity in the Rolls as we have a machine which turns them absolutely true'. He thought it might be due to the springs on one side being stiffer than on the other or the paper might have been placed to one side. [130]

More serious was Wilson's complaint about the rollers splitting. He had made a canvas-covered wooden frame to put over his machine and saw that moisture was applied to the rollers in an attempt to prevent this. [131] It was a problem probably already encountered by Watt for he replied in February 1781, 'In relation to the Iron Rolls we have not yet completed any and cannot tell how they will answer yet but are sure they will turn out much dearer than the others – I shall let you know their success and will supply you with a pair as soon as they are complete'. [132] A month later he told Wilson:

> In relation to Rolls type mettal will not answer for seven reasons which I forbear to enumerate – We make rolls perfectly true of Iron cast hollow which weigh no more than the lignum vitae & we dye them of a beautiful brown like the tea urns. We are also making some of Brass for Nabobs. [133]

This letter suggests that some rolls had been finished which does not correspond with one to Hamilton dated to September 1781 in which Watt said:

> We are now trying at the making of Iron Rolls but they will come out high & we have not yet been able to make them look well from their ugly grey colour but shall try to rust them or bronze them by some means or other, we shall also make Brass Rolls for the east and West Indies – we think that the Iron Rolls will come at £8.8. & the Brass at £10.10 but do not know yet at any certainty . [134]

The date is most likely wrong because the manufacture of presses was carefully costed in March 1781 by Pearson. He could make 36 machines at £3.3s.each. The cost for machines, boxes, paper, etc. was £4.12s.8d. which sold at £6.11s., giving an apparent profit of £1.18s.4d. Watt thought Pearson's prices were too high. He also thought that 'As there is so much loss & hindrance with the wooden rolls & I do not believe that the iron ones cost one farthing more if so much I think we should sell these at £7.7.' [135] So the change from wood to iron rolls had taken place. It is evident that not only were the presses being produced in some sort of production line process with special machine tools, but also that estimates of production costs were attempted. When Playfair was going to leave Birmingham for London, Keir told Watt, 'The manufacturing of the machines is so well regulated by Mr Playfair that it will be very easy to continue it in the same order'. [136]

Sales of Copying Machines

The partners had projected a ready sale of a thousand machines at five and a half guineas each which was far too optimistic. (See p. 193) As soon as the patent had been enrolled, they started a campaign to publicise their new invention. Boulton demonstrated the apparatus in the Houses of Parliament where it aroused such interest that the Speaker 'was often obliged to send his proper officer to fetch away from me the members to vote, and sometimes to make a House'. [137] He also went round the coffee houses and other places where samples were exhibited. Prestigious machines were sold to the Kings of England and Prussia, the Duke of Buccleugh, Lord Mansfield and others. Boulton's enthusiasm helped to launch the project. Sales were to be handled by Woodmason who would make his profits from subsequent sale of the paper. In this way, Boulton and Watt avoided setting up their own sales organisation.

A thousand copies of the proposals were ordered in March 1780 [138] when the price had to be raised to six guineas on account 'of more expence than we had expected and the necessity of giving all the little implements with it'. [139] Watt thought Boulton's methods of publicity were too profligate and told Keir:

> I have stopt printing proposals as I grudge the expence of them & the specimens and I think they may now be altered to a more convenient form and also improved in the wording – there is about 400 on hand part of which I shall send to Scotland to Liverpool & to Bristol – the giving away great numbers of them will answer no great end At least in small towns a few stuck up in public places is a much better way and some should be printed in small form for foreign letters. [140]

It must have been a tedious task to write the specimens by hand and copy them one at a time. On the prospectus was stuck a sample of the copying paper with copied onto it, 'Time Labour & Money are saved, Dispatch & Accuracy are attained, and Secrecy is preserved by this newly-invented Art of Copying Letters and other Writings'. [141] In the prospectus, Watt claimed that 'By means of this Invention, the practice of which is exceedingly easy, any person may take an exact copy of a letter or other sheet of paper written with common ink in a minute or less'. Watt wrote a second letter to Keir complaining of Boulton's tactics:

> Mr Boulton says also he wants specimens for the proposals. I know not how that is as I have repeatedly ordered loose specimens to be sent, very lately sent 60, however more shall come – I wish this matter could be managed without so much puffing, in my opinion it would puff itself if a few were out – I fear Mr B will spend too much of his time in going about shewing it to Noblemen, now a merchants 6 guineas are as good as the noblemans and he may perhaps recommend us more customers, and if while the machine is out showing to one or two noblemen 12 merchants whose time is more valuable are disappointed, we lose by the bargain. [142]

Whether it was the rise in price or the delay in delivering the machines until the middle of August because they 'take more time to make than expected', [143] or because operating them was found to be more difficult than expected, some people failed to take up their orders. Instead of the thousand expected, by the middle of June they had 'got only about 400 subscribers yet but have a certainty of a number more so soon as the presses appear'. [144] Woodmason must have been frustrated by the delays in production since Watt pointed out to him that their main concern, the steam engine business, was suffering from lack of attention:

> You desire that I would apply to this business as much as possible I assure I have done so to the great detriment of our engine business, however I was obliged last week to make a journey otherwise we run a risque of loosing the erection of a beneficial Engine besides forfeiting our credit there. [145]

Probably sales picked up because at the end of November, Watt wrote in another letter to Woodmason:

> We are happy to find that you have so good an opinion of the copying business and will be glad to prepare a larger number [of presses] if they should be wanted. As you are better able to judge of the probability of their being wanted from your hearing the sentiments of those who have used their machines but particularly from your knowing exactly how many are sold and how many are ordered we shall very willingly follow your advice in proceeding to get more machines made. We should like always to have some ready for sale, but at the same time observe that the making of so large a number as 500 will be attended with a very considerable expence and great loss if the public should not continue their present good opinion... You may depend on our being able to execute from 20 to 30 per week with our present establishment of workmen and tools. [146]

This was a period of great financial difficulty for Boulton and Watt because the steam engine business was running a considerable deficit. This letter reveals the problems of funding a new enterprise with no capital as well as the scale of production achieved through Watt's organisation of the production methods.

Smiles says that 150 machines were sold 'by the end of the year... and thirty were wanted for export'. [147] At the beginning of June 1781, Woodmason was sent a further three machines, 'making in all 489 Copying Machines sent to you'. [148] This would account for the bulk of the production with both Boulton and Watt supplying a few more directly to their friends. In May, the price had to be raised to seven guineas but 'we do not fear selling any the fewer for it as the purchasers are generally opulent & the utility of the machine is now pretty well constituted'. [149]

However for the partners in Birmingham it was not at first very profitable because Watt told Hamilton in September, 'We fixed the price of the others too low, it does not turn out so profitable as we hoped. The incidental expenses run monstrous high, however it is a gaining trade'. [150] Almost exactly a year later, he reported again:

> As to my own subject here it consists in the engine business and in my share of the Copying Business, the Balance of which in my favour is £992 but there is no money there it consists in Goods & effects Debts etc which however are daily coming in and perhaps in case of necessity some money could be raised on it. [151]

Smeaton was sent a present of a copying machine in August 1780 and, in his letter of thanks offering to pay, described his experiences using it which show the difficulties:

> I inclose you the first Specimen I took off, and was wishful to have been able to have exceeded it, by more Experience, but I don't find that upon the whole I have done it much. As it is observed in the book of Directions, the having the paper wetted to a proper degree, is a main matter; so that if by wetting it more I attempt to make it strike through more plainly I find it makes it run more into the paper; particularly after it is done, though it comes off very sharp. If some method was therefore formed by which the wetting might be adjusted, it would add much to the satisfaction of those who use the machines. I have also received a further parcill of paper... all of which I find has a small matter of size in it, so as to enable it tolerably to bear Ink. This I find if not thoroughly wetted, takes of a very faint impression; and when fully wetted is subject to want of perfect sharpness, the same as the first kind, which seems to be without any gum or size. Upon the whole, though I have constantly used the machine and find it very useful for the purpose intended, yet I

> cannot perswade my self that I have been ever able to take of copys that have struck so fairly and sharply through as the specimens I have seen done by Mr Bolton himself; and I can only impute this difference to that of wetting with fair water, instead of the liquid used by Mr Bolton; and from the whole of the Tryals I have made, it seems to me that the sharpest copys, are taken off by the thinnest paper, which is the browner sort of paper last sent, marked J. Bigg, and that which came with the machine marked Amsterdam the Pattent paper, though of a firmer texture, and easyer to manage otherways, yet by giving leave for the Ink to have more room to spread, it accordingly either spreads more, or does not get sufficiently through to the foreside.
>
> I don't mention these matters or what follows, by way of fault or obligation, but as these machines it is likely will get into hands less competent to the Execution, I do it by way of information to Messrs Bolton and Watt as to what points their customers less versed in Mechanical Matters may... [missing]. [152]

Later Smeaton purchased a press with iron rolls and had trouble with the ink. Watt pointed out to him some of the problems from which the inks and copying process suffered:

> 1 You made a wrong conclusion on thinking that the copying ink for drawings would not change by keeping. I mentioned to you that it will not continue good above 3 months in winter or 2 in summer.
> 2 Either I have neglected to shake up the bottle when I poured out the ink I gave you or you have neglected to shake it up before you used it, otherwise the original at least would have been black.
> 3 The reason the lines were not clear was either you had wetted the original too much or that you had not dryed off the copying paper sufficiently. The time it should by [sic] wet is from 3 to 4 minutes for the paper you used.
> 4 If upon shaking up the ink proves too thick dilute it with a very small quantity of strong white wine which also serves to restore its copying quality in some degree.
> 5 You have used the proper side of the paper.
> 6 If you take the impression too soon, it will be less perfect & if you let it become too wet it will be more diffuse.
> 7 China ink will not copy at all.
> 8 The cloth, oiled paper & pasteboard are glued together on purpose having found it an improvement.
> 9 There are several liquors that will spoil on a good copy but none which will repair a bad one. [153]

Although James Watt & Co. continued to make the ink, complaints from customers over both that and the quality of the paper may be one reason why Keir suggested towards the end of 1781 that Watt reveal his recipe for the ink so that people could make it fresh themselves:

> ... otherwise if the world is made to depend on our ink, or are not instructed to remedy its faults, they will be often disappointed... I am inclined it would be for our interest to resign our monopoly both of paper & ink for copying & to instruct the world, especially foreign countries, how to make both - The general use of the machine would be facilitated. [154]

While the partners could not control the quality of the ink or paper, they could guarantee that of the presses. The design of these was contained in the patent so they could still regulate that.

Watt was justified in his fears that his invention would soon be copied. In October 1784 he wrote to Matthews about a cabinetmaker called Fleming who resided in Chambers Street, London, and was making the machines. [155] In spite of such competition, James Watt & Co. carried on and in October 1794, Watt told McGrigor, 'Finding myself unable to transact business as I have done, especially with the burden of our lawsuit on my mind I have given up my share of the copying machine business to James, as Mr Boulton has given his to his son'. [156] The sons continued to develop the business with a small portable press. Although its acceptance at first may have been slow, the copying machine based on Watt's experiments became one of the indispensable pieces of office equipment throughout the nineteenth century and even into the twentieth. It may be considered as one of the earliest consumer durables based as it was on batch production. When Watt was replying to one of his correspondents in 1809, he wrote, 'It is gratifying that you find the copying-machine useful to you. It has been so much so to me for the last twenty-six years that it has been worth all the trouble I had with it, had it been attended with no other profit'. [157] His invention continued to be useful to very many people for the rest of that century at least.

Ballooning

In the early history of ballooning, we can see another example of Watt investigating this new phenomenon of natural philosophy together with his friends in Birmingham and elsewhere. [1] Their reactions reflected their characters, for Watt quickly determined mathematically the practical possibilities while Boulton enthusiastically rushed into construction of actual balloons. Black in Edinburgh took a more detached view of what could be achieved. Their activities coincided with Watt's investigations into the composition of water so his interest in ballooning may have been roused not only through the use of heat in hot air but also in hydrogen which he thought was another variety of heat as phlogiston. Therefore he may not have clearly distinguished between the different gases which the balloonists first employed.

Black had recognised the potential of hydrogen, inflammable air, in 1766 after reading the paper in which Henry Cavendish gave the results of his experiments on the specific gravity of hydrogen. [2] It occurred to him, 'as an obvious consequence of Mr. Cavendish's discovery, that if a sufficiently thin and light bladder were filled with inflammable air, the bladder and contained air would necessarily form a mass lighter than atmospheric air, and which would rise in it'. [3] Black tried with the 'allantois' of a calf and demonstrated this floating up to the ceiling to some of his friends but he never thought about any large-scale experiment. [4] It is not recorded whether Watt was one of the observers. Robison may have been there and so was encouraged to experiment, telling Watt in October 1783 about this. 'It is 17 years since I filled balls of varnish with inflammable Air, and had them rise with great rapidity, but I never could get them to dry and remain firm. They burst like Sope bubbles'. [5]

Our story really starts in France in November 1782 when Joseph Michel Montgolfier sent up a small hot-air balloon to a height of 70 ft. Shortly afterwards, he and his brother, Etienne Jacques, experimented with a larger balloon which reached an estimated height of 600 to 900 ft. The hot-air was generated by burning straw. The Académie des Sciences was invited to report on a further experiment at Annonay in the south of France. A balloon consisting of a hollow globe made of linen and paper, weighing 450 pounds and carrying 400 pounds was filled with hot air and travelled nearly 2,500 yards. Other demonstration flights followed, including one before the King and Queen at Versailles. Human flight was achieved on 15

October 1783 when J.F. Pilâtre de Rozier ascended 80 ft. in a captive Montgolfier balloon from Faubourg Saint-Antoine in Paris. This was followed on 21 November by a free flight undertaken by the same aeronaut with the Marquis d'Arlandes. They rose to a height of over 2,000 ft. and travelled nearly five miles. [6]

Considerable enthusiasm was aroused and great expectations were anticipated for this new mode of transport. Franklin saw some of these early flights in Paris and wrote to Sir Joseph Banks, President of the Royal Society in London:

> Some suppose Flying to be now invented, and since Men may be supported in the Air, nothing is wanted but some light handy Instruments to give and direct motion. Some think Progressive motion on the Earth may be advanc'd by it, and that a Running Footman or a Horse hung and suspended under such a Globe so as to leave no more of Weight pressing the Earth with their Feet, than perhaps 8 or 10 Pounds, might with a fair Wind run in a Straight Line across Countries as fast as the Wind, and over Hedges, Ditches & even Waters. It has been fancied that in time People will keep such Globes anchored in the Air, to which by Pullies they may draw up Game to be preserved in the Cool, & Water to be frozen when Ice is wanted. And that to get Money, it will be contrived to give People an extraordinary View of the Country by running them up in an Elbow Chair a Mile high for a Guinea, etc. etc. [7]

With such prospects, it is not surprising that Watt and others were also excited.

The news must have spread fast because Watt wrote to Black in September 1783:

> The whole world is full of these flying balls at present I know very little more of them than you must have seen in the papers, except that Montgolfier has found a method to make inflammable air by burning wet straw so cheap that he can fill a ball of 30 feet dia[mete]r for 1/2 a Crown, but this method of doing this is a secret yet, at least if he has communicated it, it is only to the academy, who were to send up a Ball of 0 this shape 70 long dia[mete]r, and 38 the small on last Friday and 4 or 5 criminals tyed to it. [8]

Watt received a copy of the *Journal de Paris* for 23 December 1783 which contained details of the ascent by Robert Charles in a balloon filled with hydrogen on 1 December. [9] It is possible that this paper was sent by De Luc because, in a reply to him on 2 February 1784, Watt discussed the figures for what today we would call the calorific value of straw. We can see that his interest lay in the effects of heat:

> Your letter I received by Mr. Argand... It is said that 50 lb of straw rarified the air in Mr. Mongolfier's [sic] large balloon so that it was expanded 1/3 or rather reduced it to 2/3 of its former weight and I am of opinion it was still more rarified because the rarefaction of 1/3 is deduced from the height to which it rose and it must have lost much of its heat before it rose to that height and much of the heat off the straw must have been spent on other objects, having these objections still we have 3192 pounds of air heated 243 d[egrees]. therefore each pound of straw had heated 63 lbs of air 243 d and if straw and pit coal contain the same quantity of heat 1 pound of straw would have heated 23 pounds of water the same number of degrees It would therefore appear that either straw contains an immensely greater quantity of heat than any other substance is known to do, or else that the capacity of air is to ye capacity of water as 1 to 3 instead of being 19 to 1 as Dr Crawford alleges. Or if we

> suppose that no more air was heated than was contained in the bag at the end of the operation, vixt 2115 lb then each pound of straw heated 42 lb of air and the capacity of air is to the capacity of water as 1 is to 2 N.B. I have taken the quantity of coals necessary to evaporate a given quantity of water but if we take it from Messrs Lav[oisie]r & De l'Places experiment with charcoal the coals should heat a double quantity of water and in ye later statement of the case the equilibrium of air and water should be equal, but I think the former statement the fairest as it may be reasonably ?presumed? that as much of the heat of the straw escapes as does of the coals in our furnaces. I think therefore that this subject merits further examination. It is probable there is some mistake in it such as some gas being realy [sic] formed, or a much greater quantity of straw burnt than what is said. If the thing be literally fact, it entirely demolishes the origin of heat from capacity and will I believe furnish the means of making an Engine in which heated air would be the acting power, and which would be at least on a par with steam. I therefore wish that these observations of mine may remain locked up in your own heart until I shall get some experiments made on the subject and see whether we have anything to hope from it... Making all possible allowances, I cannot see that the capacity of air for heat cannot be greater than that of water weight for weight so thus the Doctrine of capacity must fall to the ground and the expts. which have been made on it appear to be mere chimeras. But that of Latent or elementary heat shall be exalted and glorified. But let us examine what objections can be started to this expt. It may be alleged that the fluid heated is not ?atmos? but fixt air proceeding from the comb. About 1/3 of the straw is probably aqueous moisture and the remainder is earth phlogiston & fixt air. Dr. Watson gives in his analysis of 96 oz. Sallow wood, liquid 48, residuary 20, loss of weight 28, the liquid was partly oil, say about 1/3 then there was 32 oz. water liquid besides that part of the oil which is water which we shall at present suppose to be oil and we shall not take any notice of the earth of the charcoal which by the by is much less than Mr. Lr. makes it Then we have of charcoal 20, oil 16 = inflammable matter which call wholly phlogiston and 28 ?oz? of phlogisticated and inflammable air, say wholly inflammable air the whole inflammable matter forming around 2/3 of the whole weight. Therefore the 50 lb of straw would yield about 32 lbs of phlogistic matter & by Mr. Kirwan that would form 213 lbs of fixt air which is but one tenth part of the weight of the hot air and probably formed a much smaller proportion as a very great part would be reduced to water by the inflammation. Another part would be soot & all the charcoal would not be perfectly consumed – We have also made large allowances throughout. I mean to try the expt. First with a balloon of 6 feet dia. & 6 feet high, and to burn determined quantities of straw, dry shavings or splinters of wood until the air gains some certain heat and in order to determine the matter still more accurately I shall also heat it by gunpowder or by some of the mixture of nitre & charcoal or nitre & sulphur of which Mr. Lr. has tried the elementary heat – et j'espere q'une fausse route sera ferme pour toujours. I shall be happy to have your opinion on this matter as soon as you can it seems to me to open a new peeping hole into the mysteries of nature. [10]

Watt's scientific approach is well evident here, as well as his breadth of knowledge. There is his usual thoroughness in examining the problem from a variety of aspects as well as his precise observations. There is the detailed measurement of the materials involved. There is comparison with other experiments and the doubt arising when the comparisons do not tally. There is the prophetic recognition of the potential of hot-air engines and the possibility of the steam engine being eclipsed. Finally there is the excitement of new discoveries, a 'new peeping hole into the mysteries of nature'.

Watt obviously intended to carry out more experiments for he wrote again to De Luc, 'I propose as soon as I can to try the heat of the air in a balloon by an air thermometer of which expt. I shall send you an account. I shall try it by means of a mixture of nitre & charcoal the heat resulting from which is determined by Messr. L[avoisie]r & D.L. P[lace]'s experiments'. [11] He also made notes about filling a balloon with steam which he supposed would be about half the weight of air. He calculated the size of a globe 20 feet in diameter which 'would just swim in air'. He reckoned that to fill the ball and keep it up one hour would require an hundred weight of coals. [12] It is not clear how condensation of the steam would be prevented. These ideas may have been put on one side through the advent of the hydrogen balloon.

Only a few weeks after the Montgolfiers' experiments with hot-air balloons, the brothers J.A. and Robert Charles had a balloon made of silk treated with rubber. It was filled with hydrogen and released in the Champs de Mars in Paris on 27 August 1783. The advantage of using hydrogen was that the balloon did not need to be recharged in flight. The experiment was repeated by the Montgolfiers on 19 September. The Charles brothers made the first human ascent with a hydrogen balloon on 1 December. Meanwhile in that November, the first balloon to go up from British soil was released by the Italian refugee, Count Francesco Zambeccari from Cheapside in London. It was made from oiled silk so it would be less permeable. Aimé Argand demonstrated a hydrogen balloon in which Dr. John Sheldon ascended with Jean-Pierre Blanchard from Little Chelsea on 16 October 1784 while Cavendish made observations on altitudes and velocity. Then on 7 January 1785, the Channel was crossed by Blanchard and John Jeffries from Dover to twelve miles beyond Calais. [13]

In the intervening period, the Soho group had been carrying out their own experiments. Priestley's experiments of passing water vapour over iron filings reduced the cost of generating hydrogen considerably. [14] Watt told Hamilton about Smeaton's vision for equipping a person with a pair of small balloons that could barely support his weight and wings to regulate his flight but even so this would have to be with the wind. [15] On 26 October 1784, he wrote to Lind, and obviously had become rather sceptical about ballooning. Boulton had been experimenting with paper balloons while he was in Cornwall. [16]

> Mr. Boulton did idle a great deal of his time in playing with some small balloons some time ago but I hope he is now cured of the balloonomania which I look upon as a very bad disease for a man of the business & which gave me much vexations. I was guilty of being the proposer of the copper balloon but a little serious reflection, convinced me that it would not do, because the contraction & dilation it would have been subject to would have puckered & cracked it, of which I believe he was convinced. [17]

Watt's suggested solution to the problem of finding a suitable envelope to contain the hydrogen gas anticipated construction of modern all-metal airplanes.

Presumably Lind inquired further about their activities for Watt responded with the following description but unfortunately gave no date when a trial was held in Birmingham. His absence suggests that he did not look favourably on the proceedings:

> The history of Mr. Boulton's explosive balloon follows. He made a balloon of thin paper & varnished it with an oil varnish. The size was about 5 feet diar. It was filled with a mixture of about one part common air and two parts infl[ammable] air from iron. In the neck of the balloon he tied a common squib or serpent to which was

> fastened a match of about 2 feet long which was made very quick at the end next the serpent. When the balloon was filled the match was lighted & the balloon was launched. The night was very dark and nearly calm but the match being cut rather too long it was above 5 minutes before the explosion took place in which time the balloon had got above two miles from the place where it was let off. A considerable number of people was assembled to see the experiment and as they lost sight of the match soon after the balloon was let go, they expected it was gone out & that they should see or hear no more of it but when the match kindled the serpent, their murmurings were turned into joy which they expressed by a general shout which happening when the balloon exploded the noise of the explosion was very indistinctly heard, though the people who happened to be in the neighbourhood of it when it exploded said that it made a noise like thunder & almost as loud. Indeed they took the balloon for a meteor & the explosion for real thunder.
>
> Our intention of the experiment was to determine whether the Growling of thunder is owing to echos or to successive explosions but by means of that illtimed shout the question could not be solved otherwise than by the report of those who were near it who said that it growled like thunder but their observations upon it were very inaccurate and rendered more so by their being unprepared for such a Phenomenon. I was not at the place the balloon was let off from, but was at my own house, which is at least 3 miles from the place where the explosion happened. All I could observe was that the explosion was very vivid and not instantaneous, seemed to last about one second and the materials of the balloon taking fire exhibited a fine firework for a few seconds more. These are all the particulars which I remember or which are of any consequence to be known in a matter so easily repeated. [18]

The experiment does not seem to have been repeated, at least not in this form, but the scientific reasons for the trial should be noted and also Watt's reactions to them with his careful observations.

A few days later, we find Watt's final comments on ballooning in a letter to Watt junior then on the Continent. It is not clear whether this different ascent was organised by a Mr. Harper or whether he was the person in the balloon. [19]

> We have had an inflammable air balloon launched here on Tuesday last which proved a misty day with some rain but not much wind blew. It was 28 feet diameter carried up one man and went with him to 4 miles beyond Newcastle under line [sic] which may be about 50 miles in a right line. Time of the journey 1 hour & 20 minutes. The balloon was out of sight in 3 minutes and the traveller never saw the earth till he landed again but had clear sunshine he rose to the height of about 4000 feet; Bar[omete]r $26^{1}/_{4}$ thermometer 28d, the heat of the air below was about 40 d. & the barometer $28^{3}/_{8}$ inches at the ground. The balloon was filled with inflammable air from oil of vitriol & Iron at the expense of about £40. The fellow who went up was an ignorant hairdresser from Oxford, consequence made no material observations. [20]

While Watt keenly noted all the details he could, little scientific information was gained, partly owing to the Oxford hairdresser. However it was an astounding development in a short time.

It was quickly realised that ballooning could not be controlled sufficiently to yield many practical results. In addition, the expense of the hydrogen was considerable. Black had been very quick to appreciate the problems which he stated to Lind in October 1784:

> The hopes that means might be discovered for moveing these machines at pleasure & directing their course could only be formed by those who did not consider & understand their Nature – The only circumstance which enables them to subsist in the air however brisk the wind may be is that as soon as they rise they are in effect perfectly becalmed, I mean that there is not the least perceptible motion or impulse of the air against any part of them this is perfectly evident from their situation...
>
> This as I have said above is the reason why it is able to subsist in the air for it is quite unfit to sustain any considerable shock or impression of the air on it and this among other things shows the folly of attempting to give it motion at pleasure or command its motion & direction when it is up in the air for besides the impracticability of finding a power that can be applyed to produce this effect altho' the Power were found any attempt to make use of it would destroy the Balloon by impelling it against the air and exposing it to such shocks & violence as it cannot bear & further in attempting to use such a power it must necessarily happen on account of the great surface of the Balloon & the resistance the air would make to its motion thro' it, that the men & machinery by which it were moved must necessarily go foremost and the large Bladder etc would be dragged after them in a horizontal direction which would require an apparatus totally different from what has yet been contrived. [21]

Lind may have communicated Black's realistic views to Watt for the topic of ballooning does not appear in Watt's correspondence again after the letter to Watt junior in early 1785. One outcome of the activity in Birmingham was the publication by Southern in 1785 of one of the earliest books in English on ballooning. [22]

The Water Controversy

During the second half of the eighteenth century, the study of natural philosophy advanced so quickly that it became less and less possible to accept the concept of the four Aristotelian elements, earth, air, fire and water. (See Vol. I, pp. 143 – 6) Priestley wrote in 1792:

> There have been few, if any revolutions in science, so sudden, and so general, as the prevalence of what is now usually termed the new system of chemistry, or that of the Antiphlogistans, over the doctrine of Stahl, which was at one time thought to have been the greatest discovery that had ever been made in science. [1]

Watt's realization that water was not an element but some form of compound came at the turning point in the history of chemistry, leading up to John Dalton's announcement of his atomic theory in 1808. It has been seen as a key event in the battle between the supporters and opponents of the theory of phlogiston, resulting ultimately in the overthrow of earlier concepts. The revolution was profound so it is not surprising that Watt took a long time before he came to accept the new chemistry, if indeed he ever really did. Hamilton wrote to him in 1807:

> If they go on making such decompositions, they will reduce everything so as we shall not know whether there is any simple substance or not such a revolution a few years has made in Chemistry that if any of the old Chemists were to return they would know nothing of the matter & would look in vain for their old Theories which they thought so well established. [2]

While many new discoveries in chemistry were being made throughout the whole of Watt's life, Robison claimed in 1799, 'Watt was the discoverer of the Composition of Water (which discovery is the Basis of all new doctrines)'. [3] Hence it was only to be expected that Watt's thoughts might be seen later to be muddled, as the Reverend William Whewell pointed out. [4] Also, in Watt's time, there was no division between physics and chemistry as we know it today.

While it was recognised that air was some sort of mixture, gases were regarded as modifications of ordinary atmospheric air with which it was tainted or infected and were not distinct substances. Robison described steam as:

> ... the name given in our language to the visible moist vapour which arises from all bodies, which contain juices easily expelled from them by heats not sufficient for their combustion... It is distinguished from smoke by its not having been produced by combustion, by not containing any soot, and by its being condensable by cold into water, oil, inflammable spirits, or liquids composed of these. [5]

Identifying different gases began in 1754 when Black published his doctoral dissertation, *Experiments upon Magnesia Alba, Quick-lime, and some other Alcaline Substances*. (See Vol. I, p. 77) He had discovered 'fixt air' (carbon dioxide) and identified it as a distinct substance which he could quantify through the use of the balance. But he was not interested in investigating its properties and failed to advance the study of what became pneumatic chemistry. Yet others were quick to look further into this mysterious element, air, which was proving to have so many exciting forms and characteristics.

Cavendish was the next person to identify another gas which he called 'inflammable air', later termed hydrogen. He read his memoir, *Three Papers containing Experiments on factitious Airs*, to the Royal Society during 1766. The first one was entitled *Containing Experiments on Inflammable Air*. [6] He investigated some of the properties of his inflammable air by firing it when mixed with ordinary air in varying proportions and noting the loudness of the resulting explosions. He took these investigations no further at this time.

Joseph Priestley, F.R.S., 1733 – 1804.

Priestley had spent part of the autumn of 1767 reading works on air and was experimenting enthusiastically in 1771 and 1772. In June 1774, he obtained a burning glass with a lens 12 ins. diameter and set about discovering what kinds of air a variety of substances would yield. In August 1774, he isolated oxygen which he thought was 'pure air' because he was 'so fully possessed of there being no better air than common air'. He thought about a name for his new air and called it '"dephlogisticated air" (air eminently suitable for combustion, that is to receive phlogiston from combustible substances)'. [7]

Priestley had moved to Fair Hills, Birmingham, by October 1780.[8] Here, with financial support from Darwin and Boulton, he built a new laboratory and continued his investigations. Priestley was quickly accepted as a member of the Lunar Society and Watt welcomed him into his family circle. For example, in 1788, Watt sent him a note, 'I wish to lay before you all my results, and compare them at our leisure, and think this will best be done here, and if you come some day soon after dinner or take a family dinner with us, we can make any experiment you may wish to see yourself'. [9] In 1781, Priestley demonstrated what he called a 'mere random experiment, made to entertain a few philosophical friends, who had formed themselves into a private society, of which they had done me the honour to make me a member'. [10] This undoubtedly was the Lunar Society and we see the importance of its links once again. He exploded inflammable air with common air in closed glass vessels by an electric spark. [11] He noticed that, after the vessels had cooled, the inside 'became dewy' which he attributed to the mere deposition of moisture suspended in the gases. [12] His friend, John Warltire, who was studying the properties of heat, tried Priestley's experiment in copper vessels and confirmed the 'dew'. [13] Priestley informed Cavendish of the curious loss of weight which they had noticed. Cavendish decided to repeat the experiments himself which he carried out on a larger scale and observed:

> It appeared, that when inflammable and common air are exploded in a proper proportion, almost all the inflammable air, and near one-fifth of the common air, lose their elasticity, and are condensed into dew. And by this experiment it appears, that this dew is plain water, and consequently that almost all the inflammable air, and about one-fifth of the common air, are turned into pure water. [14]

Cavendish thought that these airs each contained some water which had 'condensed' to form water and made no suggestion of chemical combination between them. Also he made no attempt at this time to publish his discovery.

Meanwhile other members of the Lunar Society had taken up the study of air. Watt wrote to Darwin early in January 1781:

> I beg that you would impress on your memory the idea, that you promised to dine with sundry men of learning at my house on Monday next, and that you will realise that idea. For your encouragement there is a new book to be cut up; and it is to be determined whether or not heat is a compound of phlogiston and empyreal air, and whether a mirror can reflect the heat of the fire. [15]

Darwin replied:

> Now you must know that this said Devil has play'd me a slippery trick, and I fear prevented me from coming to join the holy men at your house...
>
> As to material philosophy, I can tell you some secrets in return for yours, viz, that atmospheric gas is composed of light and the earth of water (and aqueous earth). That water is composed of aqueous gas, which is displaced from its earth by oil of vitriol. [16]

It has been suggested that the answer to Darwin's secret gives the correct composition of water. He was pointing out that water can be broken down so is not an element and that one of its parts is a gas which is generated when oil of vitriol (sulphuric acid) acts on an earth, probably referring to a metal, to produce hydrogen. [17] If this is correct, then Darwin may have preceded both Cavendish and Watt in recognising that water was a compound and not an element.

Watt's ideas about the composition of water were linked first of all to his studies on heat. He had tumbled across the phenomenon of latent heat during his experiments to understand why the model Newcomen engine of the University of Glasgow would not work. Black explained this to him which was the beginning of a deep friendship between them. (See Vol. I, p. 78) As well as determining the amount of the extra or latent heat needed to turn water into steam, Watt investigated the relationship between the pressure of steam and the temperature both above and below atmospheric pressure. This led him to postulate in 1765 Watt's Law, that as the pressure and temperature rose, the latent heat in steam decreased while the sensible heat, or that which could be measured by the thermometer, increased so that the sum of the two always remained constant. (See Vol. I, pp. 373 – 4) These experiments explained the performance of his engine which was his main interest at this time and they did not lead into further philosophical speculations.

He carried out further experiments during the second half of 1773 and into 1774, checking his values for latent heat at various pressures. This was when he told Small that he was considering writing a book on the '"Elements of the Theory of Steam-engines", in which, however, I shall only give the enunciation of the perfect engine'. [18] In March he made the first mention about 'the quantity of air contained in water', a point which became so central later to his theory on the composition of water. [19] He continued to develop his ideas in March 1779, when he was 'trying experiment on expansion of steam by heat – when [water] is mixed with some air it begins to expand at 212 and has acquired about one fourth of its bulk at 213 but does not acquire the whole until 224 or 26'. [20] A year later, when he was explaining latent heat to Magellan, he wrote:

> ... to understand this you ought to know that all bodies can contain more of it and some others less, and that all liquids contain more heat than the solid substances by the melting of which these liquids are formed and that again when converted into vapour they contain more heat than when in the state of liquids. [21]

Like Black, he recognised the fundamental importance played by heat. For Black, the study of heat was part of chemistry and he saw how heat, and especially latent heat, determined whether a substance would be a solid or a vapour. Watt was impressed by the power of heat to change substances. For example during his work with the Delftfield pottery, he saw how heat could vitrify clay bodies as well as change the nature and colour of glazes. Heat or fire added to one substance could result in something else.

His correspondence with Magellan led to correspondence with Black about Black's original discovery of latent heat and so to further experiments on the latent heat of steam. He told Black:

> If you do not publish this Summer [1780], I will, as I will not longer run the risk of having the little pretension I have to literary fame run away with and I think I have completed the theory of the causes of the different heats at which [water] boils under various pressures and have found the true series of the Curve. [22]

A long letter followed in the spring of 1781 outlining the results of these experiments when Priestley was present but it was never published. [23] Could this have been due to Watt witnessing

Priestley firing dephlogisticated and inflammable airs and Watt recognising that his thoughts on the composition of water were more important?

Ideas about water being changed into air were not a new topic for discussion because it was noted in *Chamber's Cyclopaedia* in 1752 that there were numerous instances of such transmutation. [24] Priestley was convinced that air always contained some water. He wrote later, 'A considerable quantity of water is always produced in these decompositions of air. But this circumstance only proves that the greatest part of the weight of all kinds of air is water'. [25] He may have convinced Watt as well. As early as 1770, Watt had been puzzled by the great amount of air in his condensing apparatus and he thought that 'there was much air mixed with steam'. [26] This had led him to consider whether 'air was a modification of water'. [27] On his early engines such as the one at Bedworth, he made ample provision for the removal of air and water with triple pumps. [28]

Watt had been present at some experiments by Priestley for he told Black in March and April 1781 how Priestley had obtained 'great quantities of Dephlogisticated air from Allum and amazing quantities from Nitre by heat only' and 'excellent dephlogisticated Air from Gypsum by a violent heat'. On 8 March, Priestley found some 'nitrous air which maintain [sic] flame at a great rate but extinguishes animal life'. [29] He had produced carbon monoxide without recognising it. This caused confusion with the other inflammable air, hydrogen. Watt also saw Priestley turn water into air which he described to Boulton in December 1782. This letter begins to reveal for the first time his theories on heat and air:

> You may remember that I have often said that if water could be heated red hot, or something more, it would probably be converted into some kind of air, because steam would in that case have lost all its latent heat, and that it would have been turned wholly into sensible heat, and probably a total change of the nature of the fluid would ensue. Dr. Priestley has proved this by experiment. He took lime and chased out all the fixed air, and made it exceedingly caustic by long continued and violent heat. He added to it 2 ounces of water, and as expeditiously as possible subjected it again to a strong heat, and he obtained 2 ounces' weight of air; and, what is most surprising, a balloon which he interposed between the retort and the receiver was not sensibly moistened, nor at all heated that could be observed. The air produced was but very little worse than common air, and contained scarce any fixed air. So here is a plain account of where atmospheric air comes from. The Doctor does me justice as to the theory. [30]

Although this letter does not say so, the experiment would have been made in a porous earthenware retort.

Watt also told Black a few days later how Priestley had made this discovery 'in his usual way of Groping about'. At the same time, he gave details of his own experiments on the values of latent heat in vacuo. This seemed to confirm 'Watt's Law'. He continued by describing to Black:

> ... a theory which I have devised... showing that as {water} steam parts with its latent heat as it acquires sensible heat or is more compressed, that when it arrives at a certain point it will have no latent heat and may under proper compression/be an elastic fluid nearly as specifically heavy as water and at which point I conceive it will again change its state and become something else than steam or water. My Opinion has been that it would then become air, which many things had led me to conclude and which is confirmed by an experiment which Dr. Priestley made the other day. [31]

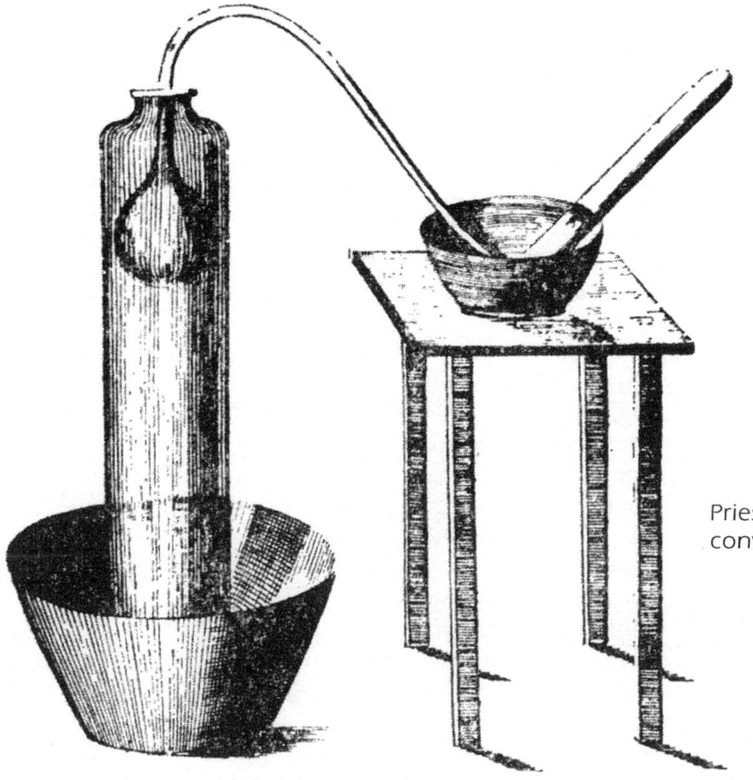

Priestley's apparatus for converting water into air.

Watt expected this change of steam into air to occur at about 1,200°F. [32]

We must remember that scientific thinking at that time was still dominated by the Aristotelian four elements which postulated that all substances were composed of different combinations of them. Watt told De Luc:

> I have observed several other processes by which I now believe air is generated from water some of which I shall mention to you when I have the pleasure of seeing you – If this process contains no deception here is an effectual account of many phenomena and one element is dismissed from the list. [33]

A little later, Watt told Smeaton, 'By the help of Dr. Priestley's experiments I have attempted to demolish 2 of the most ancient elements (air & Water) a third has been destroyed for sometime; but in return we have made 2 or 3 more'. [34] He also told Fry:

> Dr. Priestley, as you observe, converts water into air, and air into water, and I have found out the reason of all these wonders, and also what air is made of; for they are not simple elements... [Take] if you wish to be very exact, of pure air one part, of phlogiston, in a fluid form, two parts, by measure. Put them into a strong glass vessel, which admits of being shut quite close, mix them, fire them with the electric spark; they will explode, and throw out their elementary heat. Give that time to escape, and you will find the water, (equal in weight to the air), adhering to the sides of the vessel. [35]

More experiments by Priestley on turning water into air in early 1783, 'which agrees with my theory and also coincides with what I have observed in Steam Engines', [36] encouraged Watt to outline his theory in a letter to Priestley dated 26 April 1783, requesting him to deliver it to Sir Joseph Banks, President of the Royal Society. This letter covered Watt's theories on the composition of water and he may have sent it to stake a claim on his ideas as well as preparing the ground for admission to that august body. He summed up his discoveries to Hamilton:

> Dr. Priestley has made many discoveries lately in relation to the conversion of water into air; and I have from them made out what water is made of, and what air is made of... It is briefly as follows:
> Facts - 1st, Pure dry dephlogisticated air and pure dry inflammable air fired together, leave no residuum, except a small quantity of water equal to their weight.
> 2d, Pure inflammable air reduces calces of metals and is absorbed by them. The residuum, after nine-tenths was absorbed, was still inflammable air.
> 3d, All substances which produce inflammable air, are substances which contain some water firmly united to them, and have some principle which is known to attract phlogiston strongly. (Example – nitre, alum, gypsum, calces of metal, etc.)
> 4th, Porous earthen vessels imbibed with water, and slowly heated, produce air, if the process is well performed, equal in weight to the water.
> Deductions. – Pure inflammable air is phlogiston itself. Dephlogisticated air is water deprived of its phlogiston, and united to latent heat.
> Water is dephlogisticated air deprived of part of its latent heat, and united to a large dose of phlogiston... The fire gives the latent heat. [37]

But Priestley carried out further experiments which threw doubt on whether air had indeed been produced from water. This caused Watt to reconsider part of his letter and to ask that it be withdrawn, although it still remained at the Royal Society (where it is still preserved). [38]

The retort in which Priestley had heated his lime and water was made of earthenware. He announced the result of this experiment to Wedgwood on 8 December 1782 but wrote to him again on 8 January 1783 to say that this did not happen when he used a glass retort. Wedgwood pointed out that it was possible that air was coming into the earthenware retort from outside or from the fire. During March, Priestley was becoming suspicious about his water into air experiments and went to London to try some modifications with a large burning glass which confirmed to his satisfaction that air had indeed entered the retort. He was therefore surprised to receive Watt's letter of 26 April and, knowing that water had not been converted into air, replied on April 29, with a sketch of the apparatus he had used in London, 'Behold with surprise and indignation the figure of an apparatus that has utterly ruined your beautiful hypothesis, and has rendered some weeks of my labour... almost useless'. [39]

Watt answered on 2 May:

> I deny that your experiment ruins my hypothesis. It was not founded on so brittle a basis as an earthen retort nor on its converting water into air...
> I maintain my hypothesis untill it shall be shown that the water found after the explosion of pure and inflammable air had some other origin nor shall I believe that air is a child of acids or rather a modification of them untill such acids can be found after the decomp[ositio]n of it. I have many experiments to propose to you to help bring out the truth and a fair analysis made of the ? fluids. [40]

Watt was not going to be persuaded to change his mind quickly, because he told Hamilton a little later:

> As to Philosophical news Dr Priestley has by another experiment quite overturned the conversion of water into air by means of porous earthen vessels, but I shall persist in my former opinion as to the production of air from nitre etc, which confirm me in it and I hope satisfactorily to prove it or at least to account for the production on some solid basis. [41]

Perhaps Watt relied too much on Priestley's experiments at this time without carrying out many himself.

Priestley passed Watt's letter on to Banks and the experiments on changing water into air and of making water from the two airs were discussed at the Royal Society in May. There were perhaps two results. One was that, when Charles Blagden was visiting Paris later in the summer, he told Lavoisier. Blagden had been Cavendish's personal assistant and so was in a position to know about his experiments as well as Watt's letter. Watt junior pointed out much later that Cavendish never expressed anything on the theory of the composition of water, either in his diary of experiments or verbally, until after Watt's letter had been passed on to the Royal Society. So the second result may have been to prompt Cavendish to publish his experiments which he did in a paper to the Royal Society in January 1784. Certainly the anonymous writer in the *Quarterly Review* in 1854 held Blagden responsible for depriving Watt of the priority of this discovery.

In June 1783, Blagden went to Paris where he called on Lavoisier and gave him an account of Cavendish's experiments. Without delay on 24 June with Blagden present, Lavoisier, who had greater resources at his command than Watt, made an hurried attempt with De La Place to verify Cavendish's results. Next day, he gave a brief report to the Academy, stating that the combustion of inflammable air with oxygen, carried out in a closed vessel, yielded 'water in a very pure state'. [42] At a public meeting of the Academy on 12 November, Lavoisier read

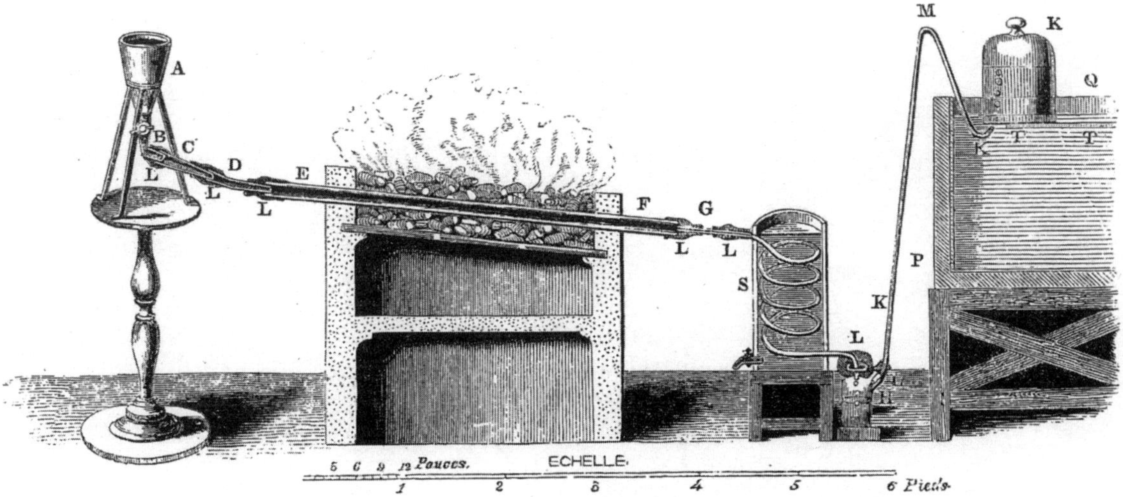

Lavoisier's apparatus for decomposing water into hydrogen.

a memoir entitled, 'On the Nature of Water and on Experiments that appear to prove that this Substance is not properly speaking an Element but can be decomposed and recombined'. The memoir was printed in the December issue of *Rozier's journal*. He concluded that water was a compound of these two airs, which were elements, with heat, or his term caloric, being generated. In this way, Lavoisier gave a closer deduction to the truth based on Cavendish's experiments and Watt's hypothesis but, in his memoir, he failed to mention either of these people. It confirmed Lavoisier in his ideas on his new theory of chemistry.

Watt quickly learnt about Lavoisier's experiments and suspected that his own hypothesis had been disclosed. On 1 December, he wrote to Kirwan:

> Mr Lav[r] has read a memoir opening a theory very similar to mine in the composition of water, indeed so similar that I cannot help suspecting that he has heard of the theory I ventured to form on ye subject as I know that some notice of it was sent to France. He does not seem to have been more lucky in his proofs of it than I have been. [43]

Kirwan confirmed that Blagden was Lavoisier's source. [44] Watt therefore rewrote his original letter of April to Priestley and sent it on 26 November to De Luc, asking him to present it to the Royal Society. [45] Since he was not a member, he could not read it himself.

On 15 January 1784, Cavendish presented a paper to the Royal Society on 'Experiments on Air'. [46] In it he stated, 'From what has been said there seems the utmost reason to think, that dephlogisticated air is only water deprived of its phlogiston, and that inflammable air, as was before said, is either phlogisticated water, or else pure phlogiston; but in all probability the former'. [47] It became a quite common and plausible concept that inflammable air and phlogiston were the same substance. Kirwan asserted that this gas was 'none other than phlogiston itself'. [48] De Luc obtained a copy of Cavendish's paper in March and informed Watt. [49] Watt's response reveals his attitude to the whole controversy and explains his subsequent behaviour:

> On the slight glance I have been able to give your extract of the paper [Cavendish's] I think his theory very different from mine; which of the two is right I cannot say; his is more likely to be so, as he has made many more experiments, and, consequently, has more facts to argue upon.
>
> I by no means wish to make any illiberal attack on Mr. C. It is barely possible he may have heard nothing of my theory; but as the Frenchman said when he found a man in bed with his wife, 'I suspect something'.
>
> As to what you say of making myself 'des jaloux', that idea would weigh little; for, were I convinced I had had foul play, if I did not assert my right, it would either be from a contempt of the modicum of reputation which could result from such a theory; from the conviction in my own mind that I was their superior; or from an indolence, that makes it easier to me to bear wrongs, than to seek redress. In point of interest, in so far as connected with money, that would be no bar; for, though I am dependent on the favour of the public, I am not a Mr. C. or his friends; and could despise the united power of the illustrious house of Cavendish, as Mr. Fox calls them. [50]

It has been suggested that Watt was an outsider and his chemical accomplishments little known to London Society. [51] He was anxious to become a member of the Royal Society and so had to remain on good terms with them.

In April 1784, Banks told Watt that his letter would be read probably on 6 May, commenting:

> How much the Royal Society & the world at large are indebted in Point of Science to the town of Birmingham. I need not declare after mentioning him [Dr. Priestley] that you are at last induced to make it the conveyance of your discovery gives I frankly confess no little pleasure. [52]

Banks was recognising the esteem in which Watt and other members of the Lunar Society were being held. On 29 April, a large meeting of fellows heard Watt's communication which was received 'with great approbation'. [53] The title was 'Thoughts on the constituent Parts of Water, and of Dephlogisticated Air; with an Account of some Experiments on that Subject'. [54] A short sequel was read on 6 May. [55] Watt wrote to Robison:

> I have lately, through the importunity of my friends, been prevailed upon to have read before the Royal Society of London, a paper containing a new hypothesis on the constituent parts of water and of dephlogisticated air, which has so far met with their approbation as to be ordered to be printed. It may seem rather bold in me to commence my publications in science by a new theory; and my natural timidity and diffidence would certainly have prevented me, if Mr. Lavoisier in France, having learned something about it from Dr. Blagden, had not adopted it as his own, and Mr. Cavendish, a year after the broaching of mine, had not published one of the same kind. [56]

Blagden oversaw the printing of Cavendish's and Watt's papers. The date of Cavendish's was given as 15 January 1783 while that of Watt's letter to Priestley as 26 April 1784, the year in both cases being wrong, giving apparent priority to Cavendish. Although Watt's paper starts, 'I send you an account of the hypothesis I have ventured to form on the probable causes of the production of water from the deflagration of a mixture of dephlogisticated and inflammable airs, in some of our friend Dr. PRIESTLEY's experiments...' the discussion on the composition of water forms only a small part. Watt also described the thermal decomposition of various nitrates including the production of dephlogisticated air from melted nitre. Because the nitre also reacted with the retort, he examined what effect there was with magnesia alba, calcareous earth, minium [a form of quicksilver] and the red calx of lead. [57] He was trying to discover what was happening to the phlogiston as part of the air thus released.

But it was the first section on the composition of water that was to stir up controversy. This had been contained in his original letter to Priestley and in another sent to Black four days earlier [58] as well as in the paper published by the Royal Society:

> Let us now consider what obviously happens in the case of the deflagration of the inflammable and dephlogisticated air. These two kinds of air unite with violence, they become red-hot, and upon cooling totally disappear. When the vessel is cooled, a quantity of water is found in it equal to the weight of the air employed. This water is then the only remaining product of the process, and water, light, and heat, are all the products, unless there be some other matter set free which escapes our senses.
>
> <u>Are we not then authorised to conclude, that water is composed of dephlogisticated air and phlogiston, deprived of part of their latent or elementary heat; that dephlogisticated or pure air is composed of water deprived of its phlogiston, and united to elementary heat and light; and that the latter are contained in a latent state, so as not to be sensible to the thermometer or to the eye; and if light be only a modification of heat, or a circumstance attending it, or a component part of the inflammable air, then pure or dephlogisticated air is composed of water deposed of its phlogiston and united to elementary heat?</u> [59]

While it can be seen that Watt no longer considered water to be a simple indivisible element, it is necessary to ask what he considered were dephlogisticated and inflammable airs, phlogiston, latent heat, elementary heat and even heat.

Watt thought that dephlogisticated air was not the element we know as oxygen but some form of water modified by heat (assuming phlogiston was a form of heat for the moment). He told De Luc, 'Dephlogisticated air is composed of water deprived of its phlogiston and united to latent or elementary heat & light'. [60] He apparently thought that what we now call oxygen was composed of water and heat. He seems to have retained the Aristotelian idea of water being the basis of other substances for he wrote the same thing in his letter to Priestley, 'I conjecture that Pure Air is composed of Dephlogisticated Water & elementary heat. I conjecture that dephlogisticated Water has a more powerful attraction for Phlogiston, than it has for its latent heat; but that it cannot perfectly unite with it, at least not to the point of saturation or to the total expulsion of the heat, unless it be red hot'. [61] Here phlogiston and latent heat appear to be different. Further it is not clear whether Watt distinguished between Priestley's pure air and ordinary air. In another letter to De Luc he wrote about Priestley's experiments. 'It does not disprove my theory... My assertion was simply, that air was water deprived of its phlogiston, and united to heat, which I grounded on the decomposition of air by inflammation with inflammable air, the residuum, or product of which, is only water and heat'. [62] He carried out further research during the latter half of 1783.

If we are confused over Watt's understanding of air, then we become more so over his ideas on heat. In the passage just quoted, Watt makes a distinction between phlogiston and heat. Perhaps he thought that, as the Aristotelian elements of air and earth were proving to have so many varieties of substances, so there would be many varieties of fire and hence heat. This was before his friend Irvine wrote in 1805:

> The invention of the thermometer superseded all the former vague and uncertain methods of estimating the degrees or qualities of caloric. It shewed that all kinds of heat were the same and differed from each other only in intensity, an opinion very dissimilar to that entertained formerly when the grandeur or baseness of the source was supposed to influence and alter the nature of the caloric itself. [63]

Watt was living during the time when this change in conception was taking place.

Further, all heat, in whatever form, would have been considered as some sort of subtle, weightless fluid, which inserted itself between the particles of air, water or earth. Watt's close friend, Robison, explained how this happened when measuring the temperature of water:

> When the sensible heat in both is in equilibrio, in a proportion depending on the nature of the two fluids [the water and fluid in the thermometer], the thermometer rises no more, because it absorbs no more heat or fire from the water; for the particles of water which are in immediate contact with the bottom, are now (by this gradual expansion of liquidity) at such distance from each other, that their laws of attraction for each other and for heat are totally changed. Each particle either no longer attracts, or perhaps it repels its adjoining particle, and now accumulates round itself a great number of the particles of heat, and forms a particle of elastic fluid, so related to the adjoining new-formed particles, as to repel them to a distance about twelve and a half times greater than their distance in the state of water. Thus a mass of elastic vapour of sensible magnitude is formed. [64]

For Watt, this elastic vapour might have been formed with latent or elementary heat or phlogiston.

Watt stated that 'pure inflammable air is phlogiston itself'. [65] Phlogiston is not mentioned in earlier experiments on latent heat. It is not clear what he thought phlogiston actually was. The probability is that he equated it with particles of fire or a substance which could be burnt but not what we consider heat or something which perhaps might be detected by a thermometer. Lavoisier, who was opposed to phlogiston, shows us the difficulty of determining what its proponents thought it was when he gave a scathing description:

> Chemists have made phlogiston a vague principle, which is not strictly defined and which consequently fits all the explanations demanded of it. Sometimes it has weight, sometimes it has not; sometimes it is free fire, sometimes it is fire combined with earth; sometimes it passes through the pores of vessels, sometimes these are impenetrable to it. It explains at once causticity and non-causticity, transparency and opacity, colour and the absence of colour. It is a veritable Proteus that changes its form every instant. [66]

While inflammable air might be phlogiston, the converse might not always be true. We have seen Priestley's confusion between hydrogen and carbon monoxide. (See p. 220) When Watt was writing to De Luc about the straw being burnt to provide the hot air for a balloon, we find the rather confused statement, 'Then we have of charcoal 20, Oil 16 = inflammable matter which call wholly phlogiston'. (See p. 213) This differed from the inflammable air. Hence phlogiston might be found in anything that would burn as some sort of weightless essence. Priestley himself admitted:

> The phlogistic theory is not without its difficulties. The chief of them is that we are not able to ascertain the weight of phlogiston, or indeed that of the oxygenous principle. But neither do any of us pretend to have weighed light, or the element of heat, though we do not doubt but that they are properly substances, capable by their addition, or abstraction, of making great changes in the properties of bodies, and of being transmitted from one substance to another. [67]

Watt had helped to convince Priestley of the truth of the phlogiston theory. He stated that 'Water is composed of pure air deprived of a great part of its latent heat and united to phlo[gisto]n'. [68] So he thought that phlogiston and latent heat were different. Latent heat he often linked with elementary heat, suggesting that these might have been the same.

Black had shown Watt the importance of latent heat in the formation of the physical world around us and how the addition of heat, which could not be detected by a thermometer, changed ice into water or water into steam. Yet it remained a puzzle how a substance could absorb heat which failed to register any increase in temperature. Watt was proud of his discovery and so continued to explore the phenomenon of latent heat for years. The reason for this interest may have been his belief that heat was a sort of basic building block for all substances. He had been confirmed in this opinion through the differences he 'had observed in the quantity of latent heat contained in steam, according as the water which produced it was evaporated under a greater or lesser pressure, which led me to conclude that in certain cases the latent heat would be changed into sensible heat'. [69] Here he suggests that latent heat and sensible heat differ. In the same letter to Priestley, he argues:

> It also seems that pure air can unite in certain degrees with phlogiston without being decomposed; and so may form fixed air, and perhaps phlogisticated air, (which seems to be proved from some of your Experiments) but that in proportion as it unites with phlogiston, it parts with some of its elementary or latent heat. [70]

Hence pure air, phlogiston and elementary or latent heat could be combined in various proportions to form our carbon dioxide and nitrogen. We have seen that he thought pure air might be formed from water, phlogiston and heat. It may be that Watt thought that, if he could accurately quantify the latent heat in steam, he would be able to determine the other types of heat.

Again, in this letter to Priestley, he suggested:

> It ought to be observed that if these Conjectures are right, the quantity of elementary or Latent heat is much greater in pure air than in Steam because the first moments after the explosion, the Water deposited by the Air remains in the form of Steam, and consequently retains the Latent heat due to that Modification of water, which is a Simple union of Water and Heat. It would be very easy to find what quantity of elementary heat enters into the Composition of air, by firing the mixture of pure and inflammable air in a Metallic Vessel immersed in another Vessel containing a known quantity of Water of a known heat, and examining how much heat the external water had gained, which being divided by the quantity of water produced by the decomposition of the air, will give the whole quantity of elementary or latent heat which it Contained, both as Air and as Steam; and if from that quantity of heat, We deduct the Latent heat of the Steam, we shall have the quantity of elementary heat which air does contain more than in Steam. [71]

Watt was trying to discover the amounts of the different types of heat in various substances. This may explain why he undertook experiments on latent heat in both 1781 and 1783 which he described later in Robison's *Mechanical Philosophy*. [72] Nowhere in these notes did he mention phlogiston or caloric. The emphasis is on latent heat. Perhaps he thought that, if he could accurately fix the amount of latent heat in one substance, in this case steam, he could discover the amount in other substances and so promulgate a cohesive theory to replace Aristotle's elements.

Watt finished his letter of April 1783 to Priestley with:

> All that I conceive to be new in this paper is the Idea, First that Dephlogisticated air is composed of Water deprived of its Phlogiston and united to a greater Quantity of Latent heat than it contains in the form of Water or of Steam; Secondly that Water is pure air deprived of part of its latent heat and united to Phlogiston. [73]

He continued to study heat for many years after submitting his papers to the Royal Society and it was only with reluctance that he accepted Lavoisier's new chemistry with its different nomenclature but even then he doubted its theory on the composition of water. He was interested in the discovery of electrolysis of water by Humphry Davy and corresponded with him in 1801. Davy pointed out that he hoped:

> ... soon to communicate... an account of some galvanic experiments which I am almost sorry to say cannot be well reconciled to the French theory & I even believe that your theory of water being the basis of all the gases will be found true at last. It

> is easy by common electricity to produce at pleasure oxygene or hydrogene from the same quantity of water without the sensible appearance of any reciprocal product. [74]

Watt persisted with his ideas that oxygen and hydrogen were not elements but were formed out of water. He thought that Davy's experiments proved this for he had 'formed an opinion that the decomposition of water was all fudge & that water united to negative gal[vanism] or elect[ricit]y formed oxygen & to positive formed Hydro[gen]'. [75] His letter to Robison in January 1804 shows that he was still searching for a solution based on the decomposition of water by electrolysis:

> It was that seeing the electricity from one end of the pile in acting upon water produced only oxygen & from the other end only hydrogen if properly managed, thus it was natural to conclude that the union of one of the electricitys with water formed oxygen & the other hydrogen & that water was not at all decomposed. Further experiments however are wanting & their illustration by someone well versed in the modern chemistry. [76]

All this shows how different were Watt's ideas on the composition of water from those we hold today.

One outcome of his paper on the composition of water was his election to the Royal Society of London. He presented a second paper in 1784 'On a method of preparing a test liquor, to show the presence of acids and alkalis in chemical mixtures'. [77] He had been looking for a good acid - alkali indicator and, after trying various substances such as a 'red infusion of violets', 'found the red cabbage (Brassica rubra) to furnish the best test from its being naturally blue, turning green with alkalies and red with acids'. This indicator was used for a considerable time. [78] In his application to the Royal Society, Watt described himself as 'inventor of the new steam engine, and author of a valuable paper in the last volume of Philosophical Transactions', without stating which. [79] His application was supported with seventeen signatures, including Smeaton, Priestley, Cavendish, Kirwan, Wedgwood and Mylne. He was elected on 24 November 1785. He travelled to London early in the New Year to be formally admitted on 12 January. The coach fare was £1.11s.6d. The admission fees for himself, Dr. Withering, Keir and Boulton were £130.4s. In addition, the porter was tipped 2s.6d. [80] Other members of the Lunar Society were also elected around this time; Wedgwood on 16 January 1783; Keir and Samuel Galton junior on 8 December 1785, thus giving recognition to the important scientific work originating in Birmingham and its surrounding regions. [81]

The question of who was the first person to recognise that water was not an element stirred up considerable controversy over the years. (See Vol. III) We have seen that Watt thought that Lavoisier had taken his ideas without recognition. His strength of feeling is revealed in a letter to Fry:

> I have had the honour, like other great men, to have had my ideas pirated. Soon after I wrote my first paper on the subject, Dr. Blagden explained my theory to M. Lavoisier at Paris; and soon after that, M. Lavoisier invented it himself and read a paper on the subject to the Royal Academy of Sciences. Since that, Mr. Cavendish has read a paper to the Royal Society on the same idea, without making the least mention of me. The one is a French financier, and the other a member of the illustrious house of Cavendish, worth above £100,000, and does not spend £1000 a year. Rich men may do mean actions. May you and I always preserve our integrity, and despise such doings. [82]

Yet Watt dined amicably with Cavendish in October 1785 and told Annie about their meeting:

> When I was in London I was received very kindly by Mr. Cavendish and Dr. Blagden, and my old friend Smeaton, who has now recovered his health and seems hearty. I dined at a turtle feast with them and the select club of the Royal Society; and never was turtle eaten with greater sobriety and temperance, or more good fellowship. I also dined at Mr. Cavendish's, who lives very elegantly, and gave us a good English dinner. [83]

Perhaps the topic of the composition of water was not raised.

Some of the later controversy might have been avoided had people taken notice of what Priestley told De Luc in December 1783 when the dispute first arose:

> I am sorry to perceive that you are intirely mistaken with respect to the concern I have in the discovery concerning the constitution of water. I really have no merit with respect to it... It was Mr. Cavendish who first found water in decomposing dephlogisticated and inflammable air. In my opinion Mr Watt first entertained the idea you mention, and the experiment of Mr. Cavendish proved the justness of it, tho Mr. Cavendish had not that idea himself. [84]

While Watt may have been the first to put his ideas in writing, those of Cavendish were the first to appear in print. But both their explanations were couched in terms of the phlogiston theory. Watt never accepted the fact that his theories were not recognised. His letter to De Luc in March 1784 gives further insights into his character:

> You may, perhaps, be surprised to find so much pride in my character. It does not seem very compatible with the diffidence that attends my conduct in general. I am diffident, because I am seldom certain that I am in the right, and because I pay respect to the opinions of others, where I think they merit it. At present, <u>je me sens un peu blessé</u>; it seems hard, that in the first attempt I have made to lay any thing before the public, I should be thus anticipated. It will make me cautious how I take the trouble of preparing anything for them another time. [85]

Perhaps this explains why he never ventured into writing a history of the steam engine. This rare instance of his venturing into 'pure science' was a failure and discouraged him from writing anything else theoretical.

Notes

1 JWP, Letter Book 1, J. Watt to J.A. De Luc, ?8 Oct. 1786.
2 For a full account, see R.L. Hills, 'James Watt and the Delftfield Pottery, Glasgow', *Proceedings of the Society of Antiquaries of Scotland*, Vol. 131, (2001), pp. 375 – 420.
3 JWP, Letter Book 6, J. Watt to R. Muirhead, 24 March 1819.
4 See Musson & Robinson, (1969), pp. 369 – 71, for a modern analysis of Watt's experiments. See also Smith & Moilliet, (1967), p. 146.
5 Robinson & McKie, (1970), p. 90, Letter 67, J. Watt to J. Black, 30 May 1780.
6 JWP, C3/6, 'Diary', 6 June 1780.
7 Robinson & McKie, (1970), pp. 93 – 4, Letter 69, J. Watt to J. Black, 9 June 1780.
8 Musson & Robinson, (1969), P. 363, J. Watt to M. Boulton, 9 July 1780.
9 JWP, W5, J. Watt to G. Hamilton, 14 Aug. 1780.
10 JWP, 4/15, G. Hamilton to J. Watt, 24 Aug. 1780.
11 Robinson & McKie, (1970), p. 95, Letter 71, J. Watt to J. Black, 15 Oct. 1780.
12 Musson & Robinson, (1969), p, 367 and Smith & Moilliet, (1967), p. 146.
13 Robinson & McKie, (1970), p. 109, Letter 79, J. Watt to J. Black, 9 April 1781. See also ibid, p. 106, Letter 77, J. Watt to J. Black, 8 March 1781, 'My Business will not probably soon permit me to think or act in the Alkali business... I resign all thought of it, sine die'.
14 JWP, Letter Book 1, J. Watt to J. Fry, 22 May 1782.
15 JWP, W5, J. Watt to J. McGrigor, 4 Feb. 1781.
16 Brewster, (1830), Vol. XVIII, p. 384.
17 JWP, JW/22, J. McGrigor to J. Watt, March 1781.
18 Muirhead, (1858), p. 278.
19 *Biographical Dictionary of the History of Technology*, (1966), p. 55.
20 Clapperton, (1967), p. 86 and Patent 4,509, enrolled 28 April 1821.
21 JWP, 3/8, J. Watt to M. Boulton, 18 April 1782.
22 Partington, (1962), Vol. III, p. 134.
23 Dickinson & Jenkins, (1927), p. 226.
24 Elton, Sir Arthur, 'William Murdock', typescript, p. 34.
25 Griffiths, (1992), p. 221.
26 Elton, Sir Arthur, 'William Murdock', typescript, p. 34.
27 Ibid.
28 Pugh & Hudson, (1985), p. 45.
29 Schofield, (1963), p. 256.
30 JWP, Letter Book 1, J. Watt to R. Kirwan, 14 Nov. 1783. See also Muirhead, (1854), Vol. II, p. 182, Letter 251, J. Watt to J.A. De Luc, 23 Nov. 1783.
31 Muirhead, (1854), Vol. II, p. 184, Letter 254, J. Watt to Magellan, 1 Jan. 1784. See JWP, C3/10, 'Common Place Book', p. 77, for Magellan's proposals.
32 JWP, Letter Book 1, J. Watt to J.A. De Luc, 11 Dec. 1785.

The Copying Machine

1 Muirhead, (1854), Vol. III, p. 27.
2 King-Hele, (1981), p. 94, Letter 78J, E. Darwin to C.F. Greville, 12 Dec. 1778. See King-Hele, (1999), pp. 151 – 55 for a more detailed account of Darwin's invention where Watt is pictured as the sour mechanical genius who was annoyed that someone else had invented a mechanical copier. The surviving papers do not support this interpretation since they show that Watt gave Darwin's machine a fair trial.
3 King-Hele, (1981), p. 96, Letter 79A, E. Darwin to C.F. Greville, 16 May 1779.
4 Arago, (1839), p. 93 has a fanciful and untrue version for the origin of Watt's copying machine.
5 JWP, C3/4, 'Journal', 14 Jan. 1779. See JWP, Letter Book 5, J. Watt to D. Brewster, 13 Dec. 1813 for Watt's description of Darwin's machine.

6 JWP, C3/4, 'Journal', 16 Jan. 1779.
7 Ibid, 13 Feb. 1779.
8 Robinson & McKie, (1970), pp. 73 – 4, Letter 57, J. Watt to J. Black, 2 Dec. 1779.
9 Muirhead, (1854), Vol. II, p. 115, Letter 175, date only 1779; JWP, W5.
10 JWP, C3/4, 'Journal', 30 March 1779.
11 Ibid, 28 April 1779.
12 JWP, W5, J. Watt to L. Henderson, 19 May 1779.
13 Robinson & McKie, (1970), p. 67, Letter 52, J. Watt to J. Black, 24 July 1778 [recte 1779].
14 JWP, W5, J. Watt to G. Hamilton, 6 May 1779.
15 Ibid, J. Watt to L. Henderson, 19 May 1779.
16 JWP, C3/5, 'Journal', 23 July 1779.
17 JWP, W5, J. Watt to P. Wilson, 18 Feb. 1781.
18 Boulton 348.140, J. Watt to M. Boulton, 27 June 1779.
19 Boulton 348.141, J. Watt to M. Boulton, 28 June 1779.
20 Roll, (1930), p. 129.
21 Boulton 348.145, J. Watt to M. Boulton, 5 July 1779.
22 Robinson & McKie, (1970), p. 68, Letter 54, J. Black to J. Watt, 25 Aug. 1779.
23 JWP, C3/5, 'Journal', 30 Aug. 1779 and 31 March 1780.
24 Ibid, 2 & 4 Nov. 1779.
25 Robinson & McKie, (1970), p. 73, Letter 57, J. Watt to J. Black, 2 Dec. 1779.
26 Muirhead, (1854), Vol. III, p. 28.
27 JWP, C3/5, 'Journal', 21 March 1780 and Muirhead, (1854), Vol. III, p. 34.
28 Muirhead, (1854), Vol. III, pp. 30 – 1.
29 I wish to thank Adam Reid, Oriel College, Oxford, for sharing with me the part of his thesis on Watt's copying process.
30 Robinson & McKie, (1970), p. 98, Letter 73, J. Watt to j. Black, 25 Oct. 1780.
31 Rees, (1819), article on INK, published c. 1810.
32 B. & W. Col., Office Letter Book 4, J. Watt to Peter ?Molari?, 21 Oct. 1780.
33 B. & W. Col., MI, 1/14, 'Journal', 11 May 1770. See also R.L. Hills & J. Reece, 'British Paper Mills; Ruthven and Methven', *The Quarterly*, British Association of Paper Historians, No. 14, March 1995, pp. 18 – 20.
34 Robinson & McKie, (1970), p. 73, Letter 57, J. Watt to J. Black, 2 Dec. 1779.
35 Rees, (1819), reprint 1972, Vol. II, p. 324, article on DYEING, c. 1808.
36 JWP, C3/4, 'Journal', 6 – 9 May 1779.
37 Hebert, (1838), Vol. II, pp. 227 & 847.
38 Ibid, (1836), Vol. I, p. 653.
39 Rees, (1819), article on GALL-nut, c. 1810.
40 Hebert, (1836), Vol. I, p. 584.
41 Rees, (1819), article on INK, c. 1811.
42 Ibid.
43 Smith & Moilliet, (1967), p. 152.
44 For early examples of press copies, see JWP, W5. See Andes, (1907), p. 122 for later comments on the copying process.
45 Boulton 348.141, J. Watt to M. Boulton, 28 June 1779.
46 JWP, C3/5, 'Journal', 6 Dec. 1779.
47 Ibid, 14 Dec. 1779.
48 Ibid, 22 Jan. 1780.
49 Muirhead, (1854), Vol. III, p. 34. See J.L. Moilliet & B.M. Smith, *A Mighty Chemist: James Keir of the Lunar Society*, privately published 1983.
50 JWP, C7/1, 'Ink Book'.
51 B. & W. Col., Office Letter Book 4, J. Watt to J. Woodmason, 16 Aug. 1780.
52 Ibid, J. Watt to J. Woodmason, 28 Aug. 1780.
53 Ibid, J. Watt to J. Woodmason, 26 May & 2 Sept. 1780.
54 JWP, 3/8, 'Receipt for Ink Powders', 4 Sept. 1780.
55 B. & W. Col., Office Letter Book 4, J. Watt to J. Woodmason, 27 Oct. 1780.
56 Ibid, J. Watt to J. Woodmason, 23 Sept. 1780.
57 B. & W. Col., Office Letter Book 4, J. Watt to J. Woodmason, 27 Oct. 1780 and JWP, W5, J. Watt to G. Hamilton, 23 Sept. 1781.
58 Hebert, (1836), Vol. I, p. 742.
59 JWP, C7/1, 'Ink Book', 22 Sept. 1780.
60 JWP, 3/8, J. Watt to J. Keir, 16 Nov. 1780.
61 JWP, C3/4, 'Journal', 13 May 1779.
62 Ibid, 16 May 1779.
63 Boulton 348.157, J. Watt to M. Boulton, Oct. 1779.

64 JWP, C3/5, 'Journal', 12 March 1780.
65 Muirhead, (1854), Vol. III, pp. 32 – 3.
66 Ibid.
67 JWP, 3/8, J. Watt to M. Boulton, 21 May 1780.
68 JWP, C7/1, 'Ink Book', July 1782.
69 JWP, C3/5, 'Journal', 3 May 1780.
70 JWP, 3/8, J. Watt to M. Boulton, 4 May 1780.
71 B. & W. Col., Office Letter Book 4, J. Watt to J. Woodmason, 23 July 1780.
72 Ibid, J. Watt to J. Woodmason, 12 Aug. 1780.
73 Ibid, J. Watt to J. Woodmason, 16 Aug. 1780.
74 Ibid, J. Watt to J. Woodmason, 2 Sept. 1780.
75 Ibid, J. Watt to J. Woodmason, 16 Sept. 1780.
76 See JWP, 3/8, 'Directions for Using the Copying Machine'.
77 JWP, 4/49, P. Wilson to J. Watt, 15 & 22 Jan. 1781.
78 JWP, W5, J. Watt to P. Wilson, 18 Feb. 1781.
79 Ibid, J. Watt to P. Wilson, 13 March 1781.
80 JWP, 3/8, J. Watt to J. Keir, 28 March 1781.
81 B. & W. Col., Office Letter Book 4, J. Woodmason to Boulton & Fothergill, 6 June 1781.
82 JWP, Letter Book 4, J. Watt to J.A. De Luc, 30 Jan. 1806.
83 Bower, (1996), pp. 11 & 15.
84 JWP, 3/8, J. Watt to J. Keir, 26 April 1780.
85 Hills, (*Exeter Papers*, 2001), p. 67.
86 JWP, 3/8, J. Watt to J. Keir, 26 April 1780.
87 JWP, W5, J. Watt to G. Hamilton, 24 April 1779 and J. Watt to A. Watt and to Capt. Marr, both 5 April 1780.
88 Balston, (1957), pp. 14 & 28.
89 B. & W. Col., Office Letter Book 4, J. Watt to E. Lea, 9 April 1780.
90 JWP, 3/8, j. Watt to J. Keir, 26 April 1780.
91 Ibid, J. Watt to J. Keir, 1 May 1780.
92 B. & W. Col., Office Letter Book 4, J. Watt to E. Lea, 3 May 1780.
93 JWP, 3/8, J. Watt to J. Keir, 16 April 1780.
94 Ibid, J. Watt to J. Keir, 26 April 1780.
95 Ibid, J. Watt to J. Keir, 1 May 1780.
96 Ibid, J. Watt to J. Lean, 9 May 1780.
97 Ibid, J. Watt to J. Keir, 30 April 1780.
98 Ibid, J. Watt to J. Keir, 26 April 1780.
99 Ibid, J. Watt to M. Boulton, 28 April 1780.
100 Ibid, J. Watt to M. Boulton, 4 May. 1780.
101 B. & W. Col., Office Letter Book 4, J. Watt to J. Woodmason, 26 May 1780.
102 Ibid, J. Watt to J. Woodmason, 23 July 1780.
103 Ibid, J. Watt to J. Woodmason, 14 Aug. 1780.
104 Ibid, J. Watt to J. Woodmason, 28 Aug. 1780.
105 Ibid, J. Watt to J. Woodmason, 28 Sept. 1780.
106 Ibid, J. Watt to J. Woodmason, 4 Oct. 1780.
107 JWP, Letter Book 1, J. Watt to M. Reveillon, 9 Dec. 1785.
108 JWP, C3/5, 'Journal', 3 May 1780.
109 Muirhead, (1854), Vol. III, p. 31.
110 JWP, C3/5, 'Journal', 18 Sept. 1779.
111 Ibid, 4 Dec. 1779.
112 Ibid, 6 Dec. 1779.
113 Ibid, 3 May 1780.
114 Ibid, 30 Dec. 1779.
115 Ibid, 21 Feb. 1780.
116 Ibid, 7 March 1780.
117 Ibid, 23 March & 13 April 1780.
118 Ibid, 7 & 23 March, 24 & 26 April, 2 & 9 May 1780.
119 JWP, 3/8, J. Watt to M. Boulton, 29 April 1780.
120 Robinson & McKie, (1970), p. 89, Letter 66, J. Watt to J. Black, 17 May 1780.
121 JWP, 3/8, J. Watt to M. Boulton, 9 May 1780.
122 B. & W. Col., Office Letter Book 4, J. Watt to 'Dear Sir', 13 June 1780.
123 JWP. 3/8, J. Watt to M. Boulton, 21 April 1780.
124 JWP, C3/5, 'Journal', 23 April 1780.
125 Ibid, 30 April 1780.
126 B. & W. Col., Office Letter Book 4, J. Watt to J. Woodmason, 28 Sept. 1780 and JWP, 4/15.23, G. Hamilton to J. Watt, 24 Aug. 1780.
127 Ibid, J. Watt to J. Woodmason, 28 Sept. 1780.
128 Ibid, J. Watt to J. Woodmason, 4 Oct. 1780.
129 Ibid, J. Watt to J. Woodmason, 10 Nov. 1780.
130 JWP, W5, J. Watt to G. Hamilton, 18 Nov. 1780.
131 JWP, 4/49, P. Wilson to J. Watt, 15 Jan. 1781.
132 JWP, W5, J. Watt to P. Wilson, 18 Feb. 1781.
133 Ibid, J. Watt to P. Wilson, 13 March 1781.
134 Ibid, J. Watt to G. Hamilton, 23 Sept. 1781.

135 JWP, 3/8, J. Watt to J. Keir and J. Watt to M. Boulton, both 24 March 1781.
136 JWP, 4/81, J. Keir to J. Watt, 12 Nov. 1781.
137 Smiles, (1878), p. 208, M. Boulton to J. Watt, 14 May 1780; Boulton Mss.
138 JWP, C3/5, 'Journal', 18 March 1780.
139 Robinson & McKie, (1970), p. 79, Letter 60, J. Watt to J. Black, 1 March 1780.
140 JWP, 3/8, J. Watt to J. Keir, 30 April 1780.
141 'Prospectus for the Copying Machine', Garret Workshop, Science Museum.
142 JWP, 3/8, J. Watt to J. Keir, 1 May 1780.
143 JWP, W5, J. Watt to J. McGrigor, 12 Aug. 1780.
144 B. & W. Col., Office Letter Book 4, J. Watt to 'Dear Sir', 26 May 1780.
145 Ibid, J. Watt to J. Woodmason, 28 Aug. 1780.
146 Ibid, J. Watt to J. Woodmason, 28 Nov. 1780. King-Hele (1999), p. 154, gives sales of 650 in the first year for £4,630. He claims that these sales prevented Boulton from being declared bankrupt, which is very doubtful as they represented only a small fraction of Boulton's debts.
147 Smiles, (1878), p. 209.
148 B. & W. Col., Office Letter Book 4, J. Watt to J. Woodmason, 4 June 1781/
149 JWP, W5, J. Watt to 'Dear Sir', 27 May 1781.
150 Ibid, J. Watt to G. Hamilton, 23 Sept. 1781/
151 JWP, Letter Book 1, J. Watt to G. Hamilton, 29 Sept. 1782.
152 Royal Society, James Watt Box PH13, J. Smeaton to J. Watt, 1 Nov. 1780.
153 JWP, Letter Book 1, J. Watt to J. Smeaton, 27 April 1783.
154 JWP, 4/81, J. Keir to J. Watt, 12 Nov. 1781.
155 JWP, Letter Book 1, J. Watt to W. Matthews, 15 Oct. 1785 [recte 1784].
156 JWP, Letter Book 2, J. Watt to J. McGrigor, 24 Oct. 1794.
157 Muirhead, (1854), Vol. III, p. 28, note, J. Watt to H.B. Wray, 27 March 1809.

Ballooning

1 Clow & Clow, (1952), pp. 151 – 64 for a more general account.
2 H. Cavendish, 'Three Papers, containing Experiments on Factitious Airs', *Philosophical Transactions*, Vol. 56, 1766.
3 Clow & Clow, (1952), p. 152.
4 See Edinburgh University, J. Black Papers, GEN 873, Vol. II, J. Black's answer to J. Lind, 8 Oct. 1784. This is also recounted in Partington, (1962), Vol. III, p. 133.
5 Robinson & McKie, (1970), p. 132, Letter 93, J. Robison to J. Watt, 22 Oct. 1783.
6 McKie, (1953), pp. 145 – 46.
7 Clark, (1983), p. 386, B. Franklin to Sir Joseph Banks, 30 Aug. 1783.
8 Robinson & McKie, (1970), p. 128, Letter 91, J. Watt to J. Black, 25 Sept. 1783. See Schofield (1963), p. 250 for A. Argand's demonstration of a balloon ascent before the King on 26 Nov. 1783.
9 Clow & Clow, (1952), p. 158.
10 JWP, Letter Book 1, J. Watt to J.A. De Luc, 2 Feb. 1784.
11 Ibid, 4 April 1784.
12 JWP, C3/6, 'Diary', undated notes at end.
13 Clow & Clow, (1952), p. 158 and Clark, (1983), p. 389.
14 Clow & Clow, (1952), p. 159 and Schofield, (1963), p, 251.
15 JWP, Letter Book 1, J. Watt to G. Hamilton, 28 Sept. 1783.
16 Schofield, (1963), p. 251.
17 JWP, Letter Book 1, J. Watt to J. Lind, 16 Dec. 1784.
18 Ibid, J. Watt to J. Lind, 16 Dec. 1784.
19 See JWP, C3/9, 'Diary', 4 Jan. 1785.
20 JWP, Letter Book 1, J. Watt to J. Watt jun., 7 Jan. 1785.
21 Edinburgh University, J. Black Papers, GEN 873, Vol. II, J. Black to J. Lind, 8 Oct. 1784.
22 Schofield, (1963), p. 252, for J. Southern, *A Treatise upon Aerostatic Machines, Containing Rules for Calculating their Powers of Ascension, etc.*, Pearson & Rollason, Birmingham, 1785.

The Water Controversy

1 Priestley, (1796), p. 19.
2 JWP, C6/6.50, G. Hamilton to J. Watt, 30 Dec.1807.
3 Robinson & McKie, (1970), p. 33, Letter 216, J. Robison to G. Black, 28 Dec. 1799.
4 Miller, (2004), p. 19.
5 Robison, (1822), Vol. II, p. 1.
6 Berry, (1960), p. 47.
7 Gibbs, (1965), pp. 119 – 22.
8 Schofield, (1963), p. 189. Watt wrote to Boulton, 12 June 1780, 'Dr. Priestley has taken Mr. Lloyds [house in Birmingham] to use'. The move did not taken place at once since Priestley was seriously ill in July and August, but by 12 October he had settled in Birmingham.
9 Royal Society, Box PH 13, J. Watt to J. Priestley, 'Thursday'.
10 Gibbs, (1965), p. 154.
11 Muirhead, (1846), pp. xxxiii & lxxxix.
12 McKie, (1952), p. 120.
13 Muirhead, (1858), p. 325, see note for Warltire's letter.
14 Quoted in McKie, (1952), p. 123. See also Crowther, (1962), p. 298.
15 Muirhead, (1854), Vol. II, p. 123, Letter 184, J. Watt to E. Darwin, 3 Jan. 1781; JWP W5.
16 King-Hele, (1981), p. 104, Letter 81 A, E. Darwin to J. Watt, 6 Jan. 1781.
17 Muirhead, (1854), Vol. I, pp. clx – clxiii and King-Hele, (1981), p. 105.
18 Muirhead, (1854), Vol. II, p. 59, Letter 120, J. Watt to W. Small, 17 Aug. 1773; Boulton 348.36.
19 Ibid, Vol. II, p. 75, Letter 132, J. Watt to W. Small, 3 March 1774; Boulton 348.41.
20 JWP, C3/4, 'Journal', 25 March 1779.
21 Robinson & McKie, (1970), p. 77, Letter 59, J. Watt to Magellan, 1 March 1780.
22 Ibid, p. 91, Letter 67, J. Watt to J. Black, 30 May 1780.
23 Ibid, pp. 103 – 5, J. Watt to J. Black, 23 Feb. to 5 March 1781 and Farey, (1827), Vol. I, p. 402.
24 Reid, (2003), p. 3.
25 Priestley, (1796), p. 35.
26 Muirhead, (1854), Vol. I, p. 99, Letter 80, J. Watt to W. Small, 16 March 1770; Boulton 348.19.
27 Reid, (2003), p. 27.
28 Partington, (1822), p. 29.
29 Robinson & McKie, (1970), J. Watt to J. Black, p. 66, Letter 77, 8 March 1781 and p. 108, Letter 79, 9 April 1781.
30 Muirhead, (1854), Vol. II, pp. 167 – 68, Letter 233, J. Watt to M. Boulton, 10 Dec. 1782. See also Smiles, (1878), pp. 303 – 4.
31 Robinson & McKie, 91970), p. 117, Letter 85, J. Watt to J. Black, 13 Dec. 1782.
32 JWP, Letter Book 1, J. Watt to J.A. De Luc, 13 Dec. 1781.
33 Ibid.
34 Ibid, J. Watt to J. Smeaton, 27 April 1783.
35 Muirhead, (1858), p. 329, J. Watt to J. Fry, 28 April 1783.
36 See Robinson & McKie, (1970), p. 122, Letter 87, J. Watt to J. Black, 3 Feb. 1783.
37 Muirhead, (1846), p. 20, J. Watt to G. Hamilton, 22 April 1783.
38 Royal Society, MM 5.38, J. Watt to J. Priestley, 26 April 1783.
39 Gibbs, (1965), p. 158.
40 JWP, Letter Book 1, J. Watt to J. Priestley, 2 May 1783 and Muirhead, (1846), p. 26.
41 Ibid, J. Watt to G. Hamilton, May or June 1783.
42 McKie, (1952), p. 125.
43 JWP, Letter Book 1, J. Watt to R. Kirwan, 1 Dec. 1783. See also Muirhead, (1846), p. 35, J. Watt to J.A. De Luc, 30 Nov. 1783.
44 Royal Society, Box PH, R. Kirwan, to J. Watt, 17 Dec. 1783.
45 Royal Society, Box PH, J. Watt to J.A. De Luc, 26 Nov. 1783.
46 *Philosophical Transactions*, Vol. LXXIV, pp. 119 – 53.
47 See Muirhead, (1858), p. 321.
48 Reid, (2003), p. 6.
49 Miller, (2004), p. 30.
50 Muirhead, (1846), pp. 47 – 9, J. Watt to J.A. De Luc, 6 March 1784.
51 *Quarterly Review*, (1854), p. 508.

52 Royal Society, Box PH 13, Sir J. Banks to J. Watt, 23 April 1784.
53 Ibid, 11 May 1784.
54 *Philosophical Transactions*, Vol. 74, 1784, p. 329 f.
55 Ibid, p. 354 f.
56 Muirhead, (1846), p. 67, J. Watt to J. Robison, 31 May 1784.
57 *Philosophical Transactions*, Vol. 74, 1784, p. 338.
58 Robinson & McKie, (1970), pp. 124 – 26, Letter 89, J. Watt to J. Black, 21 April 1783.
59 *Philosophical Transactions*, Vol. 74, 1784, p. 333.
60 JWP, Letter Book 1. J. Watt to J.A. De Luc, 26 April 1783.
61 See Schofield, (1954), p. 296 and Reid, (2003), p. 8.
62 Muirhead, (1858), J. Watt to J.A. De Luc, 18 May 1783.
63 Irvine, (1805), p. 42.
64 Robison, (1822), Vol. II, p. 11.
65 Muirhead, (1858), p. 328, J. Watt to G. Hamilton, 22 April 1783.
66 Quoted in McKie, (1952), p. 116. See also Partington, (1962), Vol. 3, p. 636.
67 Priestley, (1796), p. 42.
68 JWP, Letter Book 1, J. Watt to J.A. De Luc, 26 April 1783.
69 Royal Society, MM 5.38, J. Watt to J. Priestley, 26 April 1783.
70 Ibid.
71 Ibid.
72 Robison, (1822), Vol. II, pp. 5 – 10.
73 Royal Society, MM 5.38, J. Watt to J. Priestley, 26 April 1783.
74 JWP, 6/33.66, H. Davy to J. Watt, 29 Dec. 1801.
75 JWP, Letter Book 3, J. Watt to H. Davy, 11 Jan. 1801.
76 JWP, Letter Book 4, J. Watt to J. Robison, 24 Jan. 1804.
77 Philosophical Transactions, Vol. 74, 1784, p. 419 f.
78 Pugh & Hudson, (1985), p. 47.
79 Royal Society, Certificate of Election.
80 B. & W. Col., MI, 2/12, 'Account Book', 2 & 9 Jan. 1786.
81 Schofield, (1963), p. 236.
82 Quoted from Smiles, (1878), pp. 304 – 5, J. Watt to J. Fry, 15 May 1784.
83 Muirhead, (1854), Vol. I, p. clxxxix and Muirhead, (1858), p. 385, J. Watt to A. Watt, 31 Oct. 1785.
84 Schofield, (1963), p. 74, quoting J. Priestley to J.A. De Luc, 27 Dec. 1783.
85 Muirhead, (1846), p. 51, J. Watt to J.A. De Luc, 6 March 1784.

Epilogue

The ten years following Watt's arrival in Birmingham saw dramatic changes to his life, reputation and fortune. Boulton had taken the risk of supporting a person whose previous career can hardly be claimed to have been successful. Burdened with debt, all Watt had to offer was a patent for two new designs of steam engines, neither of which had reached even a workable stage and so were not yet a commercial proposition. And the patent would expire in nine years time. But Watt's concept of the separate condenser had arisen out of his experiments on the nature and properties of steam and so had a scientific basis which no doubt convinced Boulton that these engines would prove to show greater efficiencies than all other existing types.

Boulton thought that the gamble of supporting Watt was worth risking and his financial backing gave Watt the opportunity to concentrate on perfecting his engine. This was where Watt showed such remarkable ability, both as a scientist and as a practical engineer. He had been able to first develop a theory through studying the literature and observations on a working model of a Newcomen engine and then develop this into a full-scale machine. He applied the same techniques to other ventures, such a his copying machine where he used his knowledge of chemistry as well as his engineering experience to combine the chemistry of the ink with the mechanics of the press. The continued development of the steam engine is a further example where adaptations and modifications continued throughout this early period and beyond. Watt had a remarkable sense of determining which alteration would be an improvement and which not. His steam engines were very reliable and their simplicity masks the enormous amount of development to achieve this. Perhaps his motto should have been 'Observare, laborare et perseverare', to observe, to labour and to persevere.

Once at Birmingham, Watt could concentrate on the manufacture of his steam engines secure in the knowledge that he held a master patent through the extension of his original one of 1769 for the separate condenser up to 1800 by the 1775 Act of Parliament. The success of obtaining this Act was due first to Boulton's foresight in recognising that the period left for the patent to run was too short to return any profit and then to his influence with people of status in the land. It turned out that Boulton was correct because the original patent should have expired in 1783 when the debts of both Boulton and Watt and Watt himself had not been paid off. Had the patent not been extended, we can only surmise whether the partners would have undertaken the later development of their engines by patenting further mechanisms in the early 1780s. Probably neither Watt nor Boulton, nor indeed William Small who helped draw up the original text of the 1769 patent, ever thought of the extent to which their patent would prove to contain such all-embracing features as long as steam at low pressures continued in use. The 1775 Act extending that patent gave Boulton and Watt the security to 'perfect' not only their reciprocating engine for pumping but also the rotative beam engine which would bring them increasing profits as well as launching a new source of power crucial for the continued expansion of the Industrial Revolution in Britain.

By 1785, Boulton and Watt had erected seventy-eight reciprocating engines and some eighteen rotative ones throughout the length and breadth of Britain. While these engines may not have brought the partners fortune, they certain brought fame. Watt's theory of his perfect engine had been proved in practice. As well as the continued design improvements he had initiated to keep ahead of his rivals, Watt had established an engineering consultancy which could erect engines anywhere, even abroad. This must be the first instance of a national, let alone an international engineering company in Britain.

Watt had chosen suppliers, such as Wilkinson, capable of making parts of high standard to meet his exacting requirement as well as helping to organise some manufacturing facilities at Boulton's Soho Manufactory for small crucial parts such as the valve gear.

While at first, Watt depended on the assistance of the staff at the Soho Manufactory to launch the firm of Boulton and Watt, soon he formed his own organisation particularly for engine design. Boulton provided financial support both for Watt himself and the whole engine business. Watt's idea of charging those who ordered one of his engines a third of the cost of the fuel saved meant that Boulton and Watt received no income until after an engine had been erected and was running. Therefore Boulton had to find a considerable amount of capital to invest before seeing any return. Hence Boulton and Watt would have become heavily in debt initially even if Boulton had not used the engine premiums to bolster the finances of his loss-making Soho Manufactory. This situation might have continued far longer if Watt had not taken fright and taken control of the finances of Boulton and Watt. It was only then that the concern turned the corner and Watt's method of charging for his patent began to bring in more money as more engines were started up. But in the concluding volume we shall see that this way of charging caused great resentment, especially in Cornwall, where Boulton and Watt had erected some of their largest and hence, for them, potentially most profitable engines.

For most people, such achievements would have been more than sufficient but Watt also continued with his passion for natural philosophy. Perhaps we can understand this because it was his scientific theories of steam which had launched his steam engine, the pinnacle of his career. There was also the stimulus of his Scottish friends, Black, Roebuck and Robison, followed by the members of the Lunar Society in Birmingham. It was partly through the latter that Watt was thrust into inventing his copying machine. Here we see him launching a new sort of consumer durable with batch-production techniques similar to those he had pioneered for his scientific instrument business in Glasgow. While his steam engines were hidden away in mines, copying presses were sold through Boulton's enthusiasm to many people of great influence, starting with the King himself, so bringing recognition. Watt's ability as a natural philosopher was recognised by the Royal Society of London through his election as a member after his submission of a paper on the composition of water.

In the ten years from his arrival in Birmingham, Watt had risen from a position of comparative obscurity to being recognised as an important engineer through his addition of the separate condenser to the steam engine as well as a person capable of contributing scientific ideas and inventions. His patents were beginning to bring considerable wealth which would further enhance his standing in society. He had already laid the foundations for the next phase in the development of Boulton and Watt through the patents he took out between 1781 and 1785. These would further lessen the fuel consumption of his engines as well as turn the reciprocating type into a rotative one, thus enhancing both his fame and fortune. We can surely agree with Watt junior's assessment that these years were 'the most continuous and active exertion, and the most profusely fertile in mechanical invention, of the whole of his life'. (See p. 77) In the final volume, we will trace the development of the rotative engine, more discoveries in chemistry, his family growing up, the eventual fate of his 1769 patent and his years in retirement, full of honour and respect.

APPENDIX 1

Specification of Patent 913, January 5th, 1769 for a new method of lessening the consumption of steam and fuel in fire engines.

My method of lessening the consumption of steam and consequently fuel, in fire engines, consists of the following principles:

FIRST, that vessel in which the powers of steam are to be employed to work the engine, which is called the Cylinder in common fire engines, and which I call the Steam Vessel, must, during the whole time the engine is at work, be kept as hot as the steam that enters it; first, by inclosing it in a case of wood, or any other materials that transmit heat slowly; secondly, by surrounding it with steam or other heated bodies; and, thirdly, by suffering neither water nor any other substance colder than the steam to enter or touch it during that time.

SECONDLY, in Engines that are to be worked wholly or partially by condensation of steam, the steam is to be condensed in vessels distinct from the steam vessels or cylinders, although occasionally communicating with them: these vessels I call Condensers; and, whilst the engines are working, these condensers ought at least to be kept as cold as the air in the neighbourhood of the engines, by application of water, or other cold bodies.

THIRDLY, whatever air, or other elastic vapour, is not condensed by the cold of the condenser, and may impede the working of the engine, is to be drawn out of the steam vessels or condensers by means of pumps, wrought by the engines themselves, or otherwise.

FOURTHLY, I intend in many cases to employ the expansive force of steam to press on the pistons, or whatever may be used instead of them, in the same manner as the pressure of the atmosphere is now employed in common fire engines: in cases where cold water cannot be had in plenty, the engines may be wrought by this force of steam only, by discharging the steam into the open air after it has done its office.

FIFTHLY, where motions round an axis are required, I make the steam vessels in form of hollow rings, or circular channels, with proper inlets and outlets for the steam, mounted on horizontal axles, like wheels of a water-mill; within them are placed a number of valves, that suffer any body to go round the channel in one direction only: in these steam vessels are placed weights, so fitted to them as entirely to fill up a part or portion of their channels, yet rendered capable of moving freely in them by the means hereinafter mentioned or specified. When the steam is admitted in these engines, between these weights and the valves, it acts equally on both, so as to raise the weight to one side of the wheel, and by the re-action on the valves successively, to give a circular motion to the wheel, the valves opening in the direction in which the weights are pressed, but not in the contrary: as the steam vessel moves round, it is supplied with steam from the boiler, and that which has performed its office may either be discharged by means of condensers, or into the open air.

SIXTHLY, I intend, in some cases, to apply a degree of cold not capable of reducing the steam to water, but of contracting it considerably, so that the engines shall be worked by the alternate expansion and contraction of the steam.

LASTLY, instead of using water to render the piston, or other parts of the engines, air and steam-tight, I employ oils, wax, resinous bodies, fat of animals, quicksilver, and other metals, in their fluid state...

BE IT REMEMBERED, that the said JAMES WATT doth not intend that anything in the Fourth Article shall be understood to extend to any engine where the water to be raised enters the steam vessel itself, or any vessel having an open communication with it.

Appendix 2

'An Account of James Watt's Improvements upon the Steam or Fire Engine'. B. & W. Col., MI 8/25.

By steam is meant the vapour that arises from boiling water, which, when confined in a close vessel and heated a few degrees above the point at which water boils in the open air, becomes an elastic fluid, uniform and transparent, very much greater in bulk than the water of which it is composed, and capable of being reduced again to water when deprived of the heat which it contains.

The pressure of the atmosphere, or any other resistance equivalent to it, prevents the production of steam until the water is heated to that well known point which is called boiling water, or 212° of Fahrenheit's thermometer, but when that pressure is removed, or the water is placed in a vessel exhausted of air, steam is produced from it when considerably colder than the human blood.

On the contrary, if water be placed in a vessel, where it is pressed upon by air or steam more compressed than the atmosphere, a degree of heat above 212° is necessary for the production of steam, and the differences of the heats at which water boils under different degrees of pressure, increase in a less proportion than the pressures themselves: so that a double pressure requires less than a double increase of heat.

The inventors of the fire engine seem to have only known, that steam was a uniform fluid, unmixed with air, and capable of being reduced to water by being brought into contact with bodies colder than itself; and they do not seem to have attended to all the consequences of this property. It therefore appeared to them, that if a vessel were filled with steam, and cold water thrown into it, the steam would be reduced to water, and the greatest part of the vessel would be left empty, and its sides, top, and bottom would be pressed upon by the whole weight of the atmosphere. Consequently, if that vessel were placed about 24 feet above the surface of the water to be raised, and there were a pipe communicating between the water and the vessel, the pressure of the atmosphere would cause the water to ascend through the pipe, and fill the vessel.

The vessel being filled with water, and its return by the pipe prevented by a valve - On admission of fresh steam to counteract the pressure of the atmosphere, it might be discharged at the height to which it was raised, or might be forced through pipes to still greater heights by a hotter and consequently more compressed steam. Such were the first steam engines, the Marquis of Worcester's and Captain Savery's.

There were two great errors in the construction of these machines; the first was admitting the water, to be raised, into the same vessel with the steam; for that vessel being thereby made very cold, and the steam brought into contact with the water itself, a very great quantity of steam was condensed in giving heat to the surface of the water, and to the whole body of the vessel. The second inconvenience, was the forcing the water to the desired height by the elasticity of compressed steam: for, as the steam exerted an equal force against the sides of the steam vessel and boiler, as it did upon the surface of the water, it was found difficult to construct vessels strong enough to resist it, when the water was required to be raised to any considerable height.

The latter of these inconveniences was removed by Mr. Newcomen's invention of the Fire Engine now in common use. This machine merits a more particular description. The steam vessel is a hollow cylinder, very smooth within, open at top, shut at bottom, and firmly fixed. A plug or <u>piston</u> is fitted to it in such a manner, that it can move easily up and down, and yet be air tight; this piston is suspended by means of a chain to one end of a great lever, which is moveable upon an axis fixed near the middle of its length; to the other end of the lever is suspended the bucket or piston of a pump, which reached from the surface of the ground to the bottom of the pit, from whence the water is to be raised; and the weight of the bucket and pump rods being greater than that of the piston, it pulls down the end of the lever to which it is attached, and consequently raises the piston to the top of the cylinder. The piston being in this situation, steam is admitted from a boiler into the cylinder, and the air which it contains is extruded through a valve called the snift, made for that purpose. When the cylinder is full of steam, the passage from the boiler is shut, and cold water is injected into the cylinder, which condenses the steam, and leaves it empty, the pressure of the atmosphere, ceasing to be counterpoised by the steam within, acts upon the piston, and forces it to descend to the bottom of the cylinder; and by means of the lever, pulls up the bucket of the pump, and thereby raises the water. The injecting cock is then shut, and the steam admitted again into the cylinder, which, by counteracting the atmosphere, suffers the piston to ascend to its former place, and the operation is repeated until the desired quantity of water is raised,

By this construction it was rendered unnecessary to use steam any stronger than would counterbalance the atmosphere, and all danger of bursting the vessels was avoided: for let the depth of the pit be what it will, it is not necessary to increase the force of the steam, but to enlarge the area of the piston, until the pressure of the atmosphere upon it be equal to the weight of the column of water contained in the pump, and the steam has no other effect than to permit the ascent of the piston. Another advantage, which was unforeseen by the inventor, is, that the cylinder is not made so cold as in Savery's construction, and therefore less steam is lost in filling it after the condensation.

Such is the common steam engine, which has not been materially improved since Mr. Newcomen invented it, seventy years ago; though it be much superior to Savery's, it is still very defective, and employs much more steam, and consequently fuel than is needful, from the following causes. First, the vacuum being produced by throwing water into the cylinder to condense the steam, that water becomes hot, and being in a vessel exhausted of air, it produces a steam, which in part resists the pressure of the atmosphere upon the piston, and lessens the power of the engine; this might be remedied by throwing in so much water as would cool the whole vessel below the point, at which water boils in vacuo; but that would increase the second inconvenience, which is the destruction of steam that unavoidably happens upon attempting to fill a cold cylinder with that fluid. Those who have constructed steam engines have found, that as they make their condensation more perfect, by making the cylinder colder, they increased the consumption of steam in a greater proportion than they gained power. Though it appears that they were ignorant of the cause, they were so sensible to the effect, that they contented themselves with causing the engine to raise a load equal only to seven pounds upon the square inch of the area of the piston; whereas the pressure of the atmosphere would have raised much more, if the cylinder had been perfectly exhausted, and even that load, of seven pounds, is, in general, more than they ought to have used, if they had known the true principles upon which their engine ought to have been constructed.

But so very few experiments had been made even upon the most essential part of the subject, that the real bulk of water, when converted into steam of a given heat, remained unknown, until the author of these new improvements determined it in the year 1764. The opinions which had been entertained concerning its bulk before that time were much beyond the truth, and could by no means be deduced from the very inadequate experiments which were pretended to have been made. The boiling of water in an exhausted receiver at low heats was known; but it was neither known what these heats were, nor what progression they observed under various pressures, before he made his experiments upon that subject. These experiments pointed out the defects of the common steam engines, and the extent to which they were hurtful; but the means of avoiding them were not so obvious, and were the subject of much contemplation and experiment, of which these new improvements are the result.

These new engines are of two kinds, one of which resembles the common steam engine, and works with a cylinder, a piston, a great lever and a pump.

The cylinder or steam vessel of this engine is shut at bottom and open at top, as usual, and is included in a case of wood, or metal covered with materials which transmit heat slowly. This case is at a little distance from the cylinder, and is shut at both ends; the cover has a hole in it through which the piston rod slides, and the bottom has another by which steam from the boiler has always free entrance into this case or outer cylinder, and by the interstice between the two cylinders has access to the upper side of the piston.

To the bottom of the inner cylinder is joined a pipe, with a cock, which is opened and shut when necessary, and forms a passage to another vessel called a condenser, made of thin metal. This vessel is immersed in cold water, and is contrived in such a manner as to expose a very great surface externally to the water, and internally to the steam. It is also made air tight, and has pumps wrought by the engine which keep it always exhausted of air and water.

Both the cylinders being filled with steam, the passage is opened from the inner one to the condenser into which the steam rushes with violence by its elasticity, because that vessel is exhausted; but it no sooner enters it, than coming into contact with the cold matter of the condenser, it is reduced to water, and the vacuum still remaining, the steam continues to rush in until the inner cylinder below the piston is left empty. The steam which is above the piston, ceasing to be counteracted by that which was below it, acts upon the piston and forces it to descend to the bottom of the cylinder, and so raises the bucket of the pump by means of the lever. The passage between the inner cylinder and the condenser is then shut and another passage is opened, which permits the steam to pass from the outer cylinder, or from the boiler, into the inner cylinder under the piston, and then the superior weight of the bucket and pump rods pulls down the outer end of the lever or great beam, and raises the piston, which is suspended to the inner end of the same beam.

The advantages that accrue from this construction are, first, that the cylinder being surrounded with the steam from the boiler, is kept always uniformly as hot as the steam itself, and is therefore incapable of destroying any part of the steam which should fill it, as the common engines do. Secondly, the condenser being kept always as cold as water can be procured, and colder than the point at which it boils in vacuo, the steam is perfectly condensed, and does not oppose the descent of the piston, which is therefore forced down by the full power of the steam from the boiler, which is somewhat greater than that of the atmosphere.

In common fire engines, when they are loaded to seven pounds upon the inch, and are of a middle size, the quantity of steam which is condensed in restoring to the cylinder the heat which it had been deprived of by the former injection of cold water, is about one full of the cylinder besides what is really required to fill that vessel, so that twice the fill of the cylinder is employed to make it raise a column of water equal to seven pounds for each square inch of the piston: or to take it more simply, a cubic foot of steam raises a cubic foot of water about eight feet high, besides overcoming the friction of the engine and the resistance of the water to motion.

In the improved engine, about one full and a fourth, of the cylinder is required to fill it, because the steam is about one fourth more dense than in the common engines. This engine raises a load, equal to twelve pounds and a half upon the square inch of the piston, and each cubic foot of steam of the density of the atmosphere, raises one cubic foot of water twenty-two feet high. - These are facts proved by the actual performance of this engine, and that of the very best of many common engines which have been accurately examined; most of the common engines do not raise a foot of water six feet high by a foot of steam.

The second species of the improved engines is a machine which produces a circulating or rotative motion by the immediate action of steam; it consists of a wheel placed vertically, which has a hollow channel round its circumference, as if a tube were bent into a circle and its two ends were joined together; in this channel are placed three valves, so as to permit any body to go round the channel in one direction only, but so as to prevent even the passage of steam in the contrary direction. There is also a weight fitted to this channel, so as to be steam tight, but capable of moving easily in it, and of passing through the valves when they are open; the axis of the wheel is perforated longitudinally by three holes, which communicate with the three divisions of the channel, by as many pipes, which enter the channel near the back of the valves. By means of a hole fixed in a piece of metal, which is accurately fitted to the end of the axis, it is contrived that one of these pipes should receive steam from the boiler, while another pipe discharges the steam into the condenser after it has done its office, through another hole which is in the same fixed piece of metal, so that as the holes in the end of the axis, by the action of the wheel, come opposite to one or other of the two holes in the fixed piece of metal, they receive or discharge steam accordingly.

When the wheel is at rest, the weight naturally remains at the lowest part of it; but when steam is admitted, it acts equally upon the weight and the valve, and forces them asunder. The weight is consequently raised towards the opposite side of the wheel from the valve, until its gravity, acting upon the valve through the medium of the steam, becomes greater than the resistance or load the wheel has to overcome. As soon as that is the case the weight becomes stationary, and the valve recedes from it; and consequently the wheel, to which the valve is attached, begins to turn round in the contrary direction to that in which the weight had been raised. When a pipe passes the upper diameter of the wheel, it discharges the steam contained in the chamber to which it belongs, and another valve having then passed the weight, the steam acts upon it as it had done upon the other, and keeps up a uniform motion in the wheel.

The advantages of this particular construction are evident. It gives the means of producing rotative motions of all kinds by steam, and by the equality of its velocity is better adapted to many purposes than those engines which have a reciprocating or alternate motion. In point of power gained from a given quantity of steam, it has all the advantages of the reciprocating species, as the steam vessel is kept uniformly hot, and the steam is condensed in a distinct vessel in the same manner as in the other.

Appendix 3

Proposals to the Adventurers...

Method of Calculating tables for Wheal Maid 63 inch Cylinder: double 9 feet or 18 feet stroke:

From the time that we first offered our new-invented steam engines to the public, our conditions have uniformly been as follows:

1. To give a licence for the use of our engine during the term of our exclusive privilege granted to us by an act of parliament passed in the year 1775, and also to furnish all plans and instructions necessary for the erection of such engines.
2. The engine to be erected by the proprietors of the works for whose use it is wanted, and at their expense.
3. The profit which we require for such licence, drawings, and instructions, is to be such sums paid from time to time, during the term of our privilege, if the engine should so long work, as shall appear to be equal in value to a <u>third part of the savings of fuel</u>, which shall have been made during these respective times by the use of the engine of our invention, instead of one of the common construction of equal power.

According to these conditions, our profits are made to depend entirely on the superiority of our engine to the common one, and are always proportional to the advantages which our employers gain by the use of the engine, that is, for every pound which can become due to us, the proprietors must have gained by the use of our engine two pounds. There are other very considerable advantages which the proprietors will gain by the use of our engine instead of the common one, and in which we have no share.

The above conditions have been generally approved by the proprietors of mines who have applied to us for our engines; and we therefore propose that they shall continue to be the basis of all our future agreements.

The principal objection that lies against them is the difficulty of ascertaining what is the real consumption of fuel by common engines in performing a certain quantity of work; though this may at first sight appear an easy matter to determine, yet it is in reality more intricate than it appears, for it is necessary to ascertain the quantity of coals consumed in some given space of time, the number of strokes made by the engine during that time, the length of these strokes, the weight of the column of water raised by the engine, besides many other less material circumstances, which however do in <u>some</u> degree affect the result.

In Cornwall, where the mode of firing requires a very large quantity of fuel to be thrown in at once, and whereupon every stoppage for cleansing the boiler, or other purposes, all the coals contained in the fire place are necessarily lost: the consumption of coals for one hour or one day cannot ascertain the monthly consumption, which must therefore be determined by an account kept for some period of time, which shall include all common stoppages, and shall by that means shew the average consumption for each month or each day. Such an account was accordingly kept, during a sufficient time at Poldice mine, for the purpose of ascertaining the consumption of fuel, and the effects of two of the engines there; the result of which is explained in the following part of this paper.

This account was taken in order to adjust our premium for the engine we were to erect at that mine, and has since been acceded to as a standard of the effects of common engines, by all the other gentlemen with whom we have agreed for the erection of our engines in the county.

It may not be improper to observe, that we have had occasion to examine common engines in many parts of England, and we can venture to affirm that the effects of the coals which we have admitted in our calculations to be produced by Poldice engines, are very much greater than the average effects of coals applied to common engines.

As we continue convinced that the effects of Poldice engines have been fairly stated and deduced, we intend in all our future agreements in Cornwall, to consider them as the standard of the effects of common engines, with which we mean to compare our engines, by which means all the trouble, uncertainty, and disputes, which frequently attend trials of engines will be avoided. If it should be alleged that there are some common engines that are better than those of Poldice, we answer, That if there are such, they are not proper engines to fix a common standard by, and therefore if we were asked to compare with such extraordinary engines, we should require a larger share of the savings.

In order that those who wish to have our engines may have an opportunity of satisfying themselves of the accuracy of our rules and methods of ascertaining the profits which may become due to us, and in order to avoid all disputes about the mode of calculating such profits, we have drawn up the following explanation:

I. When the adventurers in Poldice mine proposed to substitute one engine of our construction in place of the two eastern engines of that mine; it was proposed that an account should be taken from the company's books of the quantity of coals consumed by these two engines; and of the number of strokes per minute, and the length of each stroke when the water was kept in fork, during the months of August and September 1778; and that the coals to be consumed monthly by the new engine, when erected, should be compared with those consumed by the old engines, during the abovementioned period, after making proper allowances when the new engine should go faster or slower, or when it should draw down the water from a greater depth.

II. The old engines were both in good order, one of them was lately built, and the other had undergone a thorough repair. It was agreed by all who had examined the going of the engines, that six strokes per minute was fully their average rate of working, when the water was kept in fork, and that five feet six inches was also fully the average length of their stroke. The diameter of the cylinder of the most eastern engine was fifty-six inches, and of the other sixty inches. The diameter of the pumps in both the engines was seventeen inches; and the quantity of coals consumed by both engines, during the sixty-one days of August and September, was two hundred and forty-two weys, of sixty-four bushels each; but it being alleged that part of the coals then used were not of so good a quality as usual, it was agreed to deduct twenty-two weys on that account, and to fix the consumption of both engines for these two months, at two hundred and twenty weys.

All these articles are brought into one view in the following table, and the coals are divided between the two engines, in proportion to the areas or squares of the diameters of their cylinders, and reduced to a month of thirty days.

POLDICE	Diameter of cylinder	Depth in fathoms	Diameter of Box	Strokes per minute	Length of stroke	Load per sq. ins	Bushels of coals per month of 30 days
					feet inch	pounds	
1st Eastern Engine	66 inches	33	17 inches	6	5–6	5 7/10	3954
2nd Eastern Engine	60 inches	25	17 inches	6	5–6	5 2/10	2979

III. As engines consume greater or lesser quantities of coals as they are more or less loaded, though the number of strokes they make per minute be the same, it became necessary to enquire what quantities of coals were consumed monthly by engines very lightly loaded; and it appeared both by the testimony and opinions of those persons who had experience of engines in that state, and also from calculation, that an engine with a sixty inch cylinder loaded only to one pound on each square inch of the piston, going at the rate of six strokes per minute of five feet six inches long each, would consume in thirty days 1920 bushels, that is one wey per day.

IV. It was agreed to facilitate calculation, that the increase of consumption under other loads should be proportioned to the loads in the manner which shall be explained.

And it was also agreed, for the same reason, that the consumption under other loads should be considered as proportioned to the number of strokes per minute.

The consumption by cylinders of different diameters, loaded to the same number of pounds per inch, going the same number and length of strokes per minute, and constructed equally well, will be as the squares of their diameters.

V. The two engines at Poldice consume 6924 bushels in 30 days; the squares of their diameters (60 and 66) are 3600 and 4356; the sum of these squares is 7956. To find the consumption of a 60 inch loaded to the average of these two, say, as 7956 (sum of the squares) is to 6924 bushels, so is 3600 (square 60) to 3131, the bushels which would be consumed in 30 days by a 60 inch cylinder, going six strokes per minute of five feet six inches long each, and loaded to 5 $^{4}/_{10}$ths pounds per square inch.

VI. To find what such an engine would consume at the load of five pounds on the inch, deduct 1920 (the consumption at the load of one pound on the inch, see f III.) from 3131 and there will remain 1211, which is the additional number of bushels of coals necessary monthly, when the load is increased from one pound to 5 $^{4}/_{10}$ lbs on each square inch; then as 5,4 is to 1211, so is 5 pounds to 1121 bushels, which added to 1920, gives 3041, the number of bushels such an engine would consume, when loaded to 5 pounds per inch. And 1211 divided by 4 (the number of pounds intermediate between one pound and five pounds, gives 280 bushels for the additional quantity necessary monthly for each pound on the inch increase of load; which being added to 1920 in the following manner, will shew the number of bushels consumed by a 60 inch cylinder at all loads commonly used.

1920 – – – at one pound
280
2200 – – – at two pounds
280
2480 – – – at three pounds
280
2760 – – – at four pounds
280
3040 – – – at five pounds
280
3320 – – – at six pounds
280
3600 – – – at seven pounds

VII. These calculations being confined to one particular cylinder and length of stroke, in order to make them general we shall calculate what would be consumed by a cylinder of 10 inches diameter, which will render the calculation of any other more easy. The square of 10 is 100; then, as 3600 (square of 60) is to 100, so is 1920 to 53 333/1000 bushels; and if we divide that number by 5.5 feet, the length of the stroke, we have 9.6969 for the consumption of a 10 inch cylinder with a one foot stroke at one pound per inch. And in like manner to find the increase of consumption for each pound per inch, say, as 3600 is to 100, so is 280 to 7.777; which divided by 5.5 gives 1.4141 for the increase of consumption for each pound additional load of a 10 inch cylinder with a one foot stroke.

VIII. To apply these calculations to any other given cylinder, say, as 100 (square of 10), is to the square of the cylinder required; so is 9.6969 to the number of bushels required by that cylinder with a stroke of one foot long, and loaded to one pound on the inch: Or, which is the same thing, multiply the square of the cylinder given, by the decimal .096969 (or for shortness by .097) multiply that product by the length of the stroke in feet and decimals, and it will give the monthly consumption at the load of one pound per inch with the given stroke. And to find the increase of consumption for each pound per inch, say, as 100 (square of 10) is to the square of the given cylinder, so is 1.4141 to the increase of consumption for each pound increase of load on the given cylinder, with a stroke of one foot long (or for shortness multiply the square of the cylinder given by .014) then multiply the product by the given length of the stroke, and you will have the increase of consumption by the given cylinder for each pound additional load on the square inch, with the given stroke, which being added successively to the consumption at one pound on the inch in the manner which has been shewn, will form a table for that engine.

IX. To find what fuel is saved by one of our engines, it is necessary to find the size of the common engine which would do the same work.

Our engines do actually work as briskly under a load of ten pounds and a half on the inch, as common engines do when loaded only to seven pounds on the inch; consequently, if any person wanted to erect a common engine to do the work performed by one of our engines, it would require a cylinder, the square of whose diameter exceeded the square of the diameter of ours, in the proportion of 10½ to 7, or of 1 to 1.5: therefore to find the diameter of the common engine, which would be equal in power to that proposed to be erected for you by us, Square the diameter of our cylinder, multiply the square by 10½ and divide it by 7, which will give you the square of the common cylinder equal in power, and by extraction of the square root, you will have the diameter; or to save division you may multiply the square of ours by 1.5, and it will give you the square of the common engine.

X. Thus, to find the common engine equal in power to that proposed for Wheal Maid [in ink], the diameter of which is 63 inches, and its square 3969 which multiplied by 1.5, gives 5953 the square of the common cylinder equal to it in power, the diameter of which would be 77.1 inches. To find the consumption of such common engine, the square 5953 being multiplied by .097, produces 577.441 which multiplied by 18 feet the length of the stroke, gives 10383 bushels, the monthly consumption at the load of one pound on the square inch. And again multiplying the square 5953 by .014 produces 83.342 which is multiplied by 18 feet the length of the stroke gives 1500.156 for the monthly additional consumption of coals for every pound additional load of the engine.

The square of the diameter of the cylinder, (that is, the number of circular inches it contains,) multiplied by .7854, gives the square inches of its area, which multiplied by the pounds load on the square inch, gives the total number of pounds weight, which would produce such loads per square inch, when the beam is equally divided.

XI. To find the weight of the column of water in the pumps, square the diameter of the working barrel (in inches) multiply it by the total height of the column of water (in yards) and add to it 1/44th of the product, the sum is the weight of the column of water in pounds avoirdupois. N.B. You will err very little if you add only 1/40th instead of 1/44th

To find the load raised by the cylinder when the beam is not equally divided, multiply the weight of the column of water contained in the pumps, by the length of the stroke of the pump rods, and divide that product by the length of the stroke of the cylinder piston rod; the quotient is the weight raised by the cylinder.

XII. From these calculations is formed the following table, which shows the monthly consumption for a common fire engine of a 77.1 cylinder which is equal in power to that proposed to be erected at Wheal Maid at different loads from one pound to seven pounds on each square inch, when going at the rate of six strokes per minute.

Col. 1.	Col. 2.	Col. 3.	Col. 4.
Load per ſquare inch of the piſton.	Weights of the columns of water in the pumps, and of the other loads at that end of the beam.	Load at the cylinder end of the beam when the beam is divided in the ratio of Six to Eight.	Buſhels of coals which would be conſumed in 30 days by a common engine with a cylinder of 34¼ inches diameter, making ſix ſtrokes of 8 feet long per minute.
1 pound	1231,5072 pound	923.6304 pounds	912.5 buſhels
2 pounds	2463.0144 pounds	1847.2608 pounds	1044.2 buſhels
3 pounds	3694.5216 pounds	2770.8912 pounds	1175.9 buſhels
4 pounds	4926.0288 pounds	3694.5216 pounds	1307.6 buſhels
5 pounds	6157.5360 pounds	4618.1520 pounds	1439.3 buſhels
6 pounds	7389.0432 pounds	5541.7824 pounds	1571.0 buſhels
7 pounds	8620.5504 pounds	6465.4128 pounds	1702.7 buſhels

XIII. To compute the monthly savings from the above table, having found the weight of the column of water in the pumps, see if it corresponds to any of the numbers in the table, Column 2d, in col 4th on the same line You will find the numbers of bushels which the common engine equal in power and length of stroke would have consumed at six strokes per minute when so loaded; but if the engine has made any greater or lesser number of strokes, the number of bushels must be increased or diminished accordingly; and from the number of bushels so found, you are to subtract the coals actually burnt by our engine, the remainder being divided by 3, will show the part due to us that month.

XIV. If the load of the engine corresponds to none of the numbers in the table, the shortest and easiest way will be to proceed in the following manner. Say, as 12021 pounds, the total load on the engine for each pound on the square inch, is to 1900 bushels, the encrease of consumption monthly for each pound encrease of load, so is 1000 pounds to 124.7 the monthly additional consumption for each 1000 pounds encrease of load, multiply therefore the load at the cylinder end of the beam by 124.7 the multiplier so found, and add the produce to 8383 bushels, which is the coals which would be consumed monthly, if the cylinder were loaded to 0 pounds per inch, (a) and you will have the monthly consumption at the given load, and at the rate of six strokes per minute. (Note (a) That is supposing that the engine raised no water, and its power was employed only in overcoming the vis inertiae and friction of the machine and pumps, raising the house water etc); but as the engine will seldom go exactly at that rate, divide the monthly number of strokes shewn by the counter by 8640, the strokes in a day when the engine goes 6 strokes per minute, the quotient will be the number of days the engine must have wrought at the rate of 6 strokes per minute to have made that number of strokes: then say if 30 days give bushels, what will the number of days found by this computation give? The answer will be the monthly consumption at the load and rate of going.

XV. Or for greater exactness, you may subtract the next lesser number of pounds, you find in the second column of the table, from the weight of the column of water, and say, as the load for one pound on the inch is to bushels, so is pounds the remainder so found to the additional number of bushels required for such additional load, which number so found being added to the number of bushels in the table corresponding to the above mentioned next lesser number of pounds, will give the monthly consumption required, at six strokes per minute.

In consequence of the preceding reasoning, we make the following proposal to the adventurers in

1. An engine of our construction with a cylinder of inches in diameter, capable of stroke to be erected at the charge of the proprietors of the mine, according to the directions and drawings we shall furnish you with for that purpose.
2. The consumption of coals by a common engine equal in power to the above, to be determined by the preceeding table, and to be proportioned to the depths of the pumps and to the number of strokes, in the manner which has been described.
3. The actual consumption of the new engine is to be compared monthly with the said table, according to its actual load and rate of going, and whatever smaller quantity of coals it shall appear it has burnt than what the table shows for the correspondent effect, by a common fire engine, is to be divided into three parts, the value of one of which parts we ask for our premium, and to be paid us monthly in money; the coals being valued at the price they shall cost when delivered at the engine house, the drawback of the duties being first deducted.
4. Our premium to commence from the day the engine is properly set to work to drain the mine, and to continue so long as the engine shall be used, not exceeding the term of our exclusive privilege.
5. The accounts of coals consumed, the quantity of water raised monthly, and the depths from which it shall be raised, to be entered in a book kept for that purpose at the mine, which, with the buying and debenture books of coals, are to be open for our inspection.
6. We ask a power of appointing a check on the company's coal measurer, if we shall judge it necessary.
7. We ask liberty of inspecting the engine, the pumps, and all other appurtenances of it, either by ourselves or our agents, at all reasonable times, and of directing the necessary repairs, in cases of too great consumption of fuel.
8. A power to be lodged in us of refusing improper coals, and of removing such engine men as do not do their duty, as our interest and reputation are materially concerned in the good performance of the engine.
9. The number of strokes the engine makes to be kept account of by a counter fixed on the beam.
10. Plans, drawings, and directions of all sorts for erecting, repairing, and working the engine to be furnished by us.
11. We engage that the engine shall be able to work a pump of . . . inches diameter, and . . . fathoms high, or others equivalent, at the rate of ten strokes per minute.
12. . . . pounds to be allowed for the injection or cold water pumps, and all the remaining power required to work it to be estimated as part of the load of the engine.
13. If any flat or horizontal rods are to be worked by the engine, an experiment is to be made, how much their friction and resistance to motion amounts to, and it is to be reckoned as a part of the load of the engine, and added to the weight of the column of water in the pumps.

Short Bibliography

Andes, L.E., *The Treatment of Paper for Special Purposes*, 1907, 2 English edn., Scott, Greenwood & Son, London, 1923.

Andrews, J.H., 'The Evolution of the Watt Single Acting Steam Engine', *Stationary Power, The Journal of the Stationary Engine Research Group*, 1988.

Anon, 'The Hornblower Family', *Journal of the Trevithick Society*, No. 4, 1976.

Arago, D.F.J., *Historical Eloge of James Watt*, Translated from the French with additional notes and an appendix by James Patrick Muirhead, Esq., M.A., J. Murray, London, 1839.

Balston, Thomas, *James Whatman, Father & Son*, Methuen, London, 1957.

Barker, Richard, 'John Wilkinson and the Paris Water Pipes', *Wilkinson Studies*, Vol. II, 1992.

Barton, D.B., *A History of Copper Mining in Cornwall and Devon*, D. Bradford Barton, Truro, 1961, 3 edn. 1978.

Berry, A.J., *Henry Cavendish, His Life and Scientific Work*, Hutchinson, London, 1960.

Biographical Dictionary of the History of Technology, Routledge, London, 1990.

Bootsgezel, J.J., 'William Blakey - A Rival to Newcomen', *Trans. Newcomen Society*, Vol. 16, 1935 - 6.

Bower, Peter, 'The Papers Used by John Downman', *The Quarterly, The Journal of the British Association of Paper Historians*, No. 20, Oct. 1996.

Braid, D., 'Wilkinson and New Willey Ironworks, Broseley' and 'John Wilkinson's First Patent', *Wilkinson Studies*, Vol. I, Merton Priory Press, London, 1991.

Bramah, J., *A Letter to the Rt. Hon. Sir James Eyre... on the subject of the Cause Boulton & Watt v. Hornblower & Maberley; for and infringement on Mr. Watt's Patent for An Improvement in the Steam Engine*, J. Stockdale, London, 1797.

Brewster, David, Ed., *The Edinburgh Encyclopaedia*, W. Blackwood, Edinburgh, 1830.

Brewster, David, Ed., *A System of Mechanical Philosophy*, Edinburgh, 1822.

Buckley, J.A., *The Great County Adit*, Penhellick, Camborne, 2000.

Cavendish, Henry, 'Experiments on Air', *Philosophical Transactions*, Vol. 74, 1784.

Cavendish, Henry, 'Three Papers, Containing Experiments on Factitious Airs', *Philosophical Transactions*, Vol. 56, 1766.

Clagett, Martin R., 'William Small, 1734 - 1775, Teacher, Mentor, Scientist', Ph.D. Thesis, Virginia Commonwealth University, 2003.

Clapperton, R.H., *The Paper-making Machine, Its Invention, Evolution and Development*, Pergamon Press, Oxford, 1967.

Clark, Ronald W., *Benjamin Franklin, A Biography*, Phoenix, London, 1983.

Clow, Archibald & Clow, N.L., *The Chemical Revolution, A Contribution to Social Technology*, Batchworth Press, London, 1952.

Crowther, J.G., *Scientists of the Industrial Revolution, Joseph Black, James Watt, Joseph Priestley, Henry Cavendish*, Cresset Press, London, 1962.

Curr, John, *The Coal Viewer, and Engine Builder's Practical Companion*, J. Northall, Sheffield, 1797.

Dickinson, H.W., *James Watt, Craftsman and Engineer*, Cambridge University Press, 1935.

Dickinson, H.W., *John Wilkinson, Ironmaster*, Ulverston, 1914.

Dickinson, H.W., *Matthew Boulton*, Cambridge University Press, 1936.

Dickinson, H.W., *A Short History of the Steam Engine*, Cambridge University Press, 1938.

Dickinson, H.W. & Jenkins, R., *James Watt and the Steam Engine*, Oxford, 1927, reprint Moorland, Ashbourne, 1981.

Dickinson, H.W. & Titley, Arthur, *Richard Trevithick, The Engineer and The Man*, Cambridge University Press, 1934.

Elton, Sir Arthur, 'William Murdock', typescript.

Falck, N.D., *An Account and Description of an Improved Steam-Engine*, B. Law, London, 1776.

Farey, John, *A Treatise on the Steam Engine, Historical Practical and Descriptive*, London, 1827.

Fitton R.S., *The Arkwrights, Spinners of Fortune*, Manchester University Press, 1989.

Gibbs, F.W., *Joseph Priestley, Adventurer in Science and Champion of Truth*, Nelson, London, 1965.

Goodison, Norman, *Matthew Boulton: Ormolu*, 1974, reprint Christie's, London, 2002.

Griffiths, J., *The Third Man, The Life and Times of William Murdoch, 1754 – 1834, The Inventor of Gas Lighting*, Deutsch, London, 1992.

Hadfield, Charles, *The Canals of the West Midlands*, David & Charles, Newton Abbot, 1966.

Hamilton Jenkin, A.K., *The Cornish Miner*, G. Allen & Unwin, London, 1927, 3 edn. 1962.

Harris, J.R., *The Copper King, A Biography of Thomas Williams of Llanidan*, Liverpool University Press, 1964.

Harris, J.R., *Industrial Espionage and Technology Transfer, Britain and France in the Eighteenth Century*, Ashgate, Aldershot, 1998.

Hebert, L., *The Engineer's and Mechanic's Encyclopaedia*, Kelly, London, Vol. I, 1836, Vol. II, 1838.

Hills, R.L., *The Drainage of the Fens*, Landmark, Ashbourne, 2003.

Hills, R.L., *James Watt, Vol. I, His Time in Scotland, 1736 - 1774*, Landmark, Ashbourne, 2002.

Hills, R.L., 'James Watt and his Copying Machine', *Studies in British Paper History, Vol. I, The Oxford Papers*, British Association of Paper Historians, 1966.

Hills, R.L., 'James Watt in Cornwall', *Journal of the Trevithick Society*, No. 24, 1997.

Hills, R.L., 'James Watt and the Delftfield Pottery, Glasgow', *Proceedings of the Society of Antiquaries of Scotland*, Vol. 131, 2001.

Hills, R.L., 'James Watt and Paper and Papermaking', *Studies in British Paper History, Vol. II, The Exeter Papers*, British Association of Paper Historians, 2001.

Hills, R.L., 'James Watt and his Rotary Engines', *Trans. Newcomen Society*, Vol. 70, No. 1, 1998 - 99.

Hills, R.L., 'The Origins of James Watt's Perfect Engine', *Trans. Newcomen Society*, Vol. 68, 1996-97.

Hills, R.L., *Papermaking in Britain, 1488 – 1988*, Athlone, London, 1988.

Hills, R.L., *Power from Steam, A History of the Stationary Steam Engine*, Cambridge University Press, 1989.

Hills, R.L., 'A Steam Chimera: A Review of the History of the Savery Engine', *Trans. Newcomen Society*, Vol. 58, 1986 – 87.

Hills, R.L. & Pacey, A.J., 'The Measurement of Power in Early Steam-driven Textile Mills', *Technology & Culture*, Vol. 13, No. 1, Jan. 1972.

Howard, G.S., *The New Royal Cyclopaedia and Encyclopaedia*, A. Hogg, London, 1788.

Irvine, William, *Essays Chiefly on Chemical Subjects*, J. Mawman, London, 1805.

Jenkins, Rhys, 'Jonathan Hornblower and the Compound Engine', *Trans. Newcomen Society*, Vol. 11, 1930 – 31.

King-Hele, Desmond, *Erasmus Darwin, A Life of Unequalled Achievement*, Giles de la Mare, London, 1999.

King-Hele, Desmond, Ed., *The Letters of Erasmus Darwin*, Cambridge University Press, 1981.

Lindqvist, Svante, *Technology on Trial, The Introduction of Steam Power Technology into Sweden*, Uppsala, 1984.

Lord, John, *Capital and Steam Power, 1750 - 1800*, P.S. King, London, 1923.

McKie, Douglas, *Antoine Lavoisier, Scientist, Economist, Social Reformer*, Constable, London, 1952.

Michell, F. Bice, *Michell – A Family of Cornish Engineers, 1740 – 1910*, Trevithick Society, Pendeen, 1984.

Miller, D.P., *Discovering Water, James Watt, Henry Cavendish and the Nineteenth-Century Water Controversy*, Ashgate, Aldershot, 2004.

Muirhead, J.P., *Correspondence of the late James Watt on his Discovery of the Composition of Water*, J. Murray, London, 1846.

Muirhead, James Patrick, *The Life of James Watt, with Selections from his Correspondence*, J. Murray, London, 1858.

Muirhead, James Patrick, *The Origin and Progress of the Mechanical Inventions of James Watt*, J. Murray, London, 1854.

Musson, A.E. & Robinson, E., *Science and Technology in the Industrial Revolution*, Manchester University Press, 1969.

Partington, Charles F., *An Historical and Descriptive Account of the Steam Engine*, J. Taylor, London, 1822.

Partington, J.R., *A History of Chemistry*, Macmillan, London, 4 Vols., 1961 – 64.

Philosophical Transactions.

Priestley, Joseph, *Priestley's Navigable Rivers and Canals*, 1831, reprint David & Charles, Newton Abbot, 1969.

Priestley, Joseph B., *Considerations on the Doctrine of Phlogiston, and the Decomposition of Water*, Philadelphia 1796, Ed. William Foster, Princeton University, 1929.

Pryce, W., *Mineralogia Cornubiensis; A Treatise on Minerals, Mines, and Mining*, London, 1778.

Pugh, Jennifer S. & Hudson, John, 'The Chemical Work of James Watt, F.R.S.', *Notes & Records of the Royal Society of London*, Vol. 40, No. 1, Nov. 1985.

The Quarterly Review

Rees, Abraham, *The Cyclopaedia; or Universal Dictionary of Arts, Sciences, and Literature*, London, 1819, and extracts reprint David & Charles, Newton Abbot, 1972.

Reid, Adam, 'The Composition of Water', typescript, 2003.

Richardson, A.E., *Robert Mylne, Architect and Engineer, 1733 – 1811*, Batsford, London, 1955.

Robinson, Eric, 'The Lunar Society: Its Membership and Organisation', *Trans. Newcomen Society*, Vol. 25, 1962 – 63.

Robinson, Eric & McKie, Douglas, *Partners in Science, James Watt and Joseph Black*, Constable, London, 1970.

Robinson, E. & Musson, A.E., *James Watt and the Steam Revolution*, Adams & Dart, London, 1969.

Robison, John, *A System of Mechanical Philosophy, with notes by Sir David Brewster*, Edinburgh, 1822.

Rogers, K.H., *The Newcomen Engine in the West of England*, Moonraker Press, Bradford on Avon, 1976.

Roll, Sir Eric, *An Early Experiment in Industrial Organisation, Being a History of the Firm of Boulton & Watt, 1775 – 1805*, 1930, reprint F. Cass, London, 1968.

Rolt, L.T.C. & Allen, J.S., *The Steam Engine of Thomas Newcomen*, Moorland, Hartington, 1977.

Rowe, J., *Cornwall in the Age of the Industrial Revolution*, Cornish Hillside Publications, St. Austell, 1953, 2 edn., 1993.

Schofield, Robert E., 'James Watt's Letter to Joseph Priestley, 26 April, 1783', *Annals of Science*, Vol. 10, 1954.

Schofield, Robert E., *The Lunar Society, A Social History of Provincial Science and Industry in Eighteenth Century England*, Clarendon Press, Oxford, 1963.

Schofield, Robert E., 'Still More on the Water Controversy', Chymia, Vol. 9, 1964.

Skempton, A.W., Ed., *John Smeaton, F.R.S.*, T. Telford Ltd., London, 1981.

Smeaton, John, *Reports of the Late John Smeaton, F.R.S., made on Various Occasions in the Course of his Employment as a Civil Engineer*, Longman, Hurst, Rees, Orme & Brown, London, 1812.

Smiles, Samuel, *Lives of the Engineers. The Steam-engine. Boulton and Watt*, J. Murray, London, 1878.

Smiles, Samuel, *Men of Invention and Industry*, J. Murray, London, new edn. 1890.

Smith, Barbara M.D. & Moilliet, J.L., 'James Keir of the Lunar Society', *Notes & Records of the Royal Society of London*, Vol. 22, Nos. 1 & 2, Sept. 1967.

Tann, Jennifer, 'Mr. Hornblower and His Crew: Watt Engine Pirates at the End of the Eighteenth Century', *Trans. Newcomen Society*, Vol. 51, 1979 – 80.

Tann, Jennifer, 'Riches from Copper: the Adoption of the Boulton & Watt Engine by Cornish Mine Adventurers', *Trans. Newcomen Society*, Vol. 67, 1995 – 96.

Tann, Jennifer, *The Selected Papers of Boulton & Watt, Vol. I, The Engine Partnership, 1775 - 1825*, MIT Press, Cambridge, Massachusetts, 1981.

Thomson, George, 'James Watt and the Monkland Canal', *Scottish Historical Review*, Vol. XXIX, No. 108, October 1950.

Van der Pols, K. & Verbruggen, J.A., *Steam Drainage in the Netherlands, 1770 – 1870*, Delft University Press, 1996.

Watt, James, *An Account of James Watt's Improvements upon the Steam or Fire Engine*, London, c. 1775.

Watt, James, 'On a new method of preparing a Test Liquor to show the Presence of Acids and Alkalis in Chemical Mixtures', *Philosophical Transactions*, Vol. 74, 1784.

Watt, James, 'Thoughts on the constituent Parts of Water, and of Dephlogisticated Air: with an Account of some Experiments on that Subject', *Philosophical Transactions*, Vol. 74, 1784.

Watts-Russell, P., 'A "copper revolution" cut short; the Cornish Metal Co.', (1785 – 1792), typescript, 2004.

Index People

A

Argand, Aimé, 212, 214
Arkwright, Sir Richard, 7, 113

B

Banks, Sir Joseph, 212, 222-5
Barwell, Nathan, 46
Baskerville, John, 201
Bassett, Sir Francis, 165
Bayly, Sir Nicholas, 162
Beddoes, Dr. Thomas, 8, 14-5
Bell, Thomas, 186, 204
Bentley & Co., 58-9
Bigg, John, 199, 201-4, 210
Black,Prof. Joseph, 7-8, 14-5, 21, 34, 104, 117, 130, 134, 138, 146, 155-6, 161,185, 187, 190-2, 194, 206, 211, 215-7, 219-20, 227, 239
Blackey, William, 42-5, 94
Blagden, Charles, 223-5, 229-30
Boulton, Ann, 100, 154
  Matthew, 6, 8, **13**, 13, 15-7, 19-21, 23-7, 31, 34, 38-40, 41-5, 47, 51-5, 59-64, 66-7, 71, 80-3, 85-9, 93, 96-7, 99-101, 105-110, 112, 115-7, 123-4, 128, 130-2, 137-9, 143-57, 161, 163-4, 166, 169, 171, 173-5, 185-7, 191-3, 199, 202-8, 210-1, 214, 218, 220, 229, 237-8
  Matthew Robinson, 52, 211
Boulton & Fothergill, 38, 47, 60, 78, 87, 143-4, 147-8, 150-4, 184, 200
Boulton & Watt, 47, 51, 56-7, 60, 64, 77, 94, 96-8, 100-1, 105-17, 128, 135-9, 143-5, 148-51, 153, 155, 157-8, 161-4, 166-7, 169, 173, 175, 185, 187, 209-10, 237-8
Brewster, Sir David, 18
Brindley, James, 16
Buccleugh, Duke of, 206, 208
Budge, John, 94, 99
Bull, Edward, 7, 62, 64

C

Cameron, Robert, 173
Campbell, Lord Frederick, 43
Cartwright, Richard, 6, 173
Cavendish, Henry, 211, 214, 217-9, 223-5, 229-30
Chapman, William, 146
Charles, Robert, 212, 214
Cochrane, Jean, 78
  Lord, 18, 43
  Thomas, 78
Colville, Peter, 71-2
Craig, John, 14
  William, 17, 154, 156
Cumming, Alexander, 41-42
Curr, John, 126, 169

D

Dadley, 88-9
Darwin, Dr. Erasmus, 16, 22-3, 40, 44, 80, 94, 183, 190-1, 218-9
Davy, Sir Humphry, 228-9
Dawes, John, 162
D'Heronville, Comte, 138-9
De La Place, Pierre Simon, Marquis, 188, 213-4, 223.
De Luc, Jean André, 87, 183, 188, 200, 212, 221, 224, 226-7, 230
Dudley, Thomas, 98, 100, 111-2, 114, 145

E

Edgeworth, Richard Lovell, 25
Eginton, Francis, 144
England, King of, 7, 207, 238
Ennis, Thomas, 96, 106

F

Falck, N.D., 39, 58
Farey, John, 126, 133-4, 169-70, 178
Fenton & Co., 94, 107
Fothergill, John, 6, 19, 27, 48, 87, 94, 137, 143, 145, 149-52, 154, 157
France, King of, 138, 211
Franklin, Benjamin, 22, 212
Fry, Joseph, 176, 179, 185, 221, 229

G

Gainsborough, Humphrey, 53
Garbett, Samuel,36, 45, 59, 80, 192
Gardner, John, 14, 18, 186
Green & Wyke, Liverpool, 116

H

Hamilton, Gilbert, 8, 15, 21, 47, 51, 55, 72, 77-9, 81-4, 87, 98, 106, 117, 150, 154-7, 172, 175, 183-6, 191, 197, 206-7, 209, 214, 216, 222-3
Harrison, Joseph, 36, 44, 53-5, 59, 64, 72, 112, 143
Henderson, Capt. Logan, 72, 105, 108-9, 112, 123, 128, 137, 156, 174, 191
Hornblower, Jabez Carter, 109, 128, 136-7, 173
  Jonathan, 93-4, 97-8, 100-2, 107-8, 113, 116-7, 137, 172
  Jonathan junior, 6-8, 52, 143, 154-5, 167, 171-9
  Josiah, 172
Hutton, James, 13, 34, 38, 80, 187

I

Irvine, Dr. William, 43, 186, 226

J

Jary, J.M., 127, 137-8

K

Keir, James, 20-1, 80, 89, 156-7, 183-5, 191-7, 199-202, 207-8, 210, 229
Kirwan, Richard, 188, 213, 224, 229

L

Lavoisier, Antoine L., 188, 213-4, 223-5, 227-9
Law, James, 110, 112
Lawson, James, 111
Lea, E., 201-2
Lean, John, 202, 205-6
Lind, James, 62-3, 168, 214-6
Lowe, Vere, Williams & Jennings, 144-52, 154-5

M

McDowal, Charles, 167
McGrigor, James, 6, 82-3, 89, 147, 169, 183, 186, 211
  Ann, see under Watt
Magellan, J.J., 137-8, 188, 219
Mansfield, Lord, 208
Marr, Capt., 201
  Nancy, 85
Matthews, William, 47, 148-9, 153, 155, 211
Meason, Gilbert, 72, 109, 115

Millar, Agnes,156
Betty, 156
Nancy, 79
Miller, Margaret, wife, see Watt
Montgolfier brothers, 211-2
Muirhead, James Patrick, 80, 84
Robert, 47, 78, 156, 184
Murdock, William, 52, 109-11, **109**, 116, 156, 187
Mylne, Robert, 44-5, 229

N
Neale, James, Fordyce & Downe, 16, 144
Newcomen, Thomas, 13, 32, 55, 240-11

O
Osborne, Murray, 14, 18, 186

P
Papin, Denis, 130
Paul, Capt., 98, 108, 117, 166
Pearson, James, 78, 151, 153, 157, 207
Périer, J.C., 137-9
Perrins, Isaac, 58-9, 72, 114
Playfair, William, 112, 157, 191, 200
Priestley, Dr. Joseph, 6-8, 183, 188, 214, **217**, 217-230
Prony, Baron de, 170
Pryce, William, 93-4, 99, 131

R
Rennie, John, 17, 110, 171, 178
Reynolds, Richard, 66, 101, 146
Roebuck, Dr. John, 6, 14-5, 23-7, 34, 36, 41-44, 47 66, 77, 80, 87, 136, 155, 184-5, 239
Robison, John, 8, 21-3, 34, 40, 55, 62, 211, 217, 225-6, 239

S
Savery, Thomas, 39, 45-6, 93, 240
Small, Dr. William,12, 16, 19-20, 22-3, 25-7, 31, 34, 39-40, 41, 47, 51, 56, 80, 87, 130, 150, 156, 184, 219, 237
Smeaton, John, 17, 37-8, 47, 52, 57, 60, 65, 96, 98, 103-4, 117, 123, 128, 131, 168-70, 209-10, 214, 221, 229-30
Southern, John, 8, 64, 157-8, 216
Symington, William, 72

T
Taylor, Clement, 201-4
Telford, Thomas, 17
Trevithick, Richard senior, 96, 106

V
Van Liender, J.D.H., 31, 33, 56, 61, 66, 137, 139-40

W
Warltire, John, 218
Wasborough, Matthew, 143, 171
Watt, Ann née McGrigor, 6, 63, 77, 81-5, **84**, 88, 99-100, 106, 115, 149, 152, 201
Gregory, 6, 8, **84**, 84-5
James, character, **3**, 17-8, 25-6, 40-7, 77-8, 84, 89, 94, 98, 115, 143, 145, 147-8, 154, 164-5, 183, 185, 224, 230
civil engineer, 15-7, 31, 40, 78, 112
consulting engineer, 17-8, 24, 57, 59, 72, 101, 237
finance, 14, 40, 42, 78, 81, 87-8, 105, 136, 143, 146-7, 149, 152, 154-6, 165, 184, 209, 229, 237
health, 46, 83-4, 105-6, 136, 143, 149, 155, 164, 166
instrument maker, 13-4, 123, 188, 239
marriage, 13, 77, 82-4, 88-9
merchant, 14-5, 146
natural philosopher, 14-5, 21, 34, 39, 77, 183-200, 239
partnership, 6-8, 15, 23-4, 26-7, 47, 81-2, 87-9, 143, 151-2, 155-7, 184-5
religion, 40, 46, 78, 100
James junior, 14, 79, **79**, 82-3, 85-6, 143, 158, 211, 215-6, 223, 238
James senior, 6, 14, 36, 46, 77-8
James, & Co., 192, 203, 210-1
Janet (Jessy), 6-7, 84, **84**
Margaret née Miller (Peggy), 13, 40, **79**
Margaret, daughter, 7, 14, 79, **79**, 83, 85
Wedderburn, Alexander, 39, 43
Wedgwood, Josiah, 184, 191, 222, 229
Whatman, James, 201-2
Whitehurst, John, 183, 188
Wilkinson, John, 36-7, 53, 55, 57-64, 66, 71, 79, 86, 97, 104, 113-4, 128, 137-8, 145, 167, 238
Wilson, Patrick, 192, 200, 207
Thomas, 97-100, 103, 106-8, 110-1, 115, 145, 153
Winwood, John, 173
Wiss, James, 145-7
Withering, Dr. William, 80, 85, 111, 229
Woodmason, James, 196-7, 199, 202-4, 206, 208-9
Worcester, Marquis of, 45-6, 240
Wrigley, Joshua, 168

INDEX PLACES AND SUBJECTS

A
Académie des Sciences, Paris, 8, 211, 223, 229
air and gas, 6-7, 36, 44, 52-3, 55, 60, 67, 111, 126, 134, 173, 177-8, 211-3, 217-22, 225, 228, 239
alkali, 6, 15, 80, 156-7, 183-5, 187, 196, 198, 229
Albion Corn Mill, London, 7, 154-5
Ale & Cakes engine, 108, 117, 122, 166, 171, 191
America, 7, 16, 43, 51, 96, 139-40
Anglesey, 152, 161-3

B
Balloon, 183, 211-6, 220, 277
Batavian Society, Netherlands, 7, 137, 139-40
Bath, 58, 77, 150, 179
Bedworth engine, 65-6, 68, 69, 114, 146, 169, 191, 220
Bersham Ironworks, 38, 71, 86-7, 98, 101
Birmingham, 6-7, 13-16, 19-21, 22, 25-6, 31, 34, 42, 51, 77, 79-81, 83, 96, 99, 108, 110, 112, 123, 139, 153, 167, 183, 185, 190, 196-7, 199, 211, 216, 218, 225, 229, 237
Bloomfield colliery & engine, 6, 58-60, 64-66, 69, 114, 128, 176
boiler, 32, 34, 42, 62, 66, 70, 72, 94, 97-8, 102, 106, 116, 131, 134-5, 143
boring machine, 37, 42, 53, 55, 58, 62
brass, 21, 52-3, 106, 129, 207
Bristol, 14, 43, 58, 77, 97, 99, 173, 176, 179, 205, 208
Broseley Ironworks and engine, 38, 60-1, 63, 68, 69, 79
Byker engine, 108, 127, 137

C
calculating devices, 14, 188, 189
canals, 13, 16-7, 19, 23, 73, 78, 80, 163, 169, 194
  Monkland, 15, 17, 25, 78, 89, 150
Carron Foundry, 18, 37, 57, 60
cataract mechanism, 72, 72, 108
cement, 6, 183, 186-7
Chacewater engine, 6, 94, 95, 96-8, 100, 103-8, 104, 115-6, 124, 146, 166, 168-9
Chelsea engine, 105, 111, 115-6, 123, 134, 146, 171.
chemistry, 15, 20-1, 81, 143, 192, 216-7, 219, 224, 228, 237
coal, 6, 36, 43, 51, 56, 60, 93-4, 105, 113, 115-7, 131, 138, 143, 151, 162-3, 187, 214
Coalbrookdale, 60, 97
copper, 51, 93, 96, 104, 106, 128-9, 138, 143, 147, 151, 154, 161-3, 186, 195, 214, 218
copying machine, 6, 13, 109, 143, 156-7, 183, 190-211, 237-8
Cornwall, 6, 51, 73, 84, 93-117, 123, 132, 137-8, 145, 148, 151-3, 156, 161-76, 184, 190, 196, 198, 214, 238
Cornish Metal Co., 163
Cosgarne House, Cornwall, 106, 107, 161
counter of strokes, 116-7, 164

D
Delftfield potery, 15, 146, 154, 183-4, 219
dephlogisticated air (oxygen), 218, 220, 223-6, 228-30
Dolcoath mine, 96, 116, 147
Donnington Wood engine, 108-110
drying cylinders, 6, 183, 186, 186
duty, 94, 96, 168-70

E
Edinburgh, 12, 16, 21, 43, 83, 187, 192, 204
electrolysis, 228-9
element, 216-7, 219, 221, 224, 226-9
England, 15-6, 96, 117, 139, 204
erectors, 6, 18, 52, 71-2, 98, 101, 107, 109-13, 136, 143, 156, 173, 181
experiments, 13, 24-6, 31-3, 36, 38, 45-6, 51-4, 59, 80, 87, 94, 129-30, 143, 152, 168, 176, 183-4, 191-204, 213, 218-23

F
Falmouth, 101-2, 106-8
finance, 7, 15, 18, 24-5, 27, 39, 41, 44-7, 59, 87, 107, 110, 114, 139, 143-53, 161, 166-7, 193, 200, 207, 238
fixed air (carbon dioxide), 217, 220, 228
France, 7, 26, 86, 96, 108, 137-40, 211, 224-5

G
Glasgow, 8, 13, 15, 18, 21, 25, 31, 63, 77, 79, 81, 123, 154, 183, 184
  University, 8, 13-4, 21, 130, 219
geology, 13, 105, 184
Greenock, 8, 14-5, 78-9

H
Hawksbury engine, see Bedworth.
heat, 27, 31, 33, 39, 44-6, 63, 128, 134, 177-8, 186, 211-4, 217-21, 225-8, 240-1
  latent heat, 13, 34, 51, 130, 213, 219-20, 222, 225-8
Heathfield House, Handsworth Heath, 7, 14, 81
horse & horsepower, 6, 21, 24-5, 93, 96
Hull, 57, 146, 169

I
indicator (steam engine), 7, 131, 132, 158
inflammable air (hydrogen), 211-2, 214-5, 217-20, 222-30
ink, 190-1, 194-8, 210, 237
iron, 19, 21, 52, 77, 187, 195, 198, 207, 214-5
  cast, 18, 36, 38, 52, 57, 98, 101, 128, 187, 204
  wrought, 57, 128

K
Ketley Ironworks, 66, 108, 110, 157
Kinneil, 15, 25-6, 36, 62, 128

L
lathe, 14, 109, 184, 206
Leeds, 94, 168
lens, 217, 222
Lichfield, 22, 43, 80, 191
lime, 184, 198, 220, 222
Liverpool, 116, 208
London, 6-7, 19-21, 26, 38-40, 44, 51-2, 64, 77, 93, 101, 103, 105, 123, 146, 168, 184-5, 192, 196-7, 199, 211, 214, 222
Long Benton colliery, 94, 117, 169
Lunar Society, 15, 21, 80, 106, 183-5, 188, 190-1, 196, 218, 225, 229, 239

M
Marazion mine, 166
marine (hydrochloric) acid, 185
mercury, 24, 34, 52, 54, 225
metric system, 71, 83, 188
mining, 51, 71, 73, 93-6, 131, 145, 161, 165-6

N
Neath Abbey Ironworks, 58
Netherlands, 7, 31, 136-7, 139-40
Newcastle-upon-Tyne, 93, 117, 137, 146, 168
Newhall Walk, Birmingham, 6, 81
New River Company & engine, 38, 44, 94, 168
New Willey Ironworks & engine, 6, 58, 60-8, 61, 113, 128

O
observatory, 14, 77
oil of vitriol (sulphuric acid), 184-5, 195-6, 215, 218-9

P
paper, 32, 52, 54, 186, 190-6, 198-204, 209-10, 214
Paris, 7, 43, 96, 137-9, 170, 184, 188, 212, 214, 223
Parliament, 6, 27, 36, 38-9, 41-7, 51, 66, 130, 151, 162-3, 165-6, 168, 185, 207, 237
patent, 7, 21, 25, 43, 45, 53, 93, 172, 174, 184-5, 190
  trials & infringements, 62, 86, 103, 167, 172, 175, 179
Watt's, 6-7, 20, 23, 25, 27, 31, 38-40, 42-7, 77, 81, 87, 113, 123, 130, 136-40, 156, 161, 164, 168, 176, 179, 191-3, 196, 198, 204, 207, 237-9
Parys mountain, Anglesey, 151, 162
Penrhyn Du engine, 53, 127, 135, 135-7
phlogiston, 133, 211, 213, 218, 221-2, 224-8, 230
Plymouth, 99, 106
Poldice engine, 108, 115-7, 125, 154, 161, 164, 168-71, 173
Poldory engine, 117, 166
premiums for patent, 46, 56, 63-4, 105, 107, 113-7, 136, 143, 146-7, 152, 161, 163-4, 166-7, 171, 175, 238, 24-9
press for copying, 193, 200, 204-7, 205

Q
quicksilver, see mercury.

R
Radstock engine, 173-6, 174
Redruth, 96, 105-6
Regent's Place, Harpur Hill, Birmingham, 6, 81, 81
Royal Society of London, 7, 15, 143, 171, 188, 212, 222-5, 228-30, 238
Russia, 40, 184

S
salt, 15, 184-5
scientific instruments, 14, 18, 23, 31, 77
Scorrier engine, 99, 106
Scotland, 6, 12, 15-6, 24, 40, 42, 46-7, 51, 55, 59, 71-2, 77, 79, 82, 106, 146, 155, 208
Shadwell engine, 114-6, 133, 146, 171
Sheffield, 52, 77, 126, 169
Shropshire, 97, 108
smiths, 21, 98, 101
Soho Manufactory, Birmingham, 6, 8, 13, 19, 19-21, 36, 47, 51, 55, 59, 80-1, 87, 96, 101, 109, 111, 116, 123, 137, 143-4, 152, 154, 156-7, 200, 205, 238
Soho Foundry, Birmingham, 7, 110.
Staffordshire, 97, 184
steam, 31-2, 34, 36, 39, 44, 51, 94, 126, 130, 178-9, 213-4, 217, 219-21, 227-8, 239-41
steam engine, 13-5, 17-8, 21-2, 24-7, 31, 32, 36, 37, 39, 41-5, 51-2, 53, 60, 66, 72, 77, 87, 93, 101-2, 123-36, 144, 166, 168-71, 179, 208, 213, 237-43
  air pump, 33-4, 36, 44, 52-5, 58, 62, 68, 69, 70, 99, 101, 103, 127, 136, 178-9, 220, 239
  at Soho Manufactory, 36, 38-9, 51-3, 56, 65-6, 68, 98, 128-9, 131-3, 176
  beam, 25, 36, 61, 62-3, 66, 98-9, 103, 116, 136, 173
  condenser, jet, 36, 44, 55, 58, 68, 69, 70, 136, 177-8
    separate, 13, 23, 27, 31, 33, 43-5, 56-7, 60, 64, 69, 99, 130, 135, 166, 173, 179, 237, 239, 241-2
    surface, 33, 36, 54-5, 68
  cylinder, 6, 18, 25, 27, 32-3, 36-9, 42, 45, 52, 55, 58-60, 62, 66, 71-2, 86, 98, 101-2, 104, 104, 128, 129, 135, 177, 179, 239.
  injection water, 32-3, 38, 44, 55, 68, 71, 72, 102, 111, 124, 127, 134, 177, 241.
  performance, 18, 33-4, 36, 38-9, 42, 44-5, 53-6, 59, 63, 94, 103, 105, 115, 131, 133, 168-71, 176-7, 242
  piston, 36, 44, 52, 54-5, 101, 172-3, 177, 179, 241
  steam jacket, 33, 36, 53, 62, 98, 102-4, 128, 132, 179, 242
  valve gear, 34, 366, 54, 54, 56, 62, 65-8, 67, 101, 103-4, 108-11, 123-5, 124, 125
  types of:
    double-acting, 6, 170-1
    expansive, 6, 44, 129-35, 132, 143, 170-1, 178-9, 239
    Hornblower compound, 6, 171-6
    inverted, 62-4
    Newcomen or common, 13, 18, 22, 25, 31, 33, 38-9, 42, 44-5, 52-3, 56-9, 71, 93-8, 113, 115, 123, 137-8, 168-9, 176, 219, 240-2
    rotative, 6, 42, 73, 143, 171, 237
    single-acting, 32, 32, 37, 123-36, 270
    water-returning, 22, 24, 36, 39, 52, 60, 62
  steam wheel, 13, 24, 26, 34, 35, 36, 42, 44-5, 51-2, 60, 130-1, 138-9, 239, 242-3
  Stratford-Le-Bow engine, 52, 58, 64-6, 65, 97, 115, 138
Swansea, 58, 163

T
textile mills, 7, 168, 186
tin, 6, 36, 38, 52, 128, 200
Ting Tang engine, 6, 96-8, 100-7, 102, 115-7, 124, 145
tools, 20, 23, 24, 37, 55, 59, 129, 190, 206
Torryburn engine, 66, 68, 70-2, 70, 71, 82, 108, 135
Tresavean engine, 99, 108
Truro, 93, 99, 105, 145

U
United mines, 99, 166, 174

V
vacuum, 13, 31-3, 54-5, 58, 67-8, 124, 128, 134, 168, 177-9, 187, 241

W
Wales, 97, 135
Wanlock Head, 72, 109, 115
water, 6, 15, 134, 136, 187, 195, 198, 204, 212-4, 216, 218-22, 221, 225-9, 240-1
  composition, 7, 15, 143, 183, 211, 216-30
  supply, 8, 15, 60, 73, 93, 137, 139, 171
waterwheel, 15, 19-21, 36, 42, 60, 62, 82, 93, 168
Watt's law, 219-20
Wheal Bussy, 6, 96, 98-100, 103, 107, 111, 124, 172
  Butson, 171
  Carharrack, 97
  Chance, 98, 117
  Harland, 171
  Maid, 97, 108, 117, 170, 244
  Prosper, 116
  Sparnon, 97
  Treasury, 96
  Union, 94, 106, 108, 111, 117, 145, 166
  Virgin, 6, 93, 97, 100, 107-8, 117, 128, 148, 154, 161, 164, 166, 173, 176
wood, 33, 36, 53, 101, 128, 205-7
workshop, 14, 18, 20-1, 23-4, 42, 59, 81, 86, 123, 163, 197, 200-2
Wren's Nest engine, 108, 146